Nonlinear Wave Processes in Acoustics

Cambridge Texts in Applied Mathematics

Maximum and Minimum Principles
M.J. Sewell

Solitons
P.G. Drazin and R.S. Johnson

The Kinematics of Mixing
J.M. Ottino

Introduction to Numerical Linear Algebra and Optimisation
Philippe G. Ciarlet

Integral Equations
David Porter and David S.G. Stirling

Perturbation Methods
E.J. Hinch

The Thermomechanics of Plasticity and Fracture
Gerard A. Maugin

Boundary Integral and Singularity Methods for Linearized Viscous Flow
C. Pozrikidis

Nonlinear Systems
P.G. Drazin

Stability, Instability and Chaos
Paul Glendinning

Applied Analysis of the Navier-Stokes Equations
Charles R. Doering and J.D. Gibbon

Viscous Flow
H. Ockendon and J.R. Ockendon

Similarity, Self-Similarity and Intermediate Asymptotics
G.I. Barenblatt

First Course in the Numerical Analysis of Differential Equations
A. Iserles

Complex Variables
Mark J. Ablowitz and Athanassios S. Fokas

Mathematical Models in the Applied Sciences
A. Fowler

Thinking About Ordinary Differential Equations
Robert E. O'Malley, Jr.

A Modern Introduction to the Mathematical Theory of Water Waves
R.S. Johnson

Nonlinear Wave Processes in Acoustics

K. NAUGOLNYKH

University of Colorado, ETL/NOAA, and N.N. Andreyev Acoustics Institute

L. OSTROVSKY

University of Colorado, ETL/NOAA, and Institute of Applied Physics, Russian Academy of Sciences

CAMBRIDGE UNIVERSITY PRESS
Cambridge, New York, Melbourne, Madrid, Cape Town, Singapore, São Paulo

Cambridge University Press
The Edinburgh Building, Cambridge CB2 2RU, UK

Published in the United States of America by Cambridge University Press, New York

www.cambridge.org
Information on this title: www.cambridge.org/9780521390804

First published 1998

A catalogue record for this publication is available from the British Library

Library of Congress Cataloguing in Publication data

Naugol'nykh, K. A. (Konstantin Aleksandrovich)
[Nelineĭnye volnovye prot͡sessy v akustike. English]
Nonlinear wave processes in acoustics / K. Naugolnykh, L. Ostrovsky.
p. cm. – (Cambridge texts in applied mathematics)
Includes index.
ISBN 0-521-39080-X (hb) – ISBN 0-521-39984-X (pb)
1. Nonlinear acoustics. I. Ostrovskiĭ, L. A. II. Title. III. Series.
QC244.2.N3813 1998
534–dc21 98-2925
CIP

ISBN-13 978-0-521-39080-4 hardback
ISBN-10 0-521-39080-X hardback

ISBN-13 978-0-521-39984-5 paperback
ISBN-10 0-521-39984-X paperback

Transferred to digital printing 2006

Contents

Preface

The investigation of nonlinear phenomena in acoustics has a rich history. The origins of nonlinear acoustics (NA) may be associated with the brilliant achievements in continuum mechanics by almost all the leaders of the nineteenth century in mechanical and physical science. In the first stages NA could be considered as a "weakly nonlinear" version of fluid dynamics, describing waves of small but finite amplitude when the acoustic Mach number characterizing the ratio of the medium velocity to the sound propagation velocity may be considered as a small parameter.

However paradoxical it may seem, research into weakly nonlinear perturbations in compressible media has been for a long time only rather remotely related to linear acoustics, which treated the sounding of musical instruments, acoustical properties of buildings, sound propagation in air and other, equally linear, problems. Not until the end of the 1950s was there a sharp growth of interest in nonlinear acoustic phenomena – and by no means accidentally. There was a real need to investigate the intense sounds produced by explosions, jet engines etc., in the atmosphere, the ocean and the soil. On the other hand, new sources of powerful sound and ultrasound could be used for remote sensing, for material diagnostics, and in connection with new methods of technology and surgery. It is important that even at relatively small amplitudes of the acoustic field (in the sense of smallness of the acoustic Mach number) nonlinear distortions of the wave profile often become considerable due to their cumulative character; in many cases the distances scaled in wavelengths (this scale seems to be the most natural) may be sufficiently large.

All this has stimulated various investigations of NA problems, especially in the application to ultrasound. As a result, prominent nonlinear distortions of waves in fluids and solids were observed, and considerable advances were made in theoretical models. Interesting applications appeared, associated with both the invention of original devices such as parametric arrays and new methods for acoustic diagnostics of different media.

Beginning in the 1960s, several books on various aspects of NA have been published; we mention here only those by Krasil'nikov and Zarembo (1966), Beyer (1975), Rudenko and Soluyan (1975), and Gurbatov *et al.* (1991), which are of a rather general character; many other books and reviews will be cited in subsequent chapters.

But now the spectrum of problems considered by NA is being radically widened – and not only in quantitative but also in qualitative aspects. New media and materials considered before in thermal physics, strength of materials theory, etc., are now coming within the orbit of NA. "Dispersive" NA has appeared dealing with acoustic nonlinearities in waveguides, liquids with gas bubbles and microinhomogeneous solid media. Such rather new effects for acoustics as self-focusing, "self-induced transparency", wavefront conjugation, etc., have been investigated theoretically and experimentally. All this goes far outside the framework of classical NA, which used to restrict itself to the quadratic (in amplitude) approximation (sometimes with cubic corrections) and to nondispersive (or weakly dispersive) media. The problems of modern NA seem to be becoming as various as those one used to see, for example, in plasma physics and nonlinear optics.

Such developments became one of the main stimuli for writing this book. However, another factor also played an important role. Really, modern progress in NA went in parallel with the establishment of a general theory of nonlinear waves, and one can point out a number of cases of mutual enrichment of the two disciplines. Results from NA are used in work on wave theory and vice versa.

These two circumstances affected to a great degree the contents and structure of our book. We consider here models of different media as well as the equations and behaviour of finite-amplitude waves in them. From the point of view of wave theory, the qualitative character of a wave process is determined by the joint action and competition of such factors as nonlinearity, dissipation, dispersion, refraction and diffraction. We try to present a successive consideration of the effect of these factors on the evolution and interaction of acoustic waves.

In this book we did not intend to exhaust the total field of NA, restricting ourselves to "pure acoustics", i.e. to waves of a mechanical nature, leaving aside the peculiarities of electroacoustic and magnetoacoustic processes as well as "non-wave" phenomena like "acoustic streaming", radiation pressure etc. Whenever possible we have avoided complicated mathematical procedures, and use rather simplified models or even phenomenological descriptions. Although the book is essentially of a theoretical nature we try to help readers to form a judgement about characteristic parameters of the processes considered by presenting quantitative estimates and some experimental data.

We hope that this book will be of some interest not only to specialists in acoustics but also to a wider community of mathematicians, physicists and engineers whose work is somewhat related to research on nonlinear waves in various physical systems.

Naturally, in many cases we have based our discussion on the work of Russian scientists but have tried (with the best of intentions but perhaps not always with complete success) to reflect Western advances equally well.

Any attempt to cover such a wide and actually interdisciplinary area of science inevitably involves compromises. One of them is of a technical nature: occasionally the same notation has been used for different variables; we have tried to ensure that there is no confusion and that their meaning is clear.

We are grateful to Academician A. V. Gaponov-Grekhov, who was the editor for the Russian edition of this book and who kindly stimulated and supported our work, to Professors S. N. Gurbatov, E. N. Pelinovsky, O. V. Rudenko, S. A. Rybak and A. M. Sutin for valuable discussions and to D. G. Sorokin for his contribution in preparation of the English edition. We also appreciate the help of T. A. Dunina and T. N. Orlova in the course of our work on the manuscript.

We are especially indebted to Professor D. G. Crighton for his outstanding role in improving the quality of the English edition of the book.

When we began writing this book, we were working in different locations separated by about 300 miles. While this had obvious drawbacks, it also afforded us the opportunity to consider and combine attitudes from different scientific schools in the areas under consideration. It was certainly beneficial, however, to work together on the final stages of preparation of the English version of the book in the "environment" of the NOAA Environmental Technology Laboratory (Boulder, CO, USA), and we are grateful to its director, Dr. S. Clifford, and D. Kulla of the Advanced Sensor Application Program, OSD, for making this possible. Our work at ETL was supported by the National Research Council and the Cooperative Institute for Research in Environmental Science of the University of Colorado.

References

Beyer, R.T. (1974). *Nonlinear acoustics (405 pp.)* (USA Naval Ship Syst. Command,).
Gurbatov, S.N., Malakhov, A.N., Saichev, A.I. (1991). *Nonlinear random waves in a dispersionless medium* (Manchester University Press, Manchester).
Rudenko O.V. and Soluyan S.I. (1977). *Theoretical fundamentals of nonlinear acoustics (274 pp.)* (Plenum Press, New York).
Zarembo L.K. and Krasil'nikov V.A. (1966). *Introduction to nonlinear acoustics* (Nauka, Moscow).

1

Nonlinearity, dissipation and dispersion in acoustics

1.1 Homogeneous gases and liquids

The behaviour of finite-amplitude acoustic waves may qualitatively differ, depending on the medium properties relating to nonlinearity, dispersion and dissipation. "Classical" nonlinear acoustics deals with a relatively narrow range of these properties and pertinent parameters. It mostly describes nonlinear media without dispersion (nondispersive media) but with dissipation; moreover, as a rule it is sufficient to restrict oneself to quadratic nonlinearity, i.e. second-order effects with respect to wave amplitude. Sometimes the effect of medium relaxation on nonlinear phenomena has also been taken into account.

Such models (originating from gas dynamics) may actually be applied to describe nonlinear waves in various gases, liquids and solids. At the same time we shall consider here media with internal structure such as liquids with gas bubbles and solids with dislocations, microcracks, grained structure, etc., whose properties are characterized by a complex dependence of the sound velocity and losses on frequency and, rather often, by a "nonclassical" type of nonlinearity when the "stress–strain" relation cannot be reduced to the quadratic approximation. Different models of such media have been investigated for a long time, in relation to the problems of thermal physics, the theory of elasticity, fracture mechanics, defect diagnostics, etc.; however, the nonlinear wave processes in such media, especially from the acoustical point of view, have not yet been given due attention.

In this chapter we shall consider the acoustic properties of different media with respect to intense acoustic fields, and some typical equations for these fields (mostly in the one-dimensional case, i.e. for plane waves). Meanwhile, when possible, we present quantitative estimates for the corresponding models. Of course, we were reluctant to fill the chapter with all the media models and all the types of equation (even one-dimensional) which will be met in the ensuing chapters. We shall frequently return to these problems, while for the present we discuss only relatively simple and rather typical cases.

We start by treating the equations of nonlinear acoustics concerning gases and liquids, for which the equations of continuum mechanics, taking viscosity and heat conductivity into account, are valid (Landau & Lifshitz, 1986); namely, the equation of motion

$$\rho\left(\frac{\partial \mathbf{v}}{\partial t}+\mathbf{v}\cdot\nabla\mathbf{v}\right)=-\nabla p+\eta\nabla^2\mathbf{v}+\left(\zeta+\frac{1}{3}\eta\right)\nabla\left(\nabla\cdot\mathbf{v}\right), \tag{1.1}$$

the continuity equation

$$\frac{\partial\rho}{\partial t}+\nabla\cdot(\rho\mathbf{v})=0, \tag{1.2}$$

the heat transfer equation

$$\rho T\left(\frac{\partial s}{\partial t}+\mathbf{v}\cdot\nabla s\right)=\frac{\eta}{2}\left(\frac{\partial v_i}{\partial x_k}+\frac{\partial v_k}{\partial x_i}-\frac{2}{3}\delta_{ik}\frac{\partial v_\ell}{\partial x_\ell}\right)^2+\zeta\left(\nabla\cdot\mathbf{v}\right)^2+\kappa\nabla^2 T, \tag{1.3}$$

and the equation of state

$$p=p(\rho,s). \tag{1.4}$$

Here $\mathbf{v}$ is the fluid particle velocity, ρ is density of the medium, p, s and T are pressure, entropy and temperature, respectively, η and ζ are the coefficients of shear and bulk viscosity, and κ is the heat conductivity coefficient.

Consider first small acoustical perturbations in an ideal fluid, which are governed by the linearized versions of the equation of motion

$$\rho_0\frac{\partial\mathbf{v}}{\partial t}+\nabla p'=0, \tag{1.1a}$$

continuity equation

$$\frac{\partial\rho'}{\partial t}+\rho_0\nabla\cdot\mathbf{v}=0, \tag{1.2a}$$

and adiabatic equation of state

$$p'=c_0^2\rho',\, c_0^2=\left(\frac{\partial p}{\partial\rho}\right)_s. \tag{1.4a}$$

Here $p'=p-p_0$, $\rho'=\rho-\rho_0$, and p_0, ρ_0 correspond to the equilibrium state of the medium. This set of equations can be easily transformed into a single wave equation for any scalar variable, for example, for the velocity potential φ:

$$\frac{\partial^2 \varphi}{\partial t^2} - c_0^2 \nabla^2 \varphi = 0, \tag{1.5}$$

so that $\mathbf{v} = \nabla\varphi, p' = -\rho_0 \frac{\partial \varphi}{\partial t}, \rho' = -\frac{\rho_0}{c^2}\frac{\partial \varphi}{\partial t}$, and c_0 is the sound velocity, given in (1.4*a*).

To obtain equations of nonlinear acoustics for a dissipative medium one should take into account the nonlinear, viscous and thermal effects. The corresponding terms in equations are supposed to be small so that linear acoustic relationships can be used in the course of their transformations.

Let us limit ourselves first to consideration of quadratic nonlinear effects in ideal fluid, neglecting the nonlinear terms of higher order. Then the equation of state (1.4) can be written as

$$p' = c_0^2 \rho' + \frac{1}{2}\left(\frac{\partial c^2}{\partial \rho}\right)_s \rho'^2, \tag{1.4b}$$

where the first term corresponds to the linear approximation and the second accounts for the medium nonlinearity. The formula (1.4*b*) is often presented as

$$p' = A\frac{\rho'}{\rho_0} + \frac{1}{2}B\Box\frac{\rho'}{\rho_0}, \tag{1.6}$$

and then the ratio B/A characterizes the nonlinearity of the equation of state for the medium. ρ'/ρ_0 can be considered as an acoustic Mach number M.

For gases the adiabatic equation has the Poisson form

$$\frac{p}{p_0} = \left(\frac{\rho}{\rho_0}\right)^\gamma, \tag{1.7}$$

where $\gamma = C_p/C_v$ is the adiabatic index, equal to the ratio of the specific heat at constant pressure C_p to that at constant volume, C_v. Then

$$c_0^2 = \frac{\gamma p_0}{\rho_0}, \quad \left(\frac{\partial c^2}{\partial \rho}\right)_s = \frac{c_0^2}{\rho_0}(\gamma - 1), \tag{1.8}$$

where p_0, ρ_0 are the equilibrium values of the pressure and medium density, respectively.

The equation of state for liquids is not as straightforward; one can use

Table 1.1. Measured values of B/A and $\varepsilon = 1 + B/2A$ for different media

Substance	T° C	B/A	ε
Distilled water	0	4.16	3.08
Distilled water	20	4.96	3.48
Distilled water	40	5.38	3.69
Distilled water	60	5.67	3.84
Distilled water	80	5.96	3.98
Distilled water	100	6.11	4.06
Carbon tetrachloride	30	11.54	6.77
Glycerine	20	8.80	5.40
Mercury		8.33	5.17
Ethanol		10.57	6.29
Liquid oxygen	-199	9.56	5.78
Monatomic gas	20	0.67	1.34
Diatomic gas	20	0.40	1.20
Beef liver	30	7.75	4.48
Beef heart	30	6.8–7.4	3.5–3.8
Lard		11.00	6.50
Man's fat	30	9.91	5.96
Dog's kidney	30	7.20	4.60

Tate's empirical formula

$$p = A_0 \left(\frac{\rho}{\rho_0} \right)^\gamma - A_1, \tag{1.9}$$

where A_0, A_1, γ are constants weakly dependent on the temperature and $A_0 - A_1 = p_0$, where p_0 is the hydrostatic pressure (for water $\gamma = 7$, $A_1 \sim 300\,\text{MPa}$). Hence, in the case of liquids,

$$c_0^2 = \frac{\gamma A_0}{\rho_0}, \quad \left(\frac{\partial c^2}{\partial \rho} \right)_s = \frac{c_0^2(\gamma - 1)}{\rho_0}. \tag{1.10}$$

Comparing (1.5), (1.6) and (1.10) we obtain

$$\frac{B}{A} = \gamma - 1. \tag{1.11}$$

Values of this parameter are given in Table 1.1 for several substances; the table is based on the data presented in a book by Beyer (1974), supplemented with some information on biological tissues (Dunn *et al.*, 1984; Law *et al.*, 1985).

In the quadratic approximation the set (1.1) with $\eta = 0, \zeta = 0$, (1.2) and (1.4b) may be reduced to a single equation (see, for instance, Naugol'nykh, 1971). Neglecting the dissipative effects and considering the medium state variation as adiabatic we introduce the velocity potential φ, so that $\mathbf{v} = \nabla\varphi$, and the enthalpy per unit volume W, in accordance with the thermodynamic relationship

$$W = \int_{p_0}^{P} \frac{dp}{\rho} = \frac{c^2 - c_0^2}{\gamma - 1},$$

which implies $c^2 = c_0^2 + (\gamma - 1)W$. Then, using the relations $\rho_t = c^{-2}W_t, \nabla\rho = c^{-2}\nabla p$, the continuity equation may be formulated as

$$\frac{\partial W}{\partial t} + \nabla\varphi \cdot \nabla W + c^2\nabla^2\varphi = 0. \tag{1.12}$$

Similarly, a first integral of the equation of motion (1.1) in terms of W and φ is

$$W = -\frac{\partial\varphi}{\partial t} - \frac{1}{2}(\nabla\varphi)^2.$$

Substituting the expressions for W and c^2 into (1.12) we obtain in quadratic approximation an equation describing the propagation of weakly nonlinear waves in an ideal medium (Andreev, 1955; Aanonsen *et al.*, 1984):

$$\nabla^2\varphi - \frac{1}{c_0^2}\frac{\partial^2\varphi}{\partial t^2} = \frac{1}{c_0^2}\frac{\partial}{\partial t}\left[(\nabla\varphi)^2 + \frac{1}{c_0^2}\frac{\gamma - 1}{2}\left(\frac{\partial\varphi}{\partial t}\right)^2\right]. \tag{1.13}$$

If we neglect the right-hand side, which reflects the influence of nonlinear effects, we obviously get the wave equation (1.5) representing the propagation of linear acoustic waves.

The equation (1.13) may be transformed into

$$\left(\nabla^2 - \frac{1}{c_0^2}\frac{\partial^2}{\partial t^2}\right)\phi = \frac{\varepsilon}{c_0^4}\frac{\partial}{\partial t}\left(\frac{\partial\phi}{\partial t}\right)^2, \tag{1.13a}$$

where

$$\phi = \varphi - \frac{1}{2c_0^2}\frac{\partial\varphi^2}{\partial t}, \quad \varepsilon = \frac{\gamma + 1}{2} = \frac{B}{2A} + 1.$$

The later value is usually considered as the nonlinearity parameter for fluids (see Table 1.1).

An important class of problems is associated with progressive plane waves or, more generally, with wave beams, in which a slow dependence on transverse coordinates is present. In these cases, for transformation of the small terms in the right-hand side of equation (1.13) one can use the relation

$$\partial/\partial t = -c_0 \partial/\partial x. \tag{1.14}$$

To present the result in terms of the pressure perturbation we also use the relation $p' = -\rho_0 \partial\varphi/\partial t$. The quadratic corrections to this relation cancel each other up to terms of higher order after substitution into the left-hand side of equation (1.13), due to relation (1.14), and are negligible in the nonlinear terms of the right-hand side. As a result, we get the Westervelt equation (Westervelt, 1963):

$$\nabla^2 p' - \frac{1}{c_0^2}\frac{\partial^2 p'}{\partial t^2} = -\frac{\varepsilon}{\rho_0 c_0^4}\frac{\partial^2 p'^2}{\partial t^2}. \tag{1.15}$$

For the plane wave, propagating in the x-direction, we get correspondingly the plane nonlinear wave equation:

$$\frac{\partial^2 p'}{\partial x^2} - \frac{1}{c_0^2}\frac{\partial^2 p'}{\partial t^2} = -\frac{\varepsilon}{\rho_0 c_0^4}\frac{\partial^2 p'^2}{\partial t^2}. \tag{1.15a}$$

Let us now rewrite the equation obtained in terms of density perturbations $\rho' = p'/c_0^2$, keeping in mind again that we do not need to take into account the nonlinear corrections to this equation in the course of transformations. We obtain

$$\frac{\partial^2 \rho'}{\partial t^2} - c_0^2 \frac{\partial^2 \rho'}{\partial x^2} = \frac{c_0^2 \varepsilon}{\rho_0}\frac{\partial^2 \rho'^2}{\partial x^2}. \tag{1.16}$$

Now let us proceed to the "travelling" coordinate $z = x - c_0 t$ and the "slow" variable $\tau = t$ specifying the changes at a fixed point of the wave profile travelling with velocity c_0. Since

$$\frac{\partial}{\partial x} = \frac{\partial}{\partial z}, \quad \frac{\partial}{\partial t} = -c_0\frac{\partial}{\partial z} + \frac{\partial}{\partial \tau}$$

then, neglecting the term $\partial^2\rho'/\partial\tau^2$ (due to the slowness of the profile evolution) and integrating once with respect to z, we obtain the simple-wave equation

$$\frac{\partial \rho'}{\partial \tau} + \frac{\varepsilon c_0}{\rho_0}\rho'\frac{\partial \rho'}{\partial z} = 0. \tag{1.17}$$

(The integration constant is omitted if we consider free wave propagation; it should be retained if the wave is generated by a distributed external force.)

Burgers equation. To take into account dissipative effects we start from equations (1.1–1.3) and the equation of state (1.4) in the form

$$p' = c^2\rho' + \frac{1}{2}\left(\frac{\partial c^2}{\partial \rho}\right)_s \rho'^2 + \left(\frac{\partial p}{\partial s}\right)_\rho s'. \tag{1.4c}$$

In the one-dimensional case, equations (1.1) and (1.2) can be presented as follows:

$$\frac{\partial \rho v}{\partial t} = -\frac{\partial}{\partial x}\left[\rho v^2 + p - \left(\frac{4}{3}\eta + \zeta\right)\frac{\partial v}{\partial x}\right], \tag{1.18}$$

$$\frac{\partial \rho}{\partial t} + \frac{\partial \rho v}{\partial x} = 0. \tag{1.19}$$

Differentiating (1.18) with respect to x and (1.19) with respect to t one obtains

$$\frac{\partial^2 \rho v}{\partial t \partial x} = -\frac{\partial^2}{\partial x^2}\left[\rho v^2 + p - \left(\frac{4}{3}\eta + \zeta\right)\frac{\partial v}{\partial x}\right], \tag{1.20}$$

$$\frac{\partial^2 \rho}{\partial t^2} = -\frac{\partial^2 \rho v}{\partial t \partial x}. \tag{1.21}$$

Substituting (1.20) into (1.21), subtracting term $c_0^2 \frac{\partial^2 \rho}{\partial x^2}$ from both sides of it, and using (1.4*c*) in quadratic approximation one gets

$$\begin{aligned}\frac{\partial^2 \rho}{\partial t^2} - c_0^2\frac{\partial^2 \rho}{\partial x^2} = &-\frac{\partial^2}{\partial x^2}\left[\rho_0 v^2\right] \\ &-\frac{\partial^2}{\partial x^2}\left[\frac{(\gamma-1)}{2}\frac{c_0^2}{\rho_0}\rho'^2 + \left(\frac{\partial p}{\partial s}\right)_\rho s' - \left(\frac{4}{3}\eta + \zeta\right)\frac{\partial v}{\partial x}\right].\end{aligned} \tag{1.22}$$

To transform the right-hand side of this equation we limit ourselves to consideration of plane progressive waves, so that relation $\frac{\partial}{\partial t} = -c_0 \frac{\partial}{\partial x}$ is again used. From equation (1.3) in the linear approximation one gets

$$\frac{\partial s'}{\partial t} = \frac{\kappa}{\rho_0 T_0}\frac{\partial^2 T'}{\partial x^2} = -c_0\frac{\partial s'}{\partial x}. \tag{1.23}$$

Then, using the thermodynamical relationship

$$T' = \frac{T_0 \beta c_0}{\rho_0^2 C_p} \rho',$$

where β is the thermal expansion coefficient and C_p is the specific heat at constant pressure, we obtain

$$\frac{\partial s'}{\partial x} = -\frac{\kappa \beta c_0}{\rho_0^2 c_p} \frac{\partial^2 \rho'}{\partial x^2}. \tag{1.24}$$

With the help of the thermodynamical relationship

$$\frac{\partial p}{\partial s} = \frac{C_p / C_v - 1}{\beta} \rho_0,$$

where C_v is the specific heat at constant volume, we finally obtain

$$-\left(\frac{\partial p}{\partial s}\right) \frac{\partial s'}{\partial x} = \kappa \frac{c_0}{\rho_0} \left(\frac{1}{C_v} - \frac{1}{C_p}\right) \frac{\partial^2 \rho'}{\partial x^2}. \tag{1.25}$$

After substitution of (1.25) into (1.22), using also the relationship $v = c_0 \frac{\rho'}{\rho_0}$, one gets the nonlinear equation for a dissipative medium:

$$\frac{\partial^2 \rho'}{\partial t^2} - c_0^2 \frac{\partial^2 \rho}{\partial x^2} + \frac{c_0^2}{\rho_0} \varepsilon \frac{\partial^2 \rho'^2}{\partial x^2} = \frac{c_0}{\rho_0} \theta \frac{\partial^3 \rho'}{\partial x^3}, \tag{1.26}$$

where

$$\theta = \frac{4}{3}\eta + \zeta + \kappa \left(\frac{1}{C_v} - \frac{1}{C_p}\right).$$

Proceeding again to the "travelling" coordinates $t = \tau$ and $z = x - c_0 t$ we obtain the well-known Burgers equation

$$\frac{\partial \rho'}{\partial \tau} + \frac{\varepsilon c_0}{\rho_0} \rho' \frac{\partial \rho'}{\partial z} = \frac{\theta}{2\rho_0} \frac{\partial^2 \rho'}{\partial z^2}. \tag{1.27}$$

When solving boundary (signalling) problems it is more convenient, as a rule, to deal with variables $y = t - x/c_0$ and $x' = x$, so that

$$\frac{\partial}{\partial t} = \frac{\partial}{\partial y}, \quad \frac{\partial}{\partial x} = -c_0^{-1} \frac{\partial}{\partial y} + \frac{\partial}{\partial x'}$$

and from (1.26) it follows (omitting the prime on x) that

$$\frac{\partial \rho'}{\partial x} - \frac{\varepsilon}{c_0 \rho_0} \rho' \frac{\partial \rho'}{\partial y} = \frac{\theta}{2\rho_0 c_0^3} \frac{\partial^2 \rho'}{\partial y^2}. \tag{1.28}$$

Table 1.2.

Gases	$4\pi^2\delta\cdot10^{-13}\ \frac{s^2}{cm}$	Liquids	$4\pi^2\delta\cdot10^{-17}\ \frac{s^2}{cm}$
Air	1.85	Water	23
Hydrogen	3.58	Mercury	6
Helium	2.96	Glycerine	2500
Agate	1.35	Benzene	850
Oxygen	1.68		

Here we took into account the fact that on the small right-hand side of (1.26) it is sufficient to set $\partial/\partial x = -c_0^{-1}\partial/\partial y$.

It should be noted that, while in the wave equation (1.26) the right-hand side is small compared with each term on the left-hand side, in equations (1.27) and (1.28) all the terms are, generally, of the same order (provided the nonlinear and dissipative terms are considered to be of the same order).

Notice, finally, that often one deals with the equations for the particle velocity v. Substituting $\rho' \simeq \rho_0 v/c_0$ into (1.28) (corrections to this relationship may be neglected here) we obtain the Burgers equation in a standard form:

$$\frac{\partial v}{\partial x} - \alpha v\frac{\partial v}{\partial y} = \delta\frac{\partial^2 v}{\partial y^2}, \tag{1.29}$$

where $\alpha = \varepsilon/c_0^2$, $\delta = \theta/2\rho_0 c_0^3$. Some values of the dissipative factor δ are presented in Table 1.2 (for example, see Bergman, 1942).

The Burgers equation is one of the basic equations in nonlinear acoustics; moreover, it is considered to be among the most exhaustively studied evolution equations in the theory of nonlinear waves (Whitham, 1974). It describes the propagation of an intense acoustic wave taking into consideration small but finite nonlinearity and viscous dissipation: their relative role is characterized by a dimensionless parameter, the acoustic Reynolds number

$$\mathrm{Re} = \frac{\alpha v}{\omega\delta} = \frac{2\varepsilon p'}{\omega\theta}. \tag{1.30}$$

Its value is determined both by the medium properties (through the nonlinearity parameter ε and the effective viscosity coefficient θ) and by acoustic perturbation parameters – the characteristic amplitude and frequency.

For small Reynolds numbers one may neglect the nonlinear term in

(1.29). Then the wave evolution is defined only by dissipative effects. In particular, the propagation of a harmonic wave in the form $v = v_0(x)\exp(i\omega(t - x/c_0))$ is described by a solution with exponentially decreasing amplitude

$$v_0 \sim e^{-ax},$$

where $a = \delta\omega^2$.

In Table 1.2, values of the dissipative factor δ are given for various media.

Relaxing media. For high-frequency waves in liquids and gases, relaxation processes may play an important role when the quasistatic thermodynamic relations are not satisfied exactly. In this case pressure p' may depend on ρ' at all prior times, i.e. on the "history" of the process. This is known to signify the appearance of the second (volume) viscosity. Then, the equation of state is viewed as

$$p' = c_0^2\rho' + \frac{1}{2}\left(\frac{\partial c^2}{\partial \rho}\right)_s \rho'^2 - \int_{-\infty}^{t} G(t - t')\rho'(t)\,dt, \tag{1.31}$$

where G is a kernel associated with a particular relaxation mechanism. Such mechanisms may be of many kinds (excitation of molecular degrees of freedom, impurity effects, e.g. salt in sea water, etc.). Since the real "memory time" is usually bounded by a finite value relaxation time τ, in typical cases G may be represented by the exponential form

$$G = \frac{mc_0^2}{\tau}\exp\left(-(t - t')/\tau\right), \tag{1.32}$$

where m is the relaxation parameter, equal to $(c_0^2 - c_1^2)/c_1^2$, and c_1 is the sound velocity in the high-frequency limit. The value of m is typically small, so that the corresponding term may be incorporated into the Burgers equation (1.29), which now takes the form (Rudenko & Soluyan, 1977)

$$\frac{\partial v}{\partial x} - \alpha v\frac{\partial v}{\partial y} = \mu\frac{\partial^2 v}{\partial y^2} + \frac{m}{2c_0}\frac{\partial}{\partial y}\int_{-\infty}^{y}\frac{\partial v}{\partial y'}\exp\left(-(y - y')/\tau\right)dy', \tag{1.33}$$

where $\mu = (\theta - \zeta)/2\rho_0 c_0^3$. The relaxation time τ depends on the nature of the process. For example, for sea water there are at least two pronounced relaxation processes, with typical times $\tau = 10^{-3}$ s and $\tau = 1.5 \times 10^{-5}$ s.

If we neglect the nonlinearity and shear viscosity, then in accordance with (1.33) the complex amplitude of the harmonic wave changes with distance as $\exp(-ax)$, where

$$a = \frac{\omega m(i\omega^2\tau^2 - \omega\tau)}{2c_0(1+\omega^2\tau^2)}. \tag{1.34}$$

If $\omega\tau$ is small, $a \simeq -m\omega^2\tau/2c_0$ and the wave amplitude simply decays exponentially. When $\omega\tau >> 1$, $a \simeq im\omega/2c_0$, i.e. instead of decay, a change in the wave propagation velocity appears, equal to $\Delta c_0 = -mc_0/2$. In the intermediate frequency range the relaxation effect leads to both absorption and dispersion of the wave velocity.

In the stated limiting cases the integral in (1.33) is easy to calculate. For small τ this integral is equal to $\tau\partial v/\partial y$, and that results in the Burgers equation (1.29) with the bulk viscosity

$$\zeta = \tau m \rho_0 c_0^2.$$

If, however, τ is large, the integral is equal to

$$\int_{-\infty}^{y} v(y)\,\mathrm{d}y,$$

yielding the generalized Burgers equation

$$\frac{\partial v}{\partial x} - \alpha v\frac{\partial v}{\partial y} = \mu\frac{\partial^2 v}{\partial y^2} + \frac{m}{2c_0}v. \tag{1.35}$$

It is reasonable to note that for nonequilibrium media where one or another kind of energy pumping exists (laser excitation, chemical sources, etc.) negative damping effects are possible ($m < 0$) which, in principle, may lead to various types of instability; see Chapter 2.

As could be observed in the previous discussion, the effects of various small factors such as nonlinearity, losses, etc., can be accounted for by the introduction of the corresponding additive terms into the Burgers-type evolution equation. This approach, allowing us to introduce the factors independently, will frequently be resorted to in the ensuing parts of the book.

1.2 Isotropic solids

In solids, as well as in liquids and gases, longitudinal sound waves may propagate; however, together with this phenomenon, transverse waves may also exist, due to the possible formation of strain-induced shear stresses. In the following discussion we confine ourselves to isotropic solids.

Lagrangian coordinates x_i, corresponding to the initial particle locations, are typically used to describe an elastic solid. In the linear approximation (when Eulerian and Lagrangian representations are, as a matter of fact, identical) the equation of motion of an ideal elastic medium has the form

$$\rho_0 \ddot{u}_i = \frac{\partial \sigma_{ik}}{\partial x_k} \quad (i, k = 1, 2, 3), \tag{2.1}$$

where u_i is the displacement vector component, ρ_0 is the equilibrium density, σ_{ik} is the stress tensor (repeated indices indicate summation). In the general case the relation between σ_{ik} and u_i is expressed by means of the strain tensor

$$u_{ik} = \frac{1}{2}\left(\frac{\partial u_i}{\partial x_k} + \frac{\partial u_k}{\partial x_i} + \frac{\partial u_\ell}{\partial x_i}\frac{\partial u_\ell}{\partial x_k}\right). \tag{2.2}$$

It follows from thermodynamical considerations that for adiabatic processes (Landau & Lifshitz, 1987)

$$\sigma_{ik} = \left(\frac{\partial e}{\partial u_{ik}}\right)_s, \tag{2.3}$$

where e and s are the internal energy and the entropy, respectively.

Variation of the internal energy at small deformations may be represented in terms of the strain tensor invariants:

$$e - e_0 = \frac{\lambda}{2} u_{\ell\ell}^2 + \mu u_{ik}^2, \tag{2.4}$$

where λ, μ are Lamé constants characterizing the elastic properties of a medium. Differentiating (2.4) in accordance with (2.3) we find

$$\sigma_{ik} = \lambda u_{\ell\ell}\delta_{ik} + 2\mu u_{ik}. \tag{2.5}$$

Relationship (2.5) may be represented somewhat differently if we single out deformations of uniform compression, determined by the sum of the strain tensor diagonal elements, from those associated with pure shear,

$$\sigma_{ik} = \kappa u_{\ell\ell}\delta_{ik} + 2\mu\left(u_{ik} - \frac{1}{3}\delta_{ik}u_{\ell\ell}\right). \tag{2.6}$$

Here $\kappa = \lambda + \frac{2}{3}\mu$ is the uniform compression modulus while μ is the shear modulus.

Thus, the elastic properties of an isotropic solid in the linear approximation are described by two independent constants, the Lamé constants, uniform compression and shear moduli. Other elastic moduli pairs are also used, e.g., the Young modulus $E = 9\kappa\mu/(3\kappa+\mu)$ and the shear modulus μ, or the Young modulus and Poisson ratio $\sigma = (3\kappa-2\mu)/2(3\kappa+\mu)$. The latter determines the relation between longitudinal and transverse deformations of an elastic rod.

Now consider the wave propagation equations for an elastic isotropic solid. Substituting the expression for σ_{ik} from (2.5) into equation (2.1) and making use of the linearized expression for the strain tensor (2.2), we obtain

$$\rho_0 \ddot{u}_i = (\lambda + \mu)\frac{\partial^2 u_k}{\partial x_i \partial x_k} + \mu \frac{\partial^2 u_i}{\partial x_k^2}. \tag{2.7}$$

In vector representation, this equation has the form

$$\rho_0 \ddot{\mathbf{u}} = (\lambda + \mu)\nabla(\nabla \mathbf{u}) + \mu \nabla^2 \mathbf{u}. \tag{2.8}$$

If we write down the displacement vector $\mathbf{u}$ as a sum of longitudinal (u_ℓ) and transverse ($\mathbf{u}_t$) components then equation (2.8) splits into two relationships

$$\ddot{u}_\ell - c_\ell^2 \nabla^2 u_\ell = 0, \tag{2.9}$$

$$\ddot{\mathbf{u}}_t - c_t^2 \nabla^2 \mathbf{u}_t = 0, \tag{2.10}$$

where

$$c_\ell = \sqrt{\frac{\lambda + 2\mu}{\rho}}, \quad c_t = \sqrt{\frac{\mu}{\rho}}$$

are the propagation velocities of longitudinal and transverse waves, respectively.

To obtain the equations of nonlinear acoustics for an elastic solid one has to take into account the higher-order amplitude terms. First of all it is necessary to use the complete expression for the strain tensor (2.2) (with allowance for the quadratic term). Secondly, the equation of motion, in view of the transition to the Lagrangian coordinate, may be formally written as in (2.1) (Thurston, 1966):

$$\rho_0 \ddot{u}_i = \frac{\partial P_{ik}}{\partial x_k}, \tag{2.11}$$

where P_{ik} is, generally speaking, a nonsymmetric tensor (of Lagrange or Piola–Kirchhoff type), given by

$$P_{ik} = \frac{\partial e}{\partial(\partial u_i/\partial x_k)}. \tag{2.12}$$

In the quadratic approximation, the tensor P_{ik} is expressed by the following relationship:

$$\begin{aligned}
P_{ik} =& \mu\left(\frac{\partial u_i}{\partial x_k}+\frac{\partial u_k}{\partial x_i}\right)+\left(\kappa-\frac{2\mu}{3}\right)\frac{\partial u_\ell}{\partial x_\ell}\delta_{ik} \\
&+\left(\mu+\frac{A}{4}\right)\left(\frac{\partial u_\ell}{\partial x_i}\frac{\partial u_\ell}{\partial x_k}+\frac{\partial u_k}{\partial x_\ell}\frac{\partial u_i}{\partial x_\ell}+\frac{\partial u_\ell}{\partial x_k}\frac{\partial u_\ell}{\partial x_i}\right) \\
&+\frac{1}{2}\left(\kappa-\frac{2\mu}{3}+B\right)\left[\left(\frac{\partial u_\ell}{\partial x_m}\right)^2\delta_{ik}+2\frac{\partial u_i}{\partial x_k}\frac{\partial u_\ell}{\partial x_\ell}\right] \\
&+\frac{A}{4}\frac{\partial u_k}{\partial x_\ell}\frac{\partial u_\ell}{\partial x_i}+\frac{B}{2}\left(\frac{\partial u_\ell}{\partial x_m}\frac{\partial u_m}{\partial x_\ell}\delta_{ik}+2\frac{\partial u_k}{\partial x_i}\frac{\partial u_\ell}{\partial x_\ell}\right) \\
&+C\left(\frac{\partial u_\ell}{\partial x_\ell}\right)^2\delta_{ik}.
\end{aligned} \tag{2.13}$$

This expression corresponds to the expansion (1.6) of the equation of state for fluids. Along with the moduli of uniform compression and shear, (2.13) contains three more constants – the third-order moduli A, B, C (the Landau moduli); they refer to the cubic terms of the internal energy expansion in the strain tensor invariants. Therefore, nonlinear deformations of an isotropic elastic solid are characterized, in accordance with (2.13), by five parameters (the "five-constant" theory). Substituting (2.13) into (2.11) yields

$$\rho_0\frac{\partial^2 u_i}{\partial t^2}-\mu\frac{\partial^2 u_i}{\partial x_k^2}-\left(\kappa+\frac{\mu}{3}\right)\frac{\partial^2 u_\ell}{\partial x_\ell \partial x_i}=F_i, \tag{2.14}$$

where

$$\begin{aligned}
F_i =& \left(\mu+\frac{A}{4}\right)\left(\frac{\partial^2 u_\ell}{\partial x_k^2}\frac{\partial u_\ell}{\partial x_i}+\frac{\partial^2 u_\ell}{\partial x_k^2}\frac{\partial u_i}{\partial x_\ell}+2\frac{\partial^2 u_i}{\partial x_\ell \partial x_k}\frac{\partial u_\ell}{\partial x_k}\right) \\
&+\left(\kappa+\frac{\mu}{3}+\frac{A}{4}+B\right)\left(\frac{\partial^2 u_k}{\partial x_i \partial x_k}\frac{\partial u_\ell}{\partial x_\ell}+\frac{\partial^2 u_k}{\partial x_\ell \partial x_k}\frac{\partial u_i}{\partial x_\ell}\right) \\
&-\left(\kappa+\frac{2\mu}{3}+B\right)\frac{\partial^2 u_i}{\partial x_k^2}\frac{\partial u_\ell}{\partial x_\ell} \\
&+\left(\frac{A}{4}+B\right)\left(\frac{\partial^2 u_k}{\partial x_i \partial x_k}\frac{\partial u_\ell}{\partial x_i}+\frac{\partial^2 u_\ell}{\partial x_i \partial x_k}\frac{\partial u_\ell}{\partial x_\ell}\right) \\
&+(B+2C)\left(\frac{\partial^2 u_k}{\partial x_i \partial x_k}\frac{\partial u_\ell}{\partial x_\ell}\right).
\end{aligned} \tag{2.15}$$

For the case of longitudinal wave propagation in an isotropic solid this equation reduces to

$$\frac{\partial^2 u}{\partial x^2} - \frac{1}{c_\ell^2}\frac{\partial^2 u}{\partial t^2} = -2\varepsilon_\ell \frac{\partial^2 u}{\partial x^2}\frac{\partial u}{\partial x}, \tag{2.16}$$

where the quantity

$$\varepsilon_\ell = \frac{3}{2} + (A + 3B + C)\,\rho_0^{-1} c_\ell^{-2} \tag{2.17}$$

plays the role of the nonlinear parameter for longitudinal waves.

As a rule, the values of ε_ℓ are in the range 3–10, not far from the corresponding values for liquids. After transferring to Eulerian variables the nonlinearity parameter will generally change. One should keep this in mind when comparing these results with the relationships of Section 1.1 for gases and liquids.

For a travelling wave it is easy, as in Section 1.1, to reduce the order of this equation if one goes over from displacement u to another function, the strain $s = \partial u/\partial x$. Differentiating (2.16) we obtain the following equation for nonlinear waves in isotropic solids:

$$\frac{\partial^2 s}{\partial x^2} - \frac{1}{c_\ell^2}\frac{\partial^2 s}{\partial t^2} = -\varepsilon_\ell \frac{\partial}{\partial x}\left(s\frac{\partial s}{\partial x}\right). \tag{2.18}$$

Proceeding to variables $y = t - x/c_\ell$ and $x' = x$ we have, in analogy with (1.29) at $\theta = 0$ (omitting, again, the prime on x),

$$\frac{\partial s}{\partial x} + \frac{\varepsilon_\ell}{2c_\ell} s \frac{\partial s}{\partial y} = 0. \tag{2.19}$$

This equation may be supplemented by terms describing losses. If, for example, we take into account viscous friction and heat conduction then, since the loss factor is determined by a parameter analogous to θ (Landau & Lifshitz, 1986), while losses enter additively, we obtain for s the Burgers equation of type (1.29). However, in real solids more complex loss mechanisms are frequently in action (including relaxation processes) and this leads to other dissipative terms.

Instead of A, B, C (the Landau moduli) the Murnaghan moduli ℓ, m, n are frequently used; these are simply related to the former:

$$\ell = B + C, \quad m = \frac{A}{2} + B, \quad n = A. \tag{2.20}$$

Table 1.3 presents some information on the parameters of the solid media which are often measured by nonlinear acoustics methods (Breazeale

Table 1.3.

Material	Fe armko	Glass	Polystyrene
A, 10^{10}Pa	110	42	-1.00
B, 10^{10}Pa	-158	-11.8	-0.83
C, 10^{10}Pa	123	-13.2	-1.06
ε_ℓ	-4.35	1.4	-4.00
μ, 10^9Pa	8.4	2.8	0.14
K, 10^9Pa	17.2	3.9	0.39
c_1, 10^5cm/s	5.83–5.95	5.64	2.35
c_t, 10^5cm/s	3.18–3.24	3.28	1.12

& Ford, 1965). In general the nonlinear parameters of solids and fluids are of the same order of magnitude (Breazeale, 1993).

As far as transverse waves are concerned, in the quadratic approximation nonlinear terms do not appear at all in (2.15) and the linear equation (2.10) remains valid. Transverse waves appear to be nonlinear only in the next, i.e. the third, order in amplitude.

All these results refer to an isotropic solid. To describe nonlinear effects in anisotropic media (crystals) one faces a much more cumbersome task. Even in the quadratic order we have in general to introduce 216 nonlinear constants instead of three. Nevertheless, it is worth mentioning that with the presence of symmetries this number is greatly reduced. Discussion of more complicated cases, such as those in which the propagation of longitudinal waves is accompanied by transverse deformations, together with consideration of surface waves, lies outside the framework of this book (see, however, Chapter 6).

1.3 Liquid with gas bubbles

Very complex and diverse acoustic properties are inherent in structurally inhomogeneous media. A liquid with gas bubbles is the first subject of our concern. The presence of bubbles significantly changes the sound wave propagation conditions. The resonant nature of bubble pulsations leads to phase velocity dispersion, while the high compressibility of the gas increases the medium nonlinearity. Moreover, additional losses appear in the presence of bubbles.

Nonlinear bubble dynamics. Before going into the study of the peculiarities of sound propagation let us discuss briefly the oscillations of a single bubble in an external pressure field. In the present chapter we confine our discussion to monopole, spherically symmetric pulsations, which play the main role for bubbles that are small compared to the characteristic wavelength (though later we shall see that translational motions of a bubble prove to be rather significant as well).

Due to the great compressibility of a bubble and its small dimensions the surrounding liquid may be considered as incompressible; then the velocity potential φ and the velocity $v = \partial\varphi/\partial r$ of the radial fluid flow are described by a monopole solution of the Laplace equation:

$$\varphi = -\frac{R^2\dot{R}}{r}, \quad v = \frac{R^2\dot{R}}{r^2}, \tag{3.1}$$

where r is the distance from the bubble centre and $R(t)$ is the bubble radius. Substituting into the Bernoulli integral

$$\frac{p}{\rho_0} + \frac{v^2}{2} + \frac{\partial\varphi}{\partial t} = \text{const.}, \tag{3.2}$$

where ρ_0 is the nonperturbed density of the liquid, and recovering the constant from the boundary condition at the infinity ($v \to 0, p \to p_\infty$), we find the pressure perturbations in the liquid as

$$p - p_\infty = \rho_0 \frac{R^2\ddot{R} + 2\dot{R}^2 R}{r} - \rho_0 \frac{\dot{R}^2}{2}\frac{R^4}{r^4}. \tag{3.3}$$

Setting $r = R(t)$ we find the bubble surface pressure, which has to be equal to the gas pressure P_R inside it. As a result we obtain the equation (Rayleigh, 1917)

$$R\ddot{R} + \frac{3}{2}\dot{R}^2 = \frac{1}{\rho_0}[P_R - p_\infty]. \tag{3.4}$$

The dependence of the pressure P_R on the volume (or radius) of the bubble should be defined from thermodynamical considerations. If the volume changes of the bubble are adiabatic, then

$$P_R = p_0 \left(\frac{R_0}{R}\right)^{3\gamma}, \tag{3.5}$$

where R_0 is the bubble radius at pressure P_0.

Substituting this expression into (3.3) we obtain the well-known Rayleigh equation

$$R\ddot{R} + \frac{3}{2}\dot{R}^2 = \frac{1}{\rho_0}\left[p_0 \left(\frac{R_0}{R}\right)^{3\gamma} - p_\infty\right]. \tag{3.6}$$

Linearizing (3.6) and introducing the volume perturbation of a bubble

$V' = V - V_0 = (4/3\pi)(R^3 - R_0^3)$ and adding a small dissipative term, one can easily present the equation of bubble pulsations in the form

$$\ddot{V}' + \omega_0^2 V' + \hat{f} V' = p' \frac{4\pi R_0}{\rho_0}. \qquad (3.7)$$

Here $p' = p_\infty - p_0$, $\hat{f}$ is a linear operator, and

$$\omega_0 = \left(\frac{3\gamma p_0}{\rho_0 R_0^2} \right)^{1/2} \qquad (3.8)$$

is the natural frequency of the bubble. Expression (3.8) is sometimes referred to as the Minnaert formula (Minnaert, 1933). In particular, for air bubbles ($\gamma = 1.4$) in water at atmospheric pressure it follows from (3.8) that $\omega_0 R_0 \simeq 20.5\,\mathrm{m/s}$. So, a bubble with radius 1 mm resonates at a frequency $\omega_0/2\pi = 3.26\,\mathrm{kHz}$, while at $R_0 = 10\,\mu\mathrm{m}$, $\omega_0/2\pi = 326\,\mathrm{kHz}$; for bubble radii of the order of several μm one has to take into account the surface tension.

The dissipative part $\hat{f} V'$ is specified either from the analysis of a particular loss mechanism or from phenomenological considerations; often one simply writes $f\dot{V}'$ where f is a constant factor. This is, however, only a convention of notation as f, generally speaking, depends on frequency. As far as harmonic oscillations at frequency ω are concerned the losses are characterized either by the bubble quality factor Q in accordance with the law of natural oscillation damping ($V' \sim \exp(-\omega_0 t/2Q)$) or by the loss factor $d = \omega/\omega_0 Q$ (defined as the ratio of the dissipative term in (3.7) to the term $\omega_0^2 V'$). For air bubbles in water typical values of Q are close to 10. Generally, Q and d depend on the thermal conductivity of the gas, and on the liquid viscosity and the radiation losses due to acoustic emission. An approximate expression for the loss factor accounting for all these phenomena was developed by Eller (1920) in the form $d = d_T + d_v + d_r$ where

$$d_T = 3(\gamma - 1) \left[\frac{X(\sinh X + \sin X) - 2(\cosh X - \cos X)}{X^2(\cosh X - \cos X) + 3(\gamma - 1)X(\sinh X - \sin X)} \right], \qquad (3.9)$$

where

$$X = R_0 \left(\frac{2\omega}{\chi_T} \right)^{1/2}$$

and χ_T is the temperature conductivity of the gas. Also

$$d_v = \frac{4\omega\eta}{3\gamma p_0}, \qquad (3.10)$$

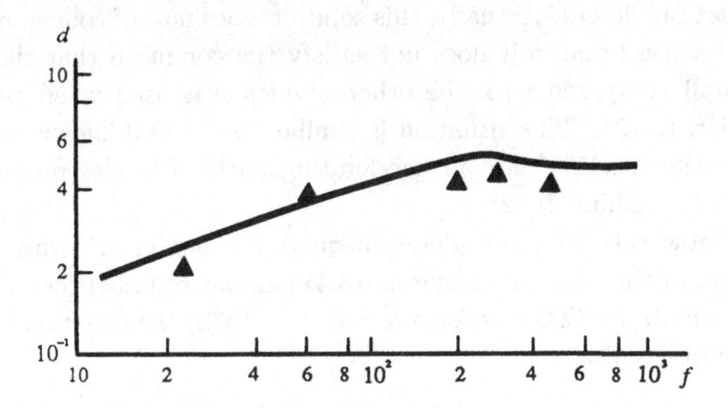

Figure 1.1. Total loss factor frequency dependence for resonant pulsations of an air bubble in water. Triangles denote experimental points.

(η is the viscosity of the liquid), and

$$d_r = \frac{\rho_0 R_0^3 \omega^3}{3\gamma p_0 c_0}, \tag{3.11}$$

where c_0 is the velocity of sound in the liquid.

The dependence of the total loss factor on frequency for a resonantly pulsating bubble in water is shown in Figure 1.1 (Devin, 1959); as is evident from the figure, this dependence offers a good description of the experimental results.

It is implied in (3.9)–(3.11) that the dissipative term in (3.7) actually has different forms depending on the loss mechanism. We demonstrate this using the example of radiation losses, which can be estimated by taking into account the liquid compressibility (finiteness of the sound velocity). In this instance the potential φ is a solution to the wave equation $\varphi = F(t - r/c_0)$. Evaluating $v = \partial\varphi/\partial r$ and substituting the boundary condition $v_{r=R} = \dot{R}$ we find the function F (at small c_0^{-1}):

$$F(t - R/c_0) = R^2\dot{R} - \frac{R^3\ddot{R}}{c_0}. \tag{3.12}$$

Allowing for this expression in (3.2) in the linear approximation it follows that, instead of (3.7),

$$\ddot{V}' + \omega_0^2 V' - \frac{R_0}{c_0}\dddot{V}' = -p'\frac{4\pi R_0}{\rho_0}. \tag{3.13}$$

For harmonic pulsations this leads directly to the result in (3.11) for d.

It is interesting to note that a "spurious" solution of equation (3.12) of the form $V' \sim \exp(c_0 t/R_0)$, rapidly increasing with time, is related to "radiation" losses. Actually, this solution does not, of course, possess physical sense because it does not satisfy the condition that the term $\dddot{V}$ is small compared with the others, which was used when deriving expression (3.12). This situation is similar to the well-known paradox of "radiation friction" for an accelerating particle in electrodynamics (Landau & Lifshitz, 1982).

If we now take into consideration quadratic nonlinear terms in the expansion of the adiabatic equation (3.5) (similar to (1.5)) then instead of (3.7) we obtain (Zabolotskaya & Soluyan, 1972) the equation of nonlinear bubble pulsation:

$$\ddot{V}' + \omega_0^2 V' - 3(\gamma+1)\beta\omega_0^2 V'^2 - \beta\left[2\ddot{V}'\dot{V}' + (\dot{V}')^2\right] + f\dot{V}' = \mathcal{X}p', \quad (3.14)$$

where $\beta = (8\pi R_0^3)^{-1}$, $\mathcal{X} = 4\pi R_0/\rho_0$, $p' = p_0 - p_\infty$.

It can be verified that the left-hand side of equation (3.14) may, to the same approximation, be converted into the standard linear form (3.7) by means of the substitution (Lopatnikov, 1980)

$$S = \left(\frac{V'}{V_0}\right)^{2/5} - 1, \quad \text{where} \quad V_0 = \frac{4}{3}\pi R_0^3. \quad (3.15)$$

The assumption of nearly adiabatic gas volume variation in a bubble used here may be violated at very low frequencies, when the length λ_T of the thermal wave in a bubble is not small compared with its radius, the thermal exchange between the gas and the surrounding liquid then becoming important.

At low frequencies, when $\omega \ll \omega_0$, $R_0/\lambda_T < 1$, $\lambda_T = \sqrt{\frac{2\chi_T}{\omega}}$, the process is nearly isothermal, and is described by the equation of state (Kobelev & Ostrovsky, 1980)

$$VP - V_0P_0 = \frac{2(\gamma-1)}{\sqrt{10}\gamma\omega_T}V_0\frac{dP}{dt}. \quad (3.16)$$

Here $\omega_T = (\sqrt{90}\chi_T)/(R_0^2)$, and V, P are the total bubble volume and pressure, respectively. The right-hand side of equation (3.16) is responsible for the deviation from isothermality.

At higher frequencies (though still much lower than ω_0) when $\omega \ll \omega_0$, but $R_0/\lambda_T \gg 1$, we have the equation of nearly adiabatic pulsations, corrected for thermal exchange:

$$\gamma \dot{V} P + V \dot{P} = 3(\gamma+1)\frac{\chi}{R_0^2} V_0 \left[P - P_0 - \frac{R_0}{\sqrt{\chi}} \int_0^t \frac{\mathrm{d}P}{\mathrm{d}t} \frac{\mathrm{d}\tau}{\sqrt{\pi(t-\tau)}} \right] . \quad (3.17)$$

Making use of (3.16) and (3.17) one can improve the bubble pulsation equation (3.14) or estimate the limits of its applicability.

Waves in a gas–liquid mixture. Now let us go over to the liquid with distributed bubbles. We characterize them by a volume gas content z, which is the fraction of the medium's volume occupied by the bubbles. If all the bubbles have the same volume V, then evidently $z = NV$, where N is the number of bubbles per unit volume. If, however, the bubbles are distributed by size, they are characterized in terms of a distribution function $n(R_0)$ such that $n(R_0)\mathrm{d}R_0$ is the number of bubbles with radii ranging from R_0 to $R_0 + \mathrm{d}R_0$, and therefore

$$z = \int_{R_0 \min}^{R_0 \max} V(R_0) n(R_0) \, \mathrm{d}R_0 . \quad (3.18)$$

Later on we shall always assume that the characteristic length of an acoustic wave λ is large compared with both the bubble radii and the average distance ℓ between the bubbles. In this case the so-called "homogeneous approximation" is valid: the liquid with gas bubbles can be considered as a homogeneous medium with averaged values of density, pressure and other variables. Thus, the average medium density is obviously equal to

$$\rho = \rho_\ell (1 - z) + \rho_g z, \quad (3.19)$$

where ρ_ℓ, ρ_g are the liquid and gas densities, respectively.

In the ensuing discussion we assume that $z \ll 1$, i.e. the condition $R_0 \ll \ell$ is fulfilled, meaning that the bubbles are small compared to the distance between them. Then taking into account that $\rho_g \ll \rho_\ell$, we write down the density perturbations as

$$\rho' = \rho'_\ell (1 - z_0) - \rho_0 z', \quad (3.20)$$

where, as above, "0" means the equilibrium state, while the prime designates perturbation.

Now we can write the hydrodynamic equations for the average density and velocity for a gas–liquid mixture. In this case we neglect the liquid nonlinearity, assuming that the main role is played by the bubble nonlinearity. Then

$$\frac{\partial \mathbf{v}}{\partial t} + \frac{1}{\rho_0}\nabla p' = 0, \tag{3.21a}$$

$$\frac{\partial \rho'}{\partial t} + \rho_0 \nabla \cdot \mathbf{v} = 0. \tag{3.21b}$$

Here ρ_0 is the equilibrium density of a gas–liquid mixture, and is close to the liquid density ρ_0. Substituting equation (3.20) and the linear relation $\rho' = p'/c_0^2$, where p' is the pressure perturbation, we derive for a plane wave

$$\frac{1}{c_0^2}\frac{\partial^2 p'}{\partial t^2} - \frac{\partial^2 p'}{\partial x^2} = \rho_0 \frac{\partial^2 z'}{\partial t^2}. \tag{3.22}$$

Together with the equation of bubble pulsations this describes wave propagation in a two-phase medium when $z \ll 1$.

Consider now in more detail the low-frequency limit, when the characteristic frequency of the field is small compared to the resonance frequency of the bubble, ω_0. In this case, we can omit all terms in (3.14) involving $\dot{V}'$ and $\ddot{V}'$ (including the dissipative one), which results in a quasistatic relation

$$V' = -\frac{4\pi R_0^3 p'}{3\gamma p_0} + \frac{2\pi(\gamma+1)R_0^3 p'^2}{3\gamma^2 p_0^2}. \tag{3.23}$$

With the aid of (3.18) and (3.23) one can find the perturbation of the bubble volume concentration:

$$z' = \int_{R_{0\,\min}}^{R_{0\,\max}} V' n(R_0)\,\mathrm{d}R_0 = z'_\ell + z'_{n\ell}, \tag{3.24a}$$

where

$$z'_\ell = -\frac{4\pi p'}{3\gamma p_0}\int_{R_{0\,\min}}^{R_{0\,\max}} R_0^3 n(R_0)\,\mathrm{d}R_0 = -\frac{z_0 p'}{\gamma p_0}, \tag{3.24b}$$

$$z'_{n\ell} = -\frac{2\pi(\gamma+1)p'^2}{3\gamma^2 p_0^2}\int_{R_{0\,\min}}^{R_{0\,\max}} R_0^3 n(R_0)\,\mathrm{d}R_0 = -\frac{(\gamma+1)z_0}{2\gamma^2 p_0^2}p'^2. \tag{3.24c}$$

Note that

$$z_0 = \frac{4\pi}{3}\int_{R_{0\,\min}}^{R_{0\,\max}} R_0^3 n(R_0)\,\mathrm{d}R_0 \tag{3.25}$$

is the equilibrium concentration of the bubbles.

Substituting these expressions into (3.22) we acquire the equation for pressure in the form

$$\left(\frac{1}{c_0^2} + \frac{z_0\rho_0}{\gamma p_0}\right)\frac{\partial^2 p'}{\partial t^2} - \frac{\partial^2 p'}{\partial x^2} = \frac{(\gamma+1)\rho_0 z_0}{2\gamma^2 p_0^2}\frac{\partial^2 p'^2}{\partial t^2}. \tag{3.26}$$

It follows that in the linear approximation the sound velocity in a gas–liquid mixture is equal to

$$c = \left(\frac{1}{c_0^2} + \frac{1}{c_1^2}\right)^{-1/2}, \quad c_1 = \frac{\gamma p_0}{\rho_0 z_0}. \tag{3.27}$$

Therefore the presence of bubbles decreases the sound velocity at low frequencies. The effect of the bubbles is substantial when $c_1^2 \lesssim c_0^2$ or $z_0 \gtrsim \bar{z}_0 = \gamma p_0/\rho_0 c_0^2$, which is quite consistent with the above condition $z_0 \ll 1$. For air bubbles in water, $\bar{z}_0 \simeq 6 \times 10^{-5}$.

Now we return to the nonlinear case. If the right-hand side of (3.26) is small one can, for a travelling wave, reduce its order again by going over to variables $y = t - x/c$, x, which yields

$$\frac{\partial p'}{\partial x} = Gp'\frac{\partial p'}{\partial y}, \quad G = \frac{(\gamma+1)\rho_0 z_0 c}{2\gamma^2 p_0^2}. \tag{3.28}$$

Let us relax to a certain degree the low-frequency restriction, and consider the term $\ddot{V}'$ in (3.14) (omitting, however, small nonlinear terms involving the derivatives of V). This will lead to the appearance of a term with $\ddot{p}'$ on the right-hand sides of (3.23) and (3.24a) (to be more exact, to the substitution of $p' - \ddot{p}'/\omega_0^2$ for p') and a term with $\partial^4 p'/\partial x^4$ in (3.26), respectively. As a result, the equation of a travelling wave takes the form

$$\frac{\partial p'}{\partial x} - Gp'\frac{\partial p'}{\partial y} - \frac{1}{2c\omega_0^2}\frac{\partial^3 p'}{\partial y^3} = 0. \tag{3.29}$$

Thus, we have arrived at the well-known Korteweg–de Vries equation, which presents a rather universal description of waves with weak nonlinearity and dispersion. If we substitute $p' = c\rho_0 v$ into (3.29) and compare to (1.19) it becomes apparent that the role of the nonlinearity parameter ε in this case is played by

$$\varepsilon = \frac{(\gamma+1)\rho_0^2 z_0 c^4}{2\gamma^2 p_0^2}. \tag{3.30}$$

This value depends nonmonotonically on the concentration: at very low z_0 we have $\varepsilon \sim z_0$, while at comparatively large z_0 (when $c_1^2 \ll c_0^2$)

$\varepsilon \simeq (\gamma+1)/2z_0$, i.e. nonlinearity increases the value of ε by a factor of z_0^{-1} as against the case of a homogeneous liquid. At $z = \bar{z}_0 = \gamma p_0/\rho_0 c_0^2$, ε attains its maximum value (Druzhinin *et al.*, 1975)

$$\varepsilon_{\max} = \frac{(\gamma+1)\rho_0 c_0^2}{8\gamma p_0}. \tag{3.31}$$

As far as air bubbles in water are concerned, $\bar{z}_0$ has the previously estimated value of 6×10^{-5}, so that $\varepsilon_{\max} \simeq 4800$. If $z_0 \ll \bar{z}_0$ then $c \approx c_0$ and (for air bubbles) $\varepsilon \simeq 3\times10^8 z_0$. For $z_0 = 10^{-6}$, say, we obtain $\varepsilon \simeq 300$. Therefore, even at quite small volume concentrations of the bubbles anomalously large nonlinearity parameter values are achieved, as compared with homogeneous media.

In (3.29) the losses have not been considered. At low frequencies the main part most frequently comes from thermal losses, and to describe these one can use (3.16) and (3.17). In the quasi-adiabatic case (and for $z_0 \ll \bar{z}_0$) equation (3.29) holds, with the addition of a relaxation term (Kobelev & Ostrovsky, 1980)

$$\frac{3(\gamma-1)c\rho_0 z_0}{2\rho_0^2 R_0^2}\left[p' - \frac{R_0}{\sqrt{\chi}}\int_0^y \frac{\partial p'}{\partial \tau}\frac{d\tau}{\sqrt{\pi(t-\tau)}}\right]. \tag{3.32}$$

Recently, Sugimoto (1989) has considered in detail the solution of the Burgers equation with a similar relaxation term in application to the theory of sound propagation in a pipe, taking into account wall friction.

For still lower frequencies, when the "quasi-isothermal" formula (3.17) can be applied, the Burgers equation may be deduced:

$$\frac{\partial p'}{\partial x} - 2\delta p'\frac{\partial p'}{\partial y} - \sqrt{\frac{2}{5}}\frac{\gamma-1}{\omega_T}\delta\frac{\partial^2 p'}{\partial y^2} = 0, \tag{3.33}$$

where $\delta = c_T z_0/2p_0$ and c_T is the isothermal sound velocity so that (for $z_0 \ll \bar{z}_0$) $y = t - x/c_T$.

Both situations, isothermal and adiabatic, can be realized within the framework of the low-frequency approximation ($\omega \ll \omega_0$). The boundary between these two cases lies in the range $\omega \sim \omega_T$. For air bubbles with radii $R_0 \simeq 10\,\mu$m in water, we have $\omega_T/2\pi \sim 16$ kHz while the resonance frequency is $\omega_0/2\pi \sim 326$ kHz.

Accordingly, even at comparatively low frequencies a medium with bubbles exhibits rather complex nonlinear and dispersive behaviour. If, however, bubbles distributed over a wide size range are present in a liquid, as is quite natural for real situations, then the medium properties will change significantly due to contributions of resonant bubbles. In addition, a vital role can be attributed to translational motion of the bubbles due to hydrodynamic (averaged) forces appearing in the sound field; new nonlinearity mechanisms are related to this motion. These questions will be tackled in later chapters of the book. It is necessary to notice that the suggestion of the spherical shape of bubbles which was taken into account is valid only at small concentrations of bubbles where one can neglect the processes of interaction. This effect can drastically change the process of pulsation (Vogel *et al.*, 1989).

1.4 Elastic medium with cavities

Using as an example the liquid with gas bubbles, we observed that the presence of microinhomogeneities ("micro" meaning that they are small compared with the wavelength) may lead to a large growth of the nonlinearity parameter even if their volume content is small. It is interesting to discuss the possibilities of similar effects in other media.

Consider briefly an isotropic solid medium with spherical cavities. Such a medium is characterized by the presence of transverse stresses connected with a nonzero shear modulus. This allows us to take into consideration empty cavities containing no gas – the subject of the opening discussion in this section. To make allowance for the gas filling the voids seems to be a simple matter.

Cavity pulsations. Let us begin with the free vibrations of a spherically symmetric cavity. First of all it is helpful to recall a well-known linear problem (Landau & Lifschitz, 1987) treating it, however, in a slightly different way.

In the linear approximation the potential ψ of displacement (but not that of velocity) defined from $\mathbf{u} = \nabla\psi$, where $\mathbf{u}$ is the displacement vector for longitudinal motions, satisfies the equation (cf. (2.29))

$$\psi_{tt} = c_\ell^2 \nabla^2 \psi, \tag{4.1}$$

where $c_\ell = \sqrt{(\lambda + 2\mu)/\rho}$ is the longitudinal wave velocity. The spherically symmetric (retarded potential) solution to (4.1) has the form

$$\psi = -\frac{F(t - r/c_\ell)}{r}, \tag{4.2}$$

where F is an arbitrary function consistent with the boundary condition

$$\sigma_{rr}|_{r=R_0} = 0. \tag{4.3}$$

R_0 is the unperturbed cavity radius and σ_{rr} is the radial component of the stress tensor which, in the linear approximation, is equal to

$$\sigma_{rr} = \lambda \nabla^2 \psi + 2\mu \psi_{rr} = \rho \psi_{tt} - 4\mu \psi_r / r \tag{4.4}$$

(here the wave equation (4.1) has been taken into account). Substituting (4.2) for ψ we consider a small cavity, i.e. we assume $kR_0 \ll 1$ where k is the characteristic wave number of the radiated longitudinal wave. Then, for the near-field zone ($kr \ll 1$) one can let $c_\ell \to \infty$ and hence (4.3) yields

$$\ddot{F} + \omega_0^2 F = 0, \tag{4.5}$$

where

$$\omega_0 = \frac{2}{R_0}\sqrt{\frac{\mu}{\rho}} \tag{4.6}$$

is the resonance frequency of the cavity. This approximation is similar to that of an incompressible liquid in the problem of the bubble vibrations. In the next approximation (in c_ℓ^{-1}) we derive from (4.2)

$$\psi = -\frac{F(t)}{r} + \frac{\dot{F}(t)}{c_\ell},$$

and

$$\ddot{F} + \omega_0^2 F - \frac{R_0}{c_\ell}\dddot{F} = 0. \tag{4.7}$$

Since the displacement is $u = \partial\psi/\partial r$, the cavity volume perturbation is equal to

$$V = 4\pi R_0^2 \frac{\partial \psi(R_0)}{\partial r} = 4\pi \left(F + \frac{R_0}{c_\ell}\dot{F} \right)$$

or, to the same degree of accuracy,

$$F = \frac{1}{4\pi}\left(V - \frac{R_0}{c_\ell}\dot{V} \right), \tag{4.8}$$

Substituting this into (4.7) and neglecting terms of order c_ℓ^{-2} we obtain the equation of elastic cavity pulsation:

$$\ddot{V} + \omega_0^2 V - \frac{R_0}{c_\ell}\dddot{V} = 0. \tag{4.9}$$

In analogy with the case of a bubble the last term in (4.7) describes the radiation losses, which are small only if $\omega_0 R_0 / c_\ell \ll 1$ or $c_t \ll c_\ell$, or $\mu \ll \lambda$ ($c_t = (\mu/\rho)^{1/2}$ is the velocity of a transverse wave). However, as will soon be demonstrated, only media satisfying these conditions may have strong nonlinearity. It is these weakly compressible media, also referred to as water- or rubber-like media, which are to be considered later in this section.

Meanwhile it should be noted that, as in the case of a bubble, the losses are described by a "radiation" term $\dddot{V}$.

Now let us turn our attention to the nonlinear problem. Suppose first that the medium is incompressible, which corresponds to $\lambda \to \infty$. Again make use of the Lagrangian coordinate r, and then $u = X - r$ where X is the physical location of a particle. With respect to incompressible media the five-constant theory cannot be applied because the elastic energy density E depends only on two invariants of the strain tensor (Lur'ye, 1980)

$$I_1 = \sum_{i=1}^{3} \nu_i^2 = \frac{r^4}{X^4} + 2\frac{X^2}{r^2}, \quad I_2 = \sum_{i=1}^{3} \nu_i^{-2} = \frac{X^4}{r^4} + 2\frac{r^2}{X^2}. \tag{4.10}$$

Here, ν_i are the elongations per unit length along the principal axes.

The equation for the vibrations of the spheres may be derived in the following way (Lur'ye, 1980). Introduce the variable $A(t) = X^3 - r^3 = 3V'/4\pi$, where V' is the cavity volume change. Due to incompressibility $A(t)$ does not depend on r: the particle displacements are such that the volume changes are the same at any radius. The particle velocity is $v = \dot{X} = \dot{A}/3X^2$, while the total kinetic energy of the particles in the medium is

$$K = \frac{1}{2}\rho \int_R^\infty v^2 \, \mathrm{d}v = \frac{1}{18}\rho A^2 4\pi \int_R^\infty \frac{\mathrm{d}X}{X^2} = \frac{2\pi}{9R}\rho \dot{A}^2, \tag{4.11}$$

where $R(t)$ is the cavity radius.

The potential energy of deformation in the Lagrangian variables is equal to

$$\Pi(A) = 4\pi \int_R^\infty E(I_1, I_2) r^2 \, \mathrm{d}r. \tag{4.12}$$

Now one can use the equation for the Lagrangian $\mathcal{L} = K - \Pi$,

$$\frac{\partial}{\partial t}\frac{\partial \mathcal{L}}{\partial A} - \frac{\partial \mathcal{L}}{\partial A} = 0, \tag{4.13}$$

and further, in accordance with (4.12),

$$\begin{aligned} \frac{\partial \Pi}{\partial A} &= 4\pi \int_R^\infty \left(\frac{\partial E}{\partial I_1}\frac{\partial I_1}{\partial A} + \frac{\partial E}{\partial I_2}\frac{\partial I_2}{\partial A} \right) r^2 \, \mathrm{d}r \\ &= \frac{16\pi}{3} A \int_R^\infty \left(\frac{\partial E}{\partial I_1} + \frac{X^2}{r^2}\frac{\partial E}{\partial I_2} \right) (X^3 + r^3) \frac{\mathrm{d}r}{X^7}. \end{aligned} \tag{4.14a}$$

The latter expression takes into account the relation (4.10) as well as the equalities $X = (A + r^3)^{1/3}$, $\partial X/\partial A = (3X^2)^{-1}$.

Then, the substitution of (4.11) and (4.14*a*) into (4.13) results in

$$\rho\left(\frac{\ddot{A}}{R} - \frac{\dot{A}^2}{6R^4}\right) + 12A\int_R^\infty \frac{\mathrm{d}r}{X^7}(X^3 + r^3)\left(\frac{\partial E}{\partial I_1} + \frac{X^2}{r^2}\frac{\partial E}{\partial I_2}\right) = 0. \quad (4.14b)$$

What follows depends on the form of the function $E(I_1, I_2)$. For incompressible media there exist various empirical approximations to this formula; we choose one of the most frequently used, i.e. the Muni potential

$$E = \frac{\mu}{4}\left[(1+\beta)(I_1 - 3) + (1-\beta)(I_2 - 3)\right], \quad (4.15)$$

where μ is the shear modulus, β being the second constant, such that $0 \le \beta \le 1$. Traditionally, we consider weakly nonlinear vibrations, when $A \ll r^3$. Then, in the quadratic approximation with respect to A, we obtain from (4.14*b*) and (4.15)

$$\rho\left(\frac{\ddot{A}}{R} - \frac{\dot{A}^2}{6R^4}\right) + \frac{4\mu}{R_0^3}A - \frac{(9+2\beta)\mu}{3R_0^6}A^2 = 0. \quad (4.16)$$

Since $R(t) \simeq R_0(1 + A/3R_0^3)$ then, in terms of the volume changes, we can write (Ostrovsky, 1991)

$$\ddot{V}' + \omega_0^2 V' - \frac{(9+2\beta)\omega_0^2}{16\pi R_0^3}V' - \frac{1}{8\pi R_0^3}(2V'\ddot{V}' + \dot{V}'^2) = 0. \quad (4.17)$$

If we now consider the small but finite compressibility of the medium then, according to (4.9), a term $(R_0/c_0)\dddot{V}$ should be added on the left-hand side of (4.17).

In order to make the transition to wave problems we propose to discuss the cavity behaviour in an external pressure field. Given the pressure P_R inside the cavity, and P_∞ at infinity, the term $3(P_R - P_\infty)$ appears on the right-hand side of (4.14*b*) and (4.16), while the right-hand side of (4.17) will contain

$$\frac{4\pi R_0}{\rho}(P_R - P_\infty).$$

Let $P_0 = 0$ (empty cavity); a generalization for the case of gas filled pore is of no difficulty (see Ostrovsky, 1991); and let P_∞ be determined by a sufficiently long wave (with wave number $k \ll r_0^{-1}$). Then the equation of cavity pulsations in an external acoustic field can be easily written in a form similar to (3.14) for a bubble:

$$\ddot{V}' + \omega_0^2 V' - \frac{r_0}{\omega}\dddot{V}' - GV'^2 - \beta_0(2V'\ddot{V}' + \dot{V}'^2) = -\chi P_\infty, \tag{4.18a}$$

where

$$G = \frac{(q+2\beta)\omega_0^2}{12\pi r_0^3}, \quad \beta_0 = \frac{1}{8\pi r_0^3}, \quad \chi = \frac{4\pi r_0}{\rho_0}. \tag{4.18b}$$

Waves in the porous medium. Let us now consider longitudinal waves in an elastic medium with pores (Ostrovsky, 1991). Under the same assumptions as for bubbles (many pores on the wavelength scale but small volume content of them) a one-dimensional wave can be described by an equation

$$\rho_o \frac{\partial^2 u}{\partial t^2} = \frac{\partial \sigma}{\partial x}, \tag{4.19}$$

where u is the longitudinal displacement and $\sigma = \sigma_{xx}$ the average stress.

Deformation of a medium layer is contributed by both the solid phase and the pores. Hence,

$$\frac{\partial u}{\partial x} = \frac{\sigma}{\lambda + z'}, \tag{4.20}$$

where z' is the perturbation of the total pore volume.

For the quasistatic (low-frequency) limit, using (4.18*a*) we readily obtain expressions for z'_ℓ and $z'_{n\ell}$ analogous to (3.24) and, as result, the "nonlinear Hooke's law" for a porous medium,

$$\frac{\sigma}{\lambda} = \frac{s}{1+g_1} - \frac{g_2 s^2}{(1+g_1)^3}. \tag{4.21}$$

Here $s = \partial u/\partial x$ is the strain tensor component and parameters $g_{1,2}$ are

$$g_1 = \frac{3\lambda z_0}{4\mu}, g_2 = \frac{3\lambda^2(q+2\beta)z_0}{64\mu^2}. \tag{4.22}$$

As a result, from (4.19) we get wave equations for u and s analogous to (2.16) and (2.17) of Section 1.2, where now the sound velocity c_ℓ is to be changed to $c_1 = c_\ell(1+g_1)^{-1/2}$, while for the nonlinearity parameter we have

$$\varepsilon = \frac{g_2}{(1+g_1)^2}. \tag{4.23}$$

This value has a maximum at some optimal pore concentration. In the case $\beta = 1$ (so-called Treloar potential) these values are equal to

$$\varepsilon_{max} = \frac{11\lambda}{64\mu}, \bar{z}_0 = \frac{4\mu}{3\lambda}, \tag{4.24}$$

which gives the optimal pore volume. Analogous consideration has been performed for a medium with parallel cylindrical pores (channels) (Ostrovsky, 1989). For low frequencies the results turn out to be almost the same as above, yielding, instead of (4.22) and (4.24), the expressions

$$g_1 = \frac{\lambda \bar{S}_0}{\mu}, g_2 = \frac{3\bar{S}_0}{2}\left(\frac{\lambda}{\mu}\right)^2 \tag{4.25}$$

and

$$\varepsilon_{max} = \frac{3\lambda}{8\mu}, \bar{S}_0 = \frac{\mu}{\lambda}, \tag{4.26}$$

where $\bar{S}_0$ is the total fraction of channels (or, which is the same, the porosity); note that there is no dependence on parameter β at all in this case.

In the "moderate frequency" approximation the Korteweg-de Vries equation similar to (3.29) can be written down for spherical pores, but not for cylindrical ones which yield a more complicated integral term in the same approximation.

Recently, a series of experiments has been performed with such rubberlike media as gelatine ($\lambda/\mu = 2.10^4$) and plastisole ($\lambda/\mu \simeq 2.10^5$) (Belyaeva & Timanin, 1991), both with spherical and cylindrical pores. Observed values of nonlinearity parameter were up to 10^4 and more, even at concentrations far from the optimal ones.

1.5 "Anomalous" nonlinearities in elastic media

In Section 2 we discussed isotropic elastic solids, which are described (in the quadratic approximation) by a "five-constant theory". However, for various elastic materials much more complicated nonlinear characteristics have been observed experimentally; and further the nonlinearity has proved to be anomalously strong (in the acoustic sense). The detailed investigation of the micromechanisms responsible for such anomalies is beyond the scope of this book; moreover, these mechanisms are frequently not altogether clear. As a rule, they are related to inhomogeneities in the material structure such as dislocations, microcracks, grains, etc.; in fact, the porous medium considered above can also be considered as belonging to this group.

Nonlinear characteristics of such media, as of an elastic medium as a whole, can be described by a relation between the stress tensor σ_{ik} and the strain tensor u_{ik}. For a plane longitudinal wave in an isotropic medium, the stress and deformation are characterized by functions $\sigma =$

σ_{xx} and $s = \partial u/\partial x$, where u is the x component of the displacement vector, i.e. the "nonlinear Hooke law" may be introduced. Then the wave equation (4.22) holds for the displacements.

In the classical theory of elasticity the relation $\sigma = \sigma(s)$ is formulated as a power series in s. For media with anomalous nonlinearity, these relations may appear rather complicated and diverse. Most intensively examined are the cases of amplitude-dependent friction: when a medium is subject to a sufficiently strong oscillating field, anomalously high energy losses are observed. Manifestations of an enhanced elastic nonlinearity (i.e. Young's modulus variations with the deformation amplitude) are also encountered. In this case, these parameters are frequently characterized by pronounced inertial behaviour as well, due to relaxation in the material. All such materials are known to show to a certain extent the hysteretic behaviour related to the irreversible motion of defects, i.e. to microplastic deformations. The effects of amplitude-dependent friction have been analysed over a wide range of frequencies ($1 - 10^6$ Hz), temperatures ($300 - 3800$°K) and deformation amplitudes ($s \sim 10^{-6} - 10^{-2}$) (Krishtal & Golovin, 1976; Rabotnov, 1987). At comparatively low frequencies such materials can be characterized by a quasistatic hysteresis curve $\sigma = \sigma(s)$ that is independent of frequency. A family of models concerning the medium inhomogeneities leading to the dependences mentioned has been discussed in recent decades. A well-known one is the Granato–Lücke string model (Granato & Lücke, 1956) which relies on consideration of the successive separation of the dislocation from the attachment points. This model leads to a dependence of the type depicted in Figure 1.2 (a). More complicated models allowing for random distribution of the point defects along dislocation lines lead to the following formulae for the internal friction coefficients and the relative variation of Young's modulus E as a function of the oscillating deformation amplitude s_0 (Hiki, 1959):

$$\delta \sim \frac{\Delta E}{E} = \frac{c_0}{s_0^{1/2}} \exp\left(-\frac{c_1}{s_0}\right), \tag{5.1}$$

where c_0 and c_1 are constants expressed in terms of the coupling force between the point defects and the dislocation and an average distance between such defects along the dislocation line. These laws have been verified experimentally.

In analytical calculations it is customary to use power series approximations with respect to each segment of a hysteresis curve $\sigma = \sigma(s)$ (Pisarenko, 1970); examples of this are given in Figure 1.2. Therefore,

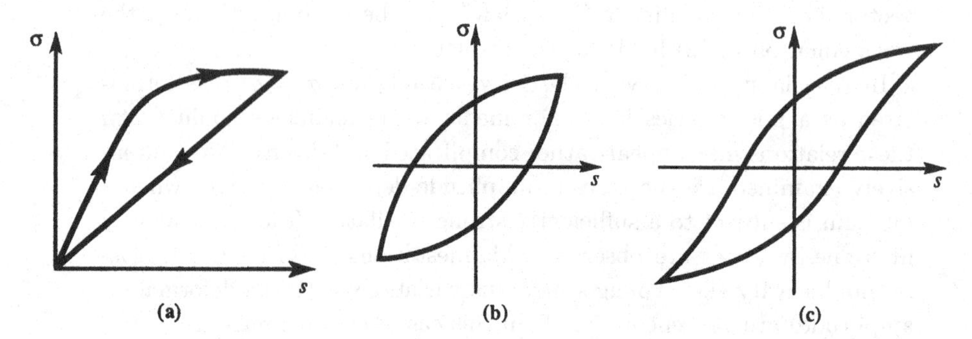

Figure 1.2. Histeresis stress–strain relation.

deformations are considered to be either reversible (as in Figure 1.2 (a)), or irreversible (as in Figure 1.2 (b, c)). One should treat as an important peculiarity the relative symmetry of the material response to compression and tension: the relation $\sigma = \sigma(s)$ may almost have odd symmetry. Here, "cubic" effects such as third harmonic generation and nonlinear frequency shift can even dominate over the "quadratic" ones, such as the second harmonic generation, which are always more pronounced for longitudinal waves within the framework of the classical five-constant theory of elasticity (Groshkov *et al.*, 1991; Nazarov, 1991).

It is worthwhile mentioning one more type of anomalously nonlinear medium: the bipolar or bilinear medium. The relation $\sigma = \sigma(s)$ for such a medium can be viewed as piecewise linear,

$$\sigma = \begin{cases} a_1 s, & s \geq 0 \\ a_2 s, & s < 0, \end{cases} \tag{5.2}$$

where a_1 and a_2 are constants (Figure 1.3). Under this relation a medium responds differently to equivalent compressions and tensions.

A detailed consideration of the "bimodular elasticity theory" models can be found in the book by Ambartsumyan (1982). Judging from the data in this book, a bimodular response is typical of a rather wide range of materials such as polymers, metals, composite systems, etc. In this case the Young's modulus under compression E^+ can be either larger or smaller than that under tension, E^-, and the differences may grow large. These data are argued against elsewhere; thus, according to Zhukov (1985), the ratio $E^+/E^- = 1$ is correct to within a few per cent (but even this level of nonlinearity is admittedly rather high for acoustics). Bimodular features are supposedly observed in earth sciences

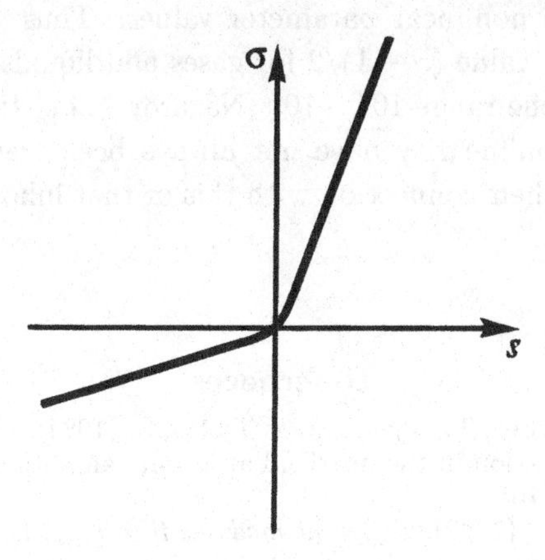

Figure 1.3. Stress–strain relation for a bilinear medium.

(Nikolaev, 1967; Myasnikov & Topalo, 1987). There is some evidence (Nelson, Blackstock, 1990) that porous absorbing materials acquire such properties at higher sound field levels.

The origin of such bimodular behaviour does not seem to be altogether clear. The following simple model (Antonets *et al.*, 1986) may be useful to gain more understanding of the above phenomenon. An elastic specimen has a slot-like defect opening in response to tension, thereby reducing the tensile rigidity of the medium compared with that under compression. It is not impossible that microdefects of this type can be responsible for bimodular characteristics of the real material.

It is interesting to note that under the model (5.2) the nonlinearity does not depend on the oscillation amplitude, being controlled by the "kink angle" alone, i.e. the ratio a_1/a_2.

Some different models of structurally inhomogeneous media can be based on a simplified representation of the medium as a chain of masses coupled by elastic forces, or else in the form of granules in contact ("grains"); the latter model will be considered in Chapter 6. More complicated models are also encountered, such as the well-known Cosserat continuum consisting of spherical particles, each sphere being characterized by both centre displacement and rotation, or the kinetic models of microcrack development leading to rather complicated elasto-plastic properties (proposed by Belyaev & Naimark in 1987). Some impulsive load processes were examined for such media, although the "nonlinear-acoustic" aspects of their theory have not been tackled as yet.

It seems important that real, natural media are sometimes characterized by very high nonlinear parameter values. Thus, the parameter ε (equivalent to the value $(\gamma + 1)/2$ for gases and liquids) in various soils attains values in the range $10^2 - 10^4$ (Nazarov *et al.*, 1988). The causes of such a high nonlinearity have not always been traceable, although one may suggest their connection with this or that inhomogeneity of the medium structure.

References

Aanonsen, S.L., Barkve, T., Tjotta, J. & Tjotta, S. (1984). Distortion and harmonic generation in the nearfield at a finite amplitude sound beam, *JASA* **75,** 744–768.

Ambartsumyan, S.A. (1982). *Different modulus theory of elasticity, p. 317* (Moscow, Nauka).

Andreev, N.N. (1955). On some values of the second order in acoustics, *Akust.Zh.* **1,** 3–11.

Antonets, V.A., Donskoy, D.M. & Sutin, A.M. (1986). Nonlinear vibrodiagnostics of enfoliation and unstikness in multilayered constractions, *Mekhanika compozit. materialov.* **5,** 936–937.

Belyaev, V.V., Naimark, O.B. (1987). Kinetical transfer in media with microcracks and destruction metals in strain waves, *PMTF* **1,** 163–171.

Belyaeva, I. Yu & Timam, E.M. (1991). Experimental investigation of nonlinear propogation in porous elastic media, *Sov. Phys. Acoust.* **37,** 1026–1029.

Bergman, L. (1942). *Der Ultraschall VIII* (Verlag C.M.B.H., Berlin).

Beyer, R.T. (1974). *Nonlinear acoustics, p. 405* (USA Naval Ship Syst. Command.,).

Breazeale, M.A. & Ford, I. (1965). Ultrasonic studies of the nonlinear behaviour of solids, *J. Appl. Phys.* **36,** 3486–3490.

Breazeale, M.A. (1993). Comparison of the nonlinear behavior of fluids and solids, *Proc. 13 ISNA, World Scientific, Singapore, New Jersey, London, Hong Kong* , 451.

Devin, C. (1959). Survey of thermal, radiation and viscous damping of pulsating air bubbles in water, *JASA* **31,** 1654–1667.

Donskoy, D.M. & Sutin, A.M. (1984). Nonlinear scattering and propagation longitudinal acoustic waves in a porous medium, *Akust. Zh.* **30,** 605–611.

Druzinin, G.A., Kryachko, V.M. & Tokman, A.S. (1975). Nonlinear phenomena at propagation of acoustic waves in porous medium, *Tes. dokl. V1 Mezd. Symp. po nelineyn. akust. M., Izd, MGU* , 166–173.

Dunn, F., Law, W. & Frizzel, L. (1984). The nonlinearity parameter B/A of biological media, *Proc. X Intern. Symp. Nonlin. Acoust. Kobe.* , 221–224.

Eller, A.J. & Flynn, H.G. (1969). Generation of subharmonics of order one-half by bubbles, *JASA* **46,** 722–727.

Kobelev, Y.A. & Ostrovsky, L.A. (1980). Models of gas liquids mixture as a nonlinear dispersive medium, in *Nonlinear Acoustics,* ed. V. Zverev & L. Ostrovsky (IAP, Gorky).

Koshkin, M.A. & Schirkevich, M.G. (1976). *Handbook of elementary physics, p. 255* (Moscow, Nauka).

Krishtal, M.A. & Golovin, S.A. (1976). Internal friction and a structure of metals, *M. Metallurgia* , 375.

Landau, L.D. & Lifshitz, E.M. (1982). *Electrodynamics of continuous media* (Moscow, Nauka).

Landau, L.D. & Lifshitz, E.M. (1986). *Hydrodynamics* (Moscow, Nauka).

Landau, L.D. & Lifshitz, E.M. (1987). *Theory of Elasticity* (Moscow, Nauka).

Law, W.K., Frizzel, L.A. & Dunn (1985). Determination of the nonlinearity parameter B/A of biological media, in *Ultrasound in Medicine and Biol.*, ed. v. 11 (307–318,).

Lopatnikov, S.L. (1980). Acoustical phase echo in a liquid with gas bubbles, *Pisma v JTF* **6, No. 10,** 623–626.

Lur'ye, A.J. (1980). *Nonlinear elasticity theory, p. 512* (Moscow, Nauka).

Minnaert, M. (1933). On musical air-bubbles and the sound of running water, *Phil. Mag.* **16,** 235–248.

Myasnikov, V.P. & Topalo, V.I. (1987). Modeling of seismic anisotropy in the lithosphere as a different moduli elastic body, *Izv. AN SSSR-Fizika Zemli.* **5,** 22–30.

Naugol'nykh, K.A. (1971). Absorption of finite amplitude waves, in *High Intensity Ultrasonic Fields,* ed. L. Rosenberg (Plenum Press, New York).

Nazarov, V.E. (1991). Effect of the structure of copper on its acoustic nonlinearity, *Fizika metallov-metallovedenie* **3,** 171–178.

Nazarov, V.E., Ostrovsky, L.A., Soustova, I.A. & Sutin, A.M. (1988). Nonlinear acoustics of micro-inhomogeneous media, *Phys. Earth and Planet Inter.* **50,** 67–73.

Nikolaev, A.V. (1967). Seismic properties of a crumbly medium, *Izv. AN SSSR Fizika Zemli* **2,** 25–31.

Ostrovsky, L.A. (1988). On the nonlinear acoustics of weakly compressible porous media, *Akust.Zh.* **34,** 908–913.

Ostrovsky, L.A. (1991). Wave processes in media with strong acoustic nonlinearity, *JASA* **90,** 3332–3338.

Pisarenko, G.S. (1970). Oscillation of mechanical systems made of imperfect materials, *Kiev. Naukova Dumka* , 379.

Rabotnov, Y.N. (1987). *Introduction to the mechanics of destructions, p. 80* (Moscow, Nauka).

Rayleigh, Lord (1917). On the pressure developed in a liquid during the collapse of a spherical cavity, *Phil. Mag.* **34,** 94–98.

Rudenko, O.V. & Soluyan, S.I. (1977). *Theoretical fundamentals of nonlinear acoustics* (Plenum Press, New York).

Sugimoto, N. (1989). "Generalized" Burgers equation and fractional analyses, in *Nonlinear Wave Motion,* ed. A. Jeffrey (Longman Sci. and Tech.,).

Thurston, R. (1966). Wave propagation in fluids and normal solids, in *Physical acoustics V.1,* ed. W. Mason (Academic Press, New York).

Vogel, A., Lauterborn, W. & Timm, R. (1989). Optical and acoustic investigations of the dynamics of laser-produced cavitation bubbles near a solid boundary, *J. Fluid Mech.* **206,** 298–338.

Westervelt, P.J. (1963). Parametric acoustic array, *JASA* **35,** 935–937.

Whitham, J. (1974). *Linear and nonlinear waves* (Wiley–Interscience, New York).

Zabolotskaya, E.A. & Soluyan, S.I. (1972). Radiation of harmonics and combination frequencies by air bubbles, *Akust. Zh.* **18,** 472–474.

Zhukov, A.M. (1985). Elasticity of materials of dilation and compression, *PMTF* **No. 4,** 128–131.

2

Simple waves and shocks in acoustics

2.1 Nonlinear effects in the presence of dissipation and dispersion

As an intense wave propagates, its profile is subject to evolution due to differences of the travel velocities of its various portions. For liquids and gases, points corresponding to a higher pressure travel faster, so that the wave steepness rises over compression portions, eventually giving rise to shock waves. As a result, periodic perturbations transform into a wave with a sawtooth profile, and an isolated pulse into a wave of triangular shape. The process of front steepening is counteracted by dissipative factors – viscosity and thermal conductivity – as well as propagation speed dispersion. These factors lead to smoothing of the wave profile. Correspondingly, the spectral composition of the wave undergoes a change: front steepening gives rise to high-frequency harmonic generation, while dissipation results in faster damping of the wave spectrum high-frequency components. Thus, the wave evolution is determined by the competing factors of nonlinearity, dissipation and dispersion.

To begin with let us estimate the relative levels of these factors in a plane wave and clarify conditions under which the nonlinearity would lead to pronounced nonlinear distortions of a profile. For this purpose we consider an example of a combined evolution equation (Korteweg–de Vries–Burgers equation, or KdVB), which allows for all three mechanisms – nonlinearity, dissipation and dispersion:

$$\frac{\partial v}{\partial x} - \alpha v \frac{\partial v}{\partial y} - \delta \frac{\partial^2 v}{\partial y^2} + \gamma \frac{\partial^3 v}{\partial y^3} = 0. \tag{1.1}$$

This equation is applicable to numerous physical situations (including acoustical ones – see Chapter 1), but we are, as yet, interested only in the general properties of its solutions.

Prescribe a boundary condition in the form of a harmonic field: $v(0,t) = v_0 \sin \omega t$, and seek a solution of (1.1) using a perturbation method, writing it as $v = v^{(1)} + v^{(2)}$ where

$$v^{(1)} = v_0 \mathrm{e}^{-ax} \sin(\omega y + px) \tag{1.2}$$

is an exact solution of the linearized (i.e. $\alpha = 0$) equation (1.1). Here $a = \delta\omega^2$, $p = \gamma\omega^3$. The correction term $v^{(2)}$ is considered to be small. Substituting (1.2) into (1.1) and assuming that in the nonlinear term we can put $v = v^{(1)}$, we obtain an inhomogeneous linear equation for $v^{(2)}$:

$$\frac{\partial v^{(2)}}{\partial x} - \delta \frac{\partial^2 v^{(2)}}{\partial y^2} + \gamma \frac{\partial^3 v^{(2)}}{\partial y^3} = \alpha \frac{v_0^2 \omega}{2} \mathrm{e}^{-ax} \sin 2(\omega y + px), \tag{1.3}$$

which, in the case considered, has the solution

$$v^{(2)} = \frac{\alpha\omega v_0^2}{4(a^2+p^2)} \Big\{ a \left[\mathrm{e}^{-2ax} \sin 2(\omega y + px) - \mathrm{e}^{-4ax} \sin 2(\omega y + 4px) \right] \\ + p \left[\mathrm{e}^{-2ax} \cos 2(\omega y + px) - \mathrm{e}^{-4ax} \cos 2(\omega y + 4px) \right] \Big\}. \tag{1.4}$$

Therefore, nonlinearity in this approximation gives rise to a small second harmonic. Let us consider two limiting cases so as to avoid cumbersome general expressions:

$\delta = 0$ (no dissipation). Then it is easy to deduce from (1.4) that

$$v^{(2)} = \frac{\alpha v_0^2}{2\gamma\omega^2} \sin 3px \sin(2\omega y + 5px). \tag{1.5}$$

Hence it is clear that dispersion suppresses nonlinear distortions: the second harmonic amplitude $v^{(2)}$ is inversely proportional to the dispersion parameter γ. In addition, the amplitude of $v^{(2)}$ oscillates with x due to interference between the "forced" waves (travelling with the velocity of the fundamental wave) and the natural free waves.

$\gamma = 0$ (no dispersion). In this case the solution has the form (Goldberg, 1957)

$$v^{(2)} = \frac{\alpha v_0^2}{2\delta\omega} \left(\mathrm{e}^{-2ax} - \mathrm{e}^{-4ax} \right) \sin 2\omega y. \tag{1.6}$$

In this instance the second harmonic amplitude grows initially to attain its maximum at $x = \ln 2/2a$, and then eventually to decay exponentially. The maximum ratio of the amplitudes $v^{(2)}$ and $v^{(1)}$ equals $Re/8$, where

$Re = \alpha v_0/\omega\delta$. Because of this, when $Re \ll 1$ this ratio is small everywhere: wave decay is accomplished prior to possible development of nonlinear effects. If, however, $Re \gg 1$, the method of perturbations yields an adequate solution only at short distances, namely, for $ax \ll Re^{-1}$. In this case (1.6) gives $v^{(2)} = \alpha v_0^2 \omega x \sin 2\omega y$. To obtain a more detailed description one needs to construct a solution, taking into account the evolution of the whole wave spectrum; this will be performed later in this chapter.

In a "dispersive" solution (1.5) the amplitude ratio $v^{(2)}/v_0$ is equal to $\alpha v_0/2\omega^2 = U$ (the so-called Ursell parameter), and if U is large, then, with the growth of x, the solution (1.5) rapidly loses its applicability, and the need for a more general approach arises again. As a result nonlinear periodic waves or solitary pulses – solitons – emerge. These questions are briefly addressed in Chapter 6.

In the general case, the second harmonic field decays in an oscillating manner, depending on the ratio of parameters δ and γ.

This simple example demonstrates that a kind of competition goes on between nonlinearity on the one hand and dissipation and dispersion on the other. Strong nonlinear distortions of a wave take place provided the corresponding parameters Re and U are large enough. In all cases nonlinearity (i.e. the Mach number value) always remains small in a local sense. It is these cases that we are going to treat later. We begin with the classical Burgers equation obtained in the preceding chapter, then we handle various cases of formation and evolution of "sawtooth" and "triangular" waves that prove especially typical of nonlinear acoustics. After that we shall proceed to more complicated instances of random signal propagation, generation of nonlinear sound, as well as to a discussion of wave propagation in "anomalous" media where the shock wave evolution laws can differ substantially from those in the classical case, from the point of view of acoustics. The last section deals with one-dimensional waves in confined systems, i.e. resonators.

2.2 Simple waves and discontinuities

As was shown in the previous chapter, the propagation of a plane sound wave in a nondispersive medium with a small quadratic nonlinearity can be described by the Burgers equation (see (1.29) of Chapter 1)

$$\frac{\partial v}{\partial x} - \alpha v \frac{\partial v}{\partial y} = \delta \frac{\partial^2 v}{\partial y^2}. \tag{2.1}$$

As is shown in Sec. 1, the qualitative behaviour of its solutions depends on the value of the acoustic Reynolds number

$$Re = \frac{\alpha v_0}{\omega \delta}, \tag{2.2}$$

where v_0 is the particle velocity perturbation amplitude and ω is the characteristic wave frequency; parameters α and δ describe nonlinear and dissipative properties of the medium. Provided $Re \ll 1$ the nonlinear term in (2.1) proves small, and the main role is played by dissipation: the wave decays before nonlinear effects develop in it, so that we need not go beyond the linear acoustics framework. If $Re \gg 1$ then, at the initial stage of the process, the dissipative term in the Burgers equation can be neglected, and we are left with the following equation:

$$\frac{\partial v}{\partial x} = \alpha v \frac{\partial v}{\partial y}, \tag{2.3}$$

which has a solution in the form of a simple wave (Riemann solution)

$$v = F(y + \alpha v x), \tag{2.4a}$$

where F is an arbitrary function defined by the boundary condition $v(0, t) = F(t)$. This solution is equivalent to a characteristic equation

$$y + \alpha v x = \varphi(v), \tag{2.4b}$$

where φ is the function inverse to F.

As $y = t - x/c_0$ and $\alpha = \varepsilon/c_0^2$, then, as is quite apparent, in terms of variables x, t the local propagation speed equals

$$c = c_0 + \varepsilon v. \tag{2.5}$$

The properties of the solution (2.4*b*) are well known (see, for example, Landau & Lifshitz, 1986): each point of the wave profile travels with its own constant speed, which depends on the value of v at this point. As a result, the trailing edge rarefaction wave (where $\partial v/\partial y < 0$) smooths out and the front edge compression wave (where $\partial v/\partial y > 0$) steepens, whilst the total profile area remains unchanged. At a certain point characterized by values x_*, y_*, v_* an "overturning" occurs, after which the wave profile becomes multivalued (Figure 2.1). These values can be deduced from the conditions $\partial x/\partial v = 0$, $\partial y/\partial v = 0$ (meaning that the profile steepness is infinite) and $\partial^2 x/\partial v^2 = 0$, $\partial^2 y/\partial v^2 = 0$ (the point (x_*, y_*) is an inflexion). In accordance with (2.4*b*) these conditions are

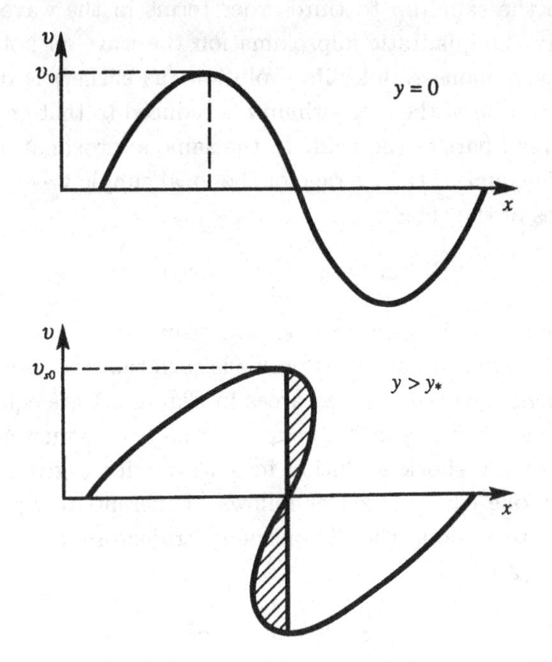

Figure 2.1. Shock front position in a simple wave profile.

reduced to the equalities

$$\alpha x = \frac{\partial \varphi}{\partial v}\bigg|_{x=x_*}, \quad \frac{\partial^2 \varphi^2}{\partial v^2}\bigg|_{x=x_*} = 0, \tag{2.6}$$

which together with (2.4b) define the values x_*, y_*, v_*.

Admittedly, if the function F in (2.4a) is not analytic, then the multivaluedness might arise exactly at the nonanalytic points (for example, at the leading point of the pulse); at these points the second condition (2.6) may not be fulfilled.

Since multivaluedness of the profile has no physical sense here, the equation (2.3) becomes inconsistent for $x > x_*$ and one needs, strictly speaking, to solve the complete Burgers equation taking dissipation into account. In the case of large Re, however, another way exists, suggesting the introduction of discontinuous solutions, i.e. *shock waves.* The appearance of a discontinuity leads, generally speaking, to reflections from it, and, consequently, to changes in the continuous part of the field: the wave ceases to be simple. Fortunately, when Mach number $M \ll 1$, i.e. in the small-nonlinearity situations that are of interest here, variations of all quantities at the shock front and in the adiabatic solution (2.4) prove to be the same up to third-order terms in the wave amplitude. Therefore, in the quadratic approximation the wave on both sides of a discontinuity remains simple, its evolution (as earlier) is described by (2.4), and the role of the discontinuity is reduced to that of "eating up" the multivalued part of the field. In the same approximation the shock velocity c_s is equal to the average of the local simple wave velocities on the two sides of the shock:

$$c_s = c_0 + \frac{\varepsilon}{2}(v_1 + v_2)\,. \tag{2.7}$$

Hence one can easily derive the known "equal areas rule": the discontinuity position in a simple-wave profile is such that the total area of the profile is unchanged (the shaded areas in Figure 2.1 are equal).

With the aid of (2.5) and (2.7) one can readily construct an analytical description of the shock evolution in a wave with a given profile. To perform this one can proceed as follows. Differentiate equation (2.4*b*) with respect to v along the discontinuity trajectory $x = x_s(y)$ where, according to (2.7),

$$\frac{\partial y}{\partial x} = -\alpha\frac{v_1 + v_2}{2},$$

which yields

$$\alpha\left(v - \frac{v_1 + v_2}{2}\right)\frac{\partial x_s}{\partial v} + \alpha x_s = \frac{\partial \varphi}{\partial v}. \tag{2.8}$$

This is valid for simple waves at either side of a discontinuity between $v = v_1$ and $v = v_2$, i.e. there are two equations (2.8). It seems simpler, however, to consider one of them, together with the relation between v_1 and v_2 (at the same point (x, t)) following directly from equations (2.4*b*):

$$\alpha x_s = \frac{\varphi(v_2) - \varphi(v_1)}{v_2 - v_1}. \tag{2.9}$$

A typical example is that of a symmetrical discontinuity where $v_1 = -v_2 = v_s$. This arises, in particular, for odd φ, when

$$\varphi(v_2) = -\varphi(v_1) = \varphi(v_s)$$

(here $c_s = c_0$), and the algebraic equation for v_s comes from (2.9),

$$\varphi(v_s) = \alpha x v_s \tag{2.10}$$

(and, of course, the same is obtained from (2.8)).

Consider first a sinusoidal boundary condition, which seems to be rather typical for acoustic problems,

$$v(0, y) = v_0 \sin \omega t. \tag{2.11}$$

Equation (2.4*b*) now takes the form

$$\omega y = \arcsin \frac{v}{v_0} + \sigma \frac{v}{v_0}, \tag{2.12}$$

where $\sigma = \alpha \omega v_0 x$ is a dimensionless coordinate (not a stress!).

This expression lends itself well to graphical analysis if, according to (2.12), we plot the relation $y(v)$ within the limits $(-\pi, \pi)$ for different values of σ (Figure 2.2); then, rotating the figure through 90° we obtain the desired function $v(y)$ (Soluyan & Khokhlov, 1961). The result for various values of σ is shown in Figure 2.3. It is evident that nonlinear distortions increase with σ until an "overlap" appears, as was stated before. The point where the discontinuity emerges can easily be found from relations (2.6): the second one in this pair produces the equality $v_* = 0$ (which is quite clear beforehand anyway due to the profile symmetry), while the first yields the coordinate of the discontinuity appearance: $\sigma_* = 1$, i.e.

$$x_* = L_N = \frac{c_0}{\varepsilon k v_0}, \quad k = \frac{\omega}{c_0}. \tag{2.13}$$

Now the meaning of the variable σ is understood, as that of a coordinate normalized to the discontinuity formation length L_N, which can be considered as a nonlinear effects development scale in a plane wave at initial amplitude v_0 and wavelength $\lambda = \frac{2\pi}{k}$. Let us present an example: an ultrasonic transducer sets up in water ($c_0 = 1.5 \times 10^3$ m/s, $\varepsilon = 4$) a field with an amplitude $v_0 = 36$ cm/s, or an acoustic Mach number $M = 2.4 \times 10^{-4}$. In such a wave $x_* = 160\lambda$ (λ is the wavelength). So, at a frequency of 1 MHz ($\lambda = 0.15$cm), we have $L_N = 24$ cm, i.e. quite moderate distances are involved here.

The further evolution of a discontinuity is described by (2.10), owing to the symmetry properties. For the case under consideration

$$\frac{v_s}{v_0} = \sin\left(\sigma \frac{v_s}{v_0}\right), \quad \sigma > 1. \tag{2.14}$$

After the shock wave formation the wave energy dissipates: although $\int_0^{2\pi/\omega} v \, dt$ is conserved as before, the value of $\int_0^{2\pi/\omega} v^2 \, dt$ diminishes. This process can be viewed as a result of the above-mentioned "eating up" of the multivalued parts of the profile. For $\sigma > 1$ the shock magnitude v_s increases, attaining its maximum at $\sigma = \pi/2$, and this

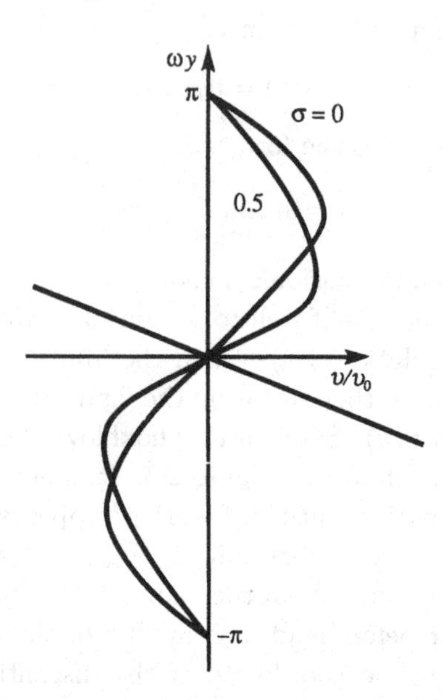

Figure 2.2. Graphical interpretation of solution (2.12).

is followed by decay afterwards. Soon after that the major part of an initial profile is "eaten up" with the exception of a domain near to the zero point. In this interval the profile is close to linear, and therefore the wave acquires a sawtooth wave shape with steep fronts and linear profiles between them (Blackstock, 1966; Soluyan & Khokhlov, 1961). For large σ we have $\sigma v_s/v_0 \ll 1$ over the whole wave, and (2.14) yields the simple asymptotic formula

$$\frac{v_s}{v_0} = \frac{\pi}{1+\sigma}. \tag{2.15}$$

Notice that this result corresponds to the weakly nonlinear (quadratic) approximation while the more exact considerations reveal the lack of symmetry of the waveform between the rarefactive and compressive phases (Rudenko & Soluyan, 1975; Inoue & Yano, 1993).

It is worth noting that if, instead of a sinusoid, we start immediately with a sawtooth wave of given initial amplitude v_{s_0}, the further wave evolution will be embodied in an analogous formula,

$$\frac{v_s}{v_{s_0}} = \frac{1}{1+\sigma}. \tag{2.16}$$

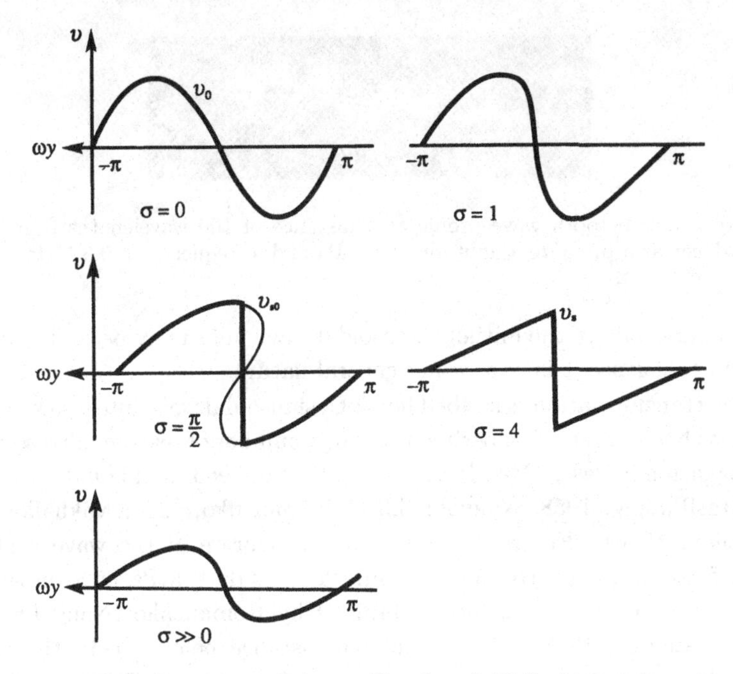

Figure 2.3. Transformation of the profile of an initially harmonic wave as it propagates.

where now $\sigma = \alpha\omega x v_{s0}$.

As is apparent from (2.15) and (2.16) the nonlinear periodic wave, as distinct from the linear one, decays by a non-exponential law. The effect of *amplitude saturation* is notable here: with the growth of the initial amplitude, the peak value v_s increases more slowly at the sawtooth wave stage, asymptotically approaching a limit v_∞ which does not depend on v_0:

$$v_\infty = \pi/\alpha\omega x. \tag{2.17}$$

This property of the nonlinear wave to "forget" its initial amplitude leads to the fact that it is impossible to transmit an acoustic power higher than the limiting value $\rho_0 c_0 v_\infty^2/3$ through a given medium layer. Thus, in water at 1 MHz, the limiting power is $120/x^2\,\mathrm{KW\,m^{-2}}$, where x is expressed in metres; note, however, that in real liquids the limiting field levels are achievable only when using very powerful sources (more frequently in focused waves).

It should be emphasized that sawtooth wave asymptotic behaviour is

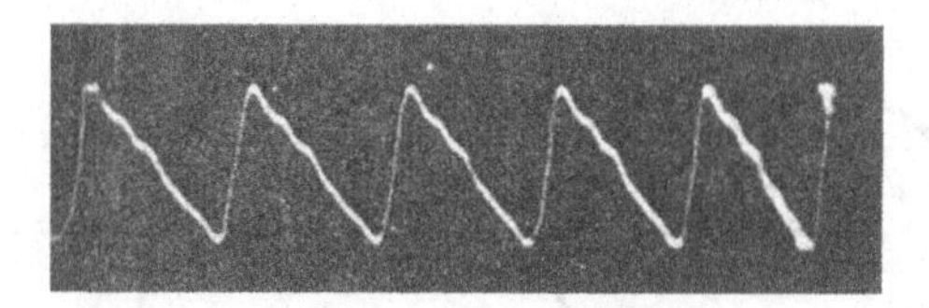

Figure 2.4. Sawtooth wave profile at a distance of 100 wavelengths from the transducer at a pressure amplitude of 8 MPa and a frequency of 0.5 MHz.

typical not only of the initially sinusoidal wave but of any periodic wave, and in this sense it is of a rather general nature.

The transformation described here of a sinusoidal wave into a sawtooth one has been observed experimentally in a number of tests on ultrasound propagation in gases (Mendousse, 1953; Werth, 1953) and liquids (Burov & Krasil'nikov, 1958; Naugol'nykh & Romanenko, 1958; Mikhailov & Shutilov, 1960). Figure 2.4 shows an oscillogram of the wave profile with frequency 0.5 MHz and pressure amplitude 8 MPa at a distance of 40 cm from the transducer, obtained by Romanenko (Naugol'nykh & Romanenko, 1958). The calculated distance before sawtooth wave formation, corresponding to $\sigma = \pi/2$, amounts to $L = \pi/2\varepsilon kM = 32\,\mathrm{cm}$ in the case under consideration.

The possibility of using the effect of nonlinear distortion to calibrate hydrophones and to measure the B/A ratio is discussed by Bacon (1990).

Spectral approach.

Let us now describe the same process in spectral terms. Formula (2.12), rewritten in the form

$$\frac{v}{v_0} = \sin\left(\omega y + \sigma\frac{v}{v_0}\right), \tag{2.18}$$

can be viewed as a transcendental equation with respect to v/v_0. The Fourier expansion of this expression is deduced using an approach that we may call the implicit argument method. To determine the coefficients of a spectral expansion

$$B_n = \frac{2}{\pi}\int_{-\pi}^{\pi} v \sin n\omega y \,\mathrm{d}(\omega y)$$

we introduce the variable $\xi = \omega y + \sigma v/v_0$; then $v = v_0 \sin\xi$, $\omega y = \xi - \sigma v/v_0$, and $\mathrm{d}(\omega y) = \mathrm{d}\xi - (\sigma/v_0)\mathrm{d}v = (1 - \sigma\cos\xi)\mathrm{d}\xi$, hence

$$B_n = \frac{2}{\pi} v_0 \int_{-\pi}^{\pi} \sin\xi(1 - \sigma\cos\xi)\sin n(\xi - \sigma\sin\xi)\mathrm{d}\xi. \tag{2.19}$$

This can be transformed as follows (Vinogradova *et al.*, 1979):

$$\begin{aligned}
\frac{\pi}{2} B_n &= -\frac{v_0}{n} \int_{-\pi}^{\pi} \sin\xi \,\mathrm{d}\cos[n(\xi - \sigma\sin\xi)] \\
&= \frac{1}{n} \int_{-\pi}^{\pi} \cos\xi \cos[n(\xi - \sigma\sin\xi)]\,\mathrm{d}\xi \\
&= \frac{v_0}{2\pi} \int_{-\pi}^{\pi} \big\{ \cos[(n+1)\xi - n\sigma\sin\xi] \\
&\qquad + \cos[(n-1)\xi - n\sigma\sin\xi] \big\}\,\mathrm{d}\xi.
\end{aligned} \tag{2.20}$$

Making use of a tabulated integral

$$\int_0^{\pi} \cos(\nu\xi - n\sigma\xi)\,\mathrm{d}\xi = J_\nu(n\sigma), \tag{2.21}$$

where J_ν is a Bessel function, by means of the recurrence relation

$$J_{n-1}(n\sigma) + J_{n+1}(n\sigma) = \frac{2}{\sigma} J_n(n\sigma)$$

we easily obtain

$$B_n = \frac{2 J_n(n\sigma)}{n\sigma} v_0. \tag{2.22}$$

The Fourier transform of an initially sinusoidal simple wave has the form

$$\frac{v}{v_0} = \sum_{n=1}^{\infty} \frac{2 J_n(n\sigma)}{n\sigma} \sin n\omega\tau. \tag{2.23}$$

This formula (or its analogue, to be more exact) was derived by Bessel when considering the Kepler problem; later on Fubini (1935) deduced it independently in connection with acoustic problems. Nowadays it is called the Bessel–Fubini formula (however, in the form (2.23) it was first written down by Blackstock, 1962).

It should be underlined that (2.23) is a special case of a more general relation describing the spectrum evolution of a simple wave of arbitrary shape. Let the wave have the form $v(y, x) = F(y + \alpha v x)$. Its spectrum is given by the standard expression

$$C(\omega, x) = \frac{1}{2\pi} \int_{-\infty}^{\infty} v(x, y) \mathrm{e}^{\mathrm{i}\omega y}\,\mathrm{d}y. \tag{2.24}$$

Proceeding to the variable $\xi = y + \alpha v x$ we get

$$C(\omega, x) = \frac{1}{2\pi} \int_{-\infty}^{\infty} v(\xi) \left(1 - \alpha x \frac{\partial v}{\partial \xi}\right) \mathrm{e}^{\mathrm{i}\omega(\xi - \alpha v x)} \, \mathrm{d}\xi.$$

Integrating by parts twice we obtain an explicit expression for the spectrum of a wave specified by a function $F(\xi)$ at $x = 0$: (Kuznetsov, 1970)

$$C(\omega, x) = \frac{1}{2\pi \mathrm{i} \alpha \omega x} \int_{-\infty}^{\infty} \left(\mathrm{e}^{-\mathrm{i}\omega\alpha F(\xi) x} - 1\right) \mathrm{e}^{\mathrm{i}\omega\xi} \, \mathrm{d}\xi. \tag{2.25}$$

Let us return now to the Bessel–Fubini formula (2.23). According to (2.23), the amplitude of the fundamental harmonic falls off monotonically, while that of any higher harmonic grows (for small σ as $(n\sigma)^{n-1}$) until it reaches the point $\sigma = 1$ where a discontinuity is formed, and solution (2.23) becomes no longer applicable. Note that at $\sigma = 1$ the first harmonic amplitude has decreased by 20 per cent, i.e. a rather small portion of the wave energy (about 10 per cent) has been transferred to the higher harmonics.

A Fourier transform can be obtained for the discontinuous stage also. Here the integral (2.19) remains valid, but with integration limits changed. Indeed, one has to exclude parts "eaten up" by the discontinuity, i.e. the integral is taken from ξ_s to π where ξ_s corresponds to the discontinuity position and, evidently, grows with time. Integrating by parts (as was done in (2.20)) and taking into account that the discontinuity is always situated at $\omega y = \xi_s - \sigma \sin \xi_s$, we obtain the expression (Blackstock, 1962)

$$B_n = \frac{2v_0}{n\pi} \left[\sin \xi_s + \frac{1}{\sigma} \int_{\xi_s}^{\pi} \cos \xi \cos n(\xi - \sigma \sin \xi) \, \mathrm{d}\xi\right], \tag{2.26}$$

where the value of ξ_s is determined by relation $v_s = v_0 \sin \xi_s$, v_s following from equation (2.14). Since equality (2.21) is fulfilled it is clear that at $\sigma = 1$, when $\xi = 0$, (2.26) coincides with (2.22). Analysis of formula (2.26) demonstrates that B_1 continues to decrease monotonically for $\sigma > 1$, while other harmonics attain their maxima, after which they decay too (Figure 2.5). At the sawtooth wave stage when σ is far in excess of unity, the harmonic amplitudes are defined by a simple expression corresponding to (2.15),

$$B_n = \frac{2v_0}{n(1+\sigma)}. \tag{2.27}$$

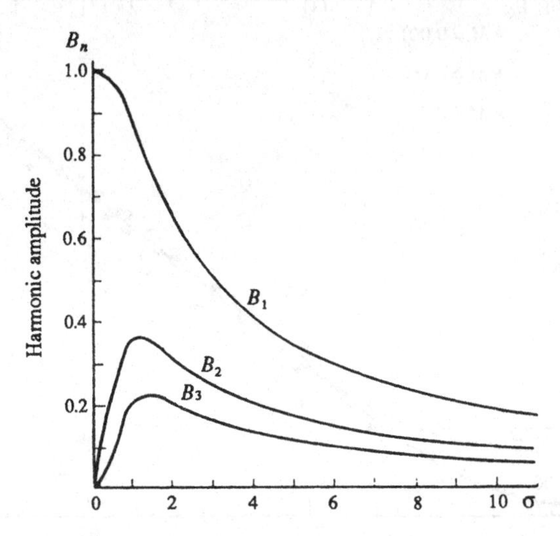

Figure 2.5. Distance dependences of the first three harmonic amplitudes of a finite-amplitude wave.

Thus, at the stage when $\sigma \gg 1$, all B_n (as the amplitude peak itself) do not depend on v_0 and decrease in inverse proportion to x.

These theoretical inferences are in fair agreement with experimental results obtained by various researchers (Thuras *et al.*, 1934; Zankel & Hiedeman, 1958; Zarembo & Krasil'nikov, 1956). Figure 2.6 demonstrates an example from a paper by Blackstock (1962) giving the range dependence of the second harmonic amplitude as measured by various authors, together with the appropriate theoretical curves. In the course of the wave dissipation the value of *Re* diminishes as x^{-1}, and the asymptotic behaviour of the wave will be different: the terminal stage will be linear, with exponential decay of the wave amplitude (see Section 2.4).

2.3 Propagation of pulses

Let us turn our attention now to the behaviour of localized pulse-like perturbations. One of the most typical cases is that of a compression pulse with a discontinuous front, propagating into an unperturbed medium (Figure 2.7*a*). The evolution of the discontinuity is described by equa-

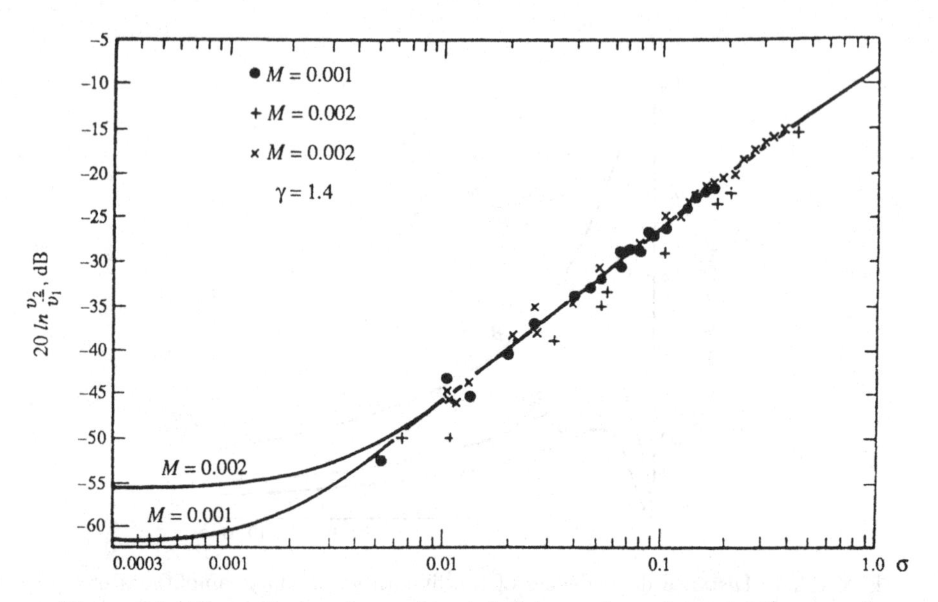

Figure 2.6. Theoretical and experimental distance dependences of the second harmonic amplitudein a near field (Blackstock, 1962), comparative data.

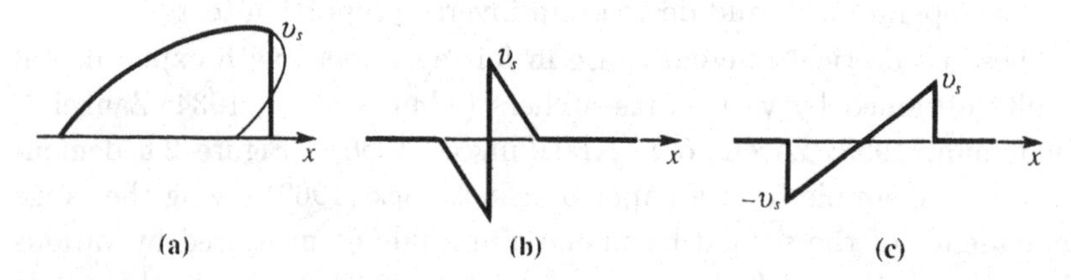

Figure 2.7. Single pulse profiles: (a) compression pulse with discontinuous front, (b) bipolar pulse, (c) N–wave.

tions (2.8) with $v_1 = 0$ (Courant & Friedrichs, 1948)

$$\frac{v}{2}\frac{\mathrm{d}x}{\mathrm{d}v} + x = \frac{1}{\alpha}\frac{\mathrm{d}\varphi}{\mathrm{d}v}, \quad v = v_2 = v_s \tag{3.1}$$

(equality (2.9) in this case gives nothing since the discontinuity does not belong to the curve $\varphi(v)$, $v_1 = 0$).

Equation (3.1) lends itself easily to integration, giving

$$x = \frac{2}{\alpha v_s^2}\int_{v_{so}}^{v_s} v\frac{\mathrm{d}\varphi}{\mathrm{d}v}\,\mathrm{d}v, \tag{3.2}$$

where v_{so} is the initial discontinuity magnitude.

For $x \gg x_*$, when the discontinuity "eats up" most of the wave profile leaving only its linear portion near the zero point, i.e. $\varphi \sim v$, equation (3.2) implies that $v_s \sim x^{-1/2}$, rather than x^{-1} as was observed for a sawtooth wave. A slower discontinuity damping rate in this situation is easy to understand. A triangular pulse with a discontinuous front and a continuously extending length is formed; the velocity difference between a simple wave and a discontinuity and, hence, the energy decay rate, is less, compared with that of a wave with an unchanged length (as was the case for a sawtooth periodic wave).

From the discussion above it follows that a triangular profile is formed as the "old-age" behaviour of any unipolar pulse of finite length. If at $t = 0$ a triangular pulse with amplitude v_0 and length ℓ_0 is given ($\varphi(v)$ being a linear function) then it follows from (3.2) that the amplitude v_s and duration ℓ of this pulse change according to

$$v_s = v_0(1 + \varepsilon v_0 x / c_0 \ell_0)^{-1/2}, \tag{3.3}$$

$$\ell = \ell_0(1 + \varepsilon v_0 x / c_0 \ell_0)^{1/2}. \tag{3.4}$$

The latter formula can most easily be deduced from (3.3) using the condition of area conservation, $v_s \ell = v_0 \ell_0$, that results from the above-mentioned "equal areas rule", which holds true both for (2.3) and for the complete Burgers equation (Figure 2.7*a*). In this case the pulse amplitude at long range continues to depend on the initial amplitude although this dependence is weaker (square root) than that of the linear case.

Among other examples, a pulse with an exponential tail is of practical interest:

$$v(0, y) = v_0 \exp(-y/\tau), \quad y = t - \frac{x}{c_0}, \tag{3.5}$$

approximating a typical head wave initiated by an explosion in water. Evolution of this pulse is described by the formulae (Fridman, 1980)

$$\begin{aligned} v_s &= 2v_0 \left(1 + \sqrt{1 + \frac{2\varepsilon v_0 x}{c_0 \ell}}\right)^{-1}, \\ T_s &= (T/\mathrm{e}) \left[1 + (\mathrm{e} - 1)\sqrt{1 + 2\varepsilon v_0 x / c_0 \ell_0}\right], \end{aligned} \tag{3.6}$$

where $\ell_0 = c_0 T$ and T_s is the pulse duration at the amplitude level $v_0 \mathrm{e}^{-1}$. The asymptotic shape of the pulse is again triangular, and one obtains, for large x, $v_s \sim x^{-1/2}$, $T_s \sim x^{1/2}$.

If two discontinuous pulses of the same polarity propagate in the same direction, then the discontinuity that travels second always overtakes the one travelling first, yielding a single pulse with the same "triangular" asymptotics. (Generally speaking, this inference proves to be unconditionally valid only in the discontinuous approximation.)

Somewhat more complicated is the case of alternating pulses. Let us discuss the behaviour of a pulse consisting of two triangles with different signs separated by a discontinuity (Figure 2.7*b*)

$$v(0,y) = \begin{cases} v_{0-}(y-y_0)/(y_0-y_2), & y_2 < y < 0 \\ v_{0+}(y_1-y)/(y_1-y_0), & 0 < y < y_1, \end{cases} \tag{3.7}$$

and $v = 0$ outside this interval.

Utilizing (2.8) and (2.9) one can demonstrate that the peak velocity values in a wave, for $v > 0$ and $v < 0$, are given by (Fridman, 1976)

$$v_{\pm} = \frac{R}{S}\left(\frac{1+RS}{1-R^2}\right)\left[\left(\frac{R+S}{1+RS}\right)\left(\frac{Z+R^{-1}}{Z+R}\right)^{\pm 1/2} - 1\right], \tag{3.8}$$

where $Z = xR$,

$$S = \left|\frac{v_{0+}T_+}{v_{0-}T_-}\right|^{1/2}, \quad R = \left|\frac{v_{0-}T_+}{v_{0+}T_-}\right|^{1/2}$$

are the area and profile shape ratios respectively for the pulses at $t = 0$ and $T_{\pm} = |(y_{1,2} - y_0)|$ are the initial lengths of the positive and negative phases (respectively, $RS = T_+/T_-$ and $R/S = v_-/v_+$).

On the basis of (3.8) a conclusion can be drawn about the special case $S = 1$ when the wave asymptotic behaviour is presented by a symmetric wave corresponding to a period in a sawtooth profile; the amplitude of such a discontinuity as $x \to \infty$ decreases as x^{-1}. If, however, $S \neq 1$, the pulse with smaller area will always be absorbed by that with larger area, which is followed by the formation of a triangular pulse considered above.

It seems instructive to mention also the case of the "N"-wave, the alternating pulse with zero total area and two discontinuities at the leading and trailing edges, respectively (Figure 2.7*c*). Such a pulse is, in particular, typical of the "sonic boom" induced in the atmosphere by a supersonic aircraft. In such a wave the positive and negative parts behave, in fact, independently. The areas of each are conserved and the long-distance amplitudes and the lengths equalize so that the pulse acquires a symmetric "N-shape". However, at very large distances these two parts are coupled by viscous effects, the area of each part decreases and the symmetric N-wave ultimately decays in a linear manner (see equations (4.14) and (4.15)). A review of investigations of the propagation of N-waves of different shape is presented in the paper of Nakamura (1984).

Thus, for a pulse of finite duration, three different types of asymptotic behaviour are possible (for $Re \gg 1$): a triangular unipolar pulse and two types of pulse with zero area of which one retains its length while the other spreads indefinitely.

The pulse evolution also manifests itself in the spectrum change. Let us examine, as an example, the Gaussian pulse propagation:

$$v(y) = v_0 \exp(-b^2 y^2).$$

Using (2.25) with respect to the simple wave spectrum, one obtains (Pelinovsky, 1976)

$$C(\omega, x) = C(\omega, 0) \sum_{m=0}^{\infty} \frac{\left(-\mathrm{i}\omega x v_0/c_0^2\right)^m}{m!(m+1)^{3/2}} \exp\left[\frac{m\omega^2}{(m+1)4b^2c_0^2}\right], \tag{3.9}$$

where $C(\omega, 0)$ is the spectrum of the initial profile.

The early stage of the process (near to a source) is described by the first terms in this sum:

$$|C(\omega, x)/C(\omega, 0)| = 1 + \frac{\omega^2 x^2 v_0^2}{c_0^4} \exp\left(\frac{\omega^2}{4b^2}\right) \times \left[1 - 1.6 \exp\left(-\frac{\omega^2}{12b^2}\right)\right]. \tag{3.10}$$

This solution shows that within the low-frequency range where $\omega < \omega_1$ and

$$\omega_1^2 = 18\, b^2 \ln \frac{4}{3}$$

the expression in the parentheses is negative, becoming positive for $\omega > \omega_1$, which reflects the energy transfer from the low-frequency range to the high-frequency range.

Nowadays the investigations of nonlinear pulse propagation and determination of the relative effects of turbulence and molecular relaxation on the weak shock formation are stimulated by the sonic boom problem (Pierce, 1993).

2.4 A medium with finite dissipation; some solutions of the Burgers equation

A discontinuous approximation, valid for Re $\to \infty$, seems to demonstrate most clearly the peculiarities of the evolution of finite amplitude acoustic waves. However, in order to describe such an evolution more completely, one has to return to the Burgers equation in its complete form (2.1).

Integrating (2.1) with respect to y we can readily verify that the value $\int_{-\infty}^{\infty} v\,dy$ is an integral of motion, i.e. for any localized perturbation (for which the integral is finite) the profile area is invariant. The same holds true for a periodic wave provided the integral is taken over the period. This feature (reflecting both momentum and mass conservation) is known to assist frequently in producing a qualitative description of the wave evolution. The energy of the latter is of course not conserved.

It has already been mentioned that at low *Re* the Burgers equation becomes the linear equation of diffusion type. It is, however, of much more interest that such a transition is feasible in the general case as well, if the following substitution is undertaken:

$$v = \frac{2\delta}{\alpha}\frac{\partial}{\partial y}\ln\zeta. \tag{4.1}$$

Then after a single integration (2.1) takes the form†

$$\frac{\partial \zeta}{\partial x} = \delta \frac{\partial^2 \zeta}{\partial y^2}. \tag{4.2}$$

This important transformation was suggested by Hopf (1950) and Cole (1951). It provides a unique chance of obtaining the exact solution to a nonlinear wave problem for a dissipative medium.‡ Note that due to the linearity of equation (4.2) the sum of the solutions to this equation will also be a solution of the Burgers equation. Should the boundary condition $v(\sigma,\tau) = v_0(\tau)$ be imposed, the solution will take the form

$$\zeta(x,y) = \frac{1}{\sqrt{4\pi\delta x}}\int_{-\infty}^{\infty} \exp\left[-\frac{(y-y')^2}{4\delta x} - \frac{1}{2\delta}\int_0^{y'} v_0(y')\,dy'\right] dy'. \tag{4.3}$$

The integral here converges if the initial perturbation satisfies the condition $\int_0^y v_0(y')\,dy' \le \text{const}\,.\,y\ (y \to \infty)$.

† Actually, the term $A(x)\zeta$ is also present on the right-hand side of (4.2) where $A(x)$ is an arbitrary function. This addition, however, leads only to the multiplication of ζ by $\exp\int A\,dx$ which, according to (4.1), does not influence the desired solution for v.

‡ The majority of the current exact methods to solve nonlinear evolution equations (such as the inverse scattering method) refer to nondissipative systems.

The solutions of the Burgers equation are treated thoroughly in the literature (Karpman, 1973; Rudenko *et al.*, 1977). Here we shall give a few results that clarify the physical situation. Of course, it is not always convenient to use the transformation (4.1) since the simple boundary or initial conditions in the Burgers equation (2.1) can lead to rather complicated integrals in (4.3), so that it may be better to handle equation (2.1) directly.

Stationary shock wave and periodic solutions.

Let us consider self-similar solutions of the Burgers equation that depend on some specified combinations of the variables x and y. The simplest of the latter is a stationary travelling wave of type $v = v(\eta)$, where $\eta = y + bx$, and b is a constant. Substitution into (2.1) results in an equation in ordinary derivatives,

$$\delta \frac{dv}{d\eta} = \frac{\alpha}{2} v^2 + bv + A, \tag{4.4}$$

where A is an integration constant. This equation has, generally speaking, two equilibrium points v_1 and v_2. The solution which is referred to as a Taylor shock (for example, see Crighton, 1986) has the form of a transition between these points:

$$2v = (v_1 + v_2) + (v_2 - v_1)\text{th}[\mathcal{X}(y + bx)], \tag{4.5}$$

where

$$\mathcal{X} = \frac{\alpha}{4\delta}(v_2 - v_1), \quad b = -\frac{\alpha}{2}(v_1 + v_2). \tag{4.6}$$

It is essentially the same shock wave as that interpreted earlier as a discontinuity; its speed coincides with that of a discontinuity and its characteristic duration is $\tau \sim \mathcal{X}^{-1}$ (i.e. its spatial thickness is $\ell_s \sim c_0/\mathcal{X} \sim 4\nu_{eff}/\varepsilon v_0$) where $\nu_{eff} = \delta c_0^3$ presents an effective value of the kinematic viscosity, and v_0 is the velocity jump at the shock front). Consider an example: the shock front thickness in water ($\nu_{eff} = 3 \times 10^{-2}\,\text{cm}^2/\text{s}$) with a pressure jump 1 MPa amounts to $\ell_s = 10^{-3}\,\text{cm}$. It is evident that the discontinuous approximation has meaning only if ℓ_s is small compared with the external field scale, i.e. the characteristic wavelength λ. Thus, for example, for water at a pressure amplitude of 0.1 MPa (1 atmosphere) one obtains $\ell_s \simeq 5 \times 10^{-3}\,\text{cm}$ and at a frequency of, say, 1 MHz ($\lambda \sim 1.5\,\text{mm}$), the profile of such a wave can be considered as discontinuous.

It is to be noted that the ratio λ/ℓ_s has the order of the acoustic

Reynolds number for the wave. This means, for example, that for a periodic wave the discontinuity amplitude at long distances decreases, whereas λ is not changed, so that the discontinuous approximation always appears only as an intermediate asymptotic description and eventually becomes inadequate: the asymptotic behaviour of a periodic wave for $x \to \infty$ is ultimately described by a linear approximation ($Re \ll 1$).

A more detailed description of the periodic wave evolution can be obtained by making use of the following solution (Khokhlov & Soluyan, 1964; Blackstock, 1964):

$$\frac{v}{v_0} = \frac{1}{1+\sigma}\left[-\omega y + \pi \operatorname{th}(\omega y/\Delta)\right], \quad -\pi \leq \omega y \leq \pi, \tag{4.7}$$

where

$$\Delta = \frac{2\omega\delta(1+\sigma)}{\alpha\pi v_0} = \frac{2}{\pi Re}(1+\sigma) \tag{4.8}$$

and σ and the Reynolds number Re are related to v_0 as before. It may be interesting to note that this solution, consisting of a shock of type (4.7) (with $v_2 = v_1$) and a rarefaction wave with linear profile, proves to be an exact solution of the Burgers equation, as can easily be confirmed by direct substitution. When $Re \gg 1$ the solution (4.7) can (approximately) produce a damped periodic wave with period 2π that is, at first, close to a sawtooth wave form, but as it decays develops a smoother profile (Figure 2.3). Such a periodic solution can be readily expressed as a Fourier series, and at high Re it coincides with an exact solution in Fourier form obtained by Fay (1931):

$$\frac{v}{v_0} = \frac{2}{Re}\sum_{n=1}^{\infty}\frac{\sin n\omega y}{\operatorname{sh}\frac{n(1+\sigma)}{2Re}}. \tag{4.9}$$

It is evident that for $Re \gg n(1+\sigma)$, the harmonic amplitudes are proportional to n^{-1} and depend similarly on x, which corresponds to the formula (2.27) deduced for a discontinuous sawtooth wave. It is, however, clear from (4.9) that this formula holds only for a restricted number of harmonics, up to $n \lesssim Re/(1+\sigma)$. At large n, when

$$Re < n(1+\sigma),$$

the harmonic amplitudes decay exponentially, with the decrement proportional to n (and not to n^2 as in the linear case). The shock profile broadens with distance, and when $Re \ll (1+\sigma)$ the wave transforms into an exponentially decaying sinusoid (in this instance (4.9) remains valid while (4.7) becomes, of course, inadequate).

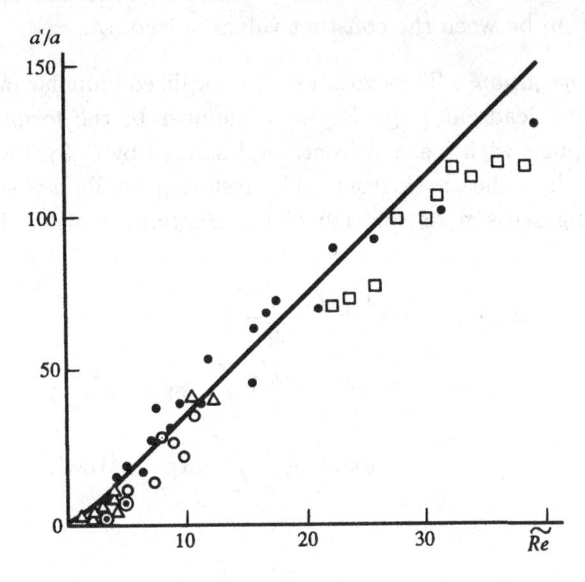

Figure 2.8. The first harmonic absorption coefficient dependence on the current value of the Reynolds number, equation (4.10). Symbols represent experimental data.

In fact, the solution (4.9) describes the asymptotic behaviour of any alternating wave. When $Re \gg 1$, such a wave goes through a sawtooth stage and then is converted to a decaying sinusoid with the basic frequency, and with amplitude no longer depending on the initial amplitude v_0, this dependence being lost at the sawtooth stage.

It is sometimes helpful to introduce a local absorption coefficient as regards the first harmonic amplitude $v_1(x)$ (Naugol'nykh, 1958):

$$a' = -\frac{1}{v_1}\frac{\mathrm{d}v_1}{\mathrm{d}x} = a\sqrt{1 + \widetilde{Re}^2}\,. \tag{4.10}$$

Here a is the absorption coefficient for a linear wave at the basic frequency, and $\widetilde{Re} = Re/(1+\sigma)$ is the current value of the Reynolds number. The value of a' changes during propagation. At the sawtooth stage where $\widetilde{Re}$ is high, (4.9) leads to $a' = a\widetilde{Re}$, which is far in excess of a, a' being proportional to the wave amplitude. This result has received experimental verification in many studies (Figure 2.8). At large x, when the linear stage is reached, $a' \to a$.

It is also worth saying that the stationary solution (4.5) appears as the asymptotic behaviour for an arbitrary boundary condition in the form of a transition between the constant values v_1 and v_2.

Pulse propagation. The evolution of a localized unipolar pulse of arbitrary shape leads at large Reynolds number to the formation of a triangular pulse with a sharp front, as described by (3.3), (3.4). Dissipation broadens the shock front. The resulting profile is described by the following self-similar solution of the Burgers equation (Karpman, 1973):

$$\begin{aligned} v(x,y) &= -\frac{2\delta}{\alpha}\frac{\mathrm{d}}{\mathrm{d}y}F\left(\frac{y}{\sqrt{4\delta x}}\right), \\ F(z) &= \frac{1}{\sqrt{\pi}}\Big[\exp\left(-\frac{A}{4\delta}\right)\int_{-\infty}^{z}\exp(-\eta^2)\,\mathrm{d}\eta \\ &\quad + \exp\left(\frac{A}{4\delta}\right)\int_{z}^{\infty}\exp(-\eta^2)\,\mathrm{d}\eta\Big], \end{aligned} \tag{4.11}$$

where

$$A = \int_{-\infty}^{\infty} v(y')\,\mathrm{d}y',$$

which, being the constant pulse area, is easily proved to be conserved. The shock front thickness is

$$\ell_s = c_0\tau \approx \sqrt{\alpha A x}\cdot Re^{-1}, \quad Re = \alpha A/\delta, \tag{4.12}$$

and grows with the wave propagation range as $\sqrt{x}$, but the pulse duration increases according to the same law as well (see (3.4)). Because of this the pulse shape expressed in the coordinates $y/\sqrt{\alpha A x}$, $v\sqrt{\alpha A x}$ remains invariant (Figure 2.9). The ratio of the pulse duration $\ell = \sqrt{\alpha A x}$ to the shock front thickness also remains constant, being equal, according to (4.12), to $\ell/\ell_s \sim Re$, so that the importance of nonlinearity is not diminished with propagation for such a pulse, as distinct from, for example, the sawtooth wave.

At sufficiently large Reynolds number values, (4.11) yields

$$v \sim \begin{cases} y/\alpha x & 0 < y < \sqrt{\alpha A x}, \\ 0 & \text{elsewhere}, \end{cases}$$

which agrees with the discontinuous solution (3.3), (3.4).

Consider now the solution to the Burgers equation describing N-wave propagation. For (4.2) it has the following form:

$$\zeta = -\left[1 + \sqrt{\frac{\ell}{x}}\exp\left(\frac{-y^2}{4\delta x}\right)\right], \tag{4.13}$$

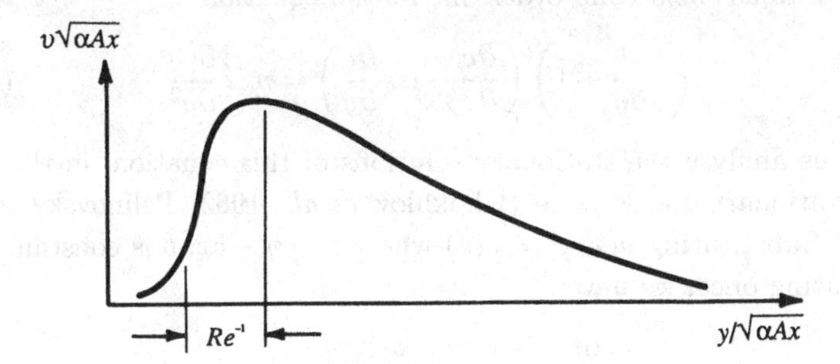

Figure 2.9. Asymptotic pulse profile corresponding to the self-similar solution of the Burgers equation.

where ℓ is the length of the N-wave, or, in terms of v (Lighthill, 1956; Whitham, 1977):

$$v = \frac{y}{\alpha x}\left[\frac{\sqrt{\ell/x}\exp\left(-y^2/4\delta x\right)}{1+\sqrt{\ell/x}\exp\left(-y^2/4\delta x\right)}\right]. \tag{4.14}$$

The area of the positive profile phase is given as

$$\int_0^\infty v\,dy = \frac{2\delta}{\alpha}\ln\left(1+\sqrt{\frac{\ell}{x}}\right). \tag{4.15}$$

Denote the value of the integral (4.15) at the initial point x_0 as A_0 and introduce the Reynolds number $Re_0 = \alpha A_0/2\delta$ characterizing the nonlinearity of the N-wave. For high Re the solution (4.14) has a simple form corresponding to the discontinuous N-wave:

$$v \sim \begin{cases} y/\alpha x, & -\sqrt{\alpha A x} < y < \sqrt{\alpha A x}, \\ 0, & \text{elsewhere.} \end{cases}$$

Shock waves in a medium with relaxation.

It has already been shown in Chapter 1 that dissipation can be related to relaxation processes. To this end consider equation (1.33) of Chapter 1 (omitting the shear viscosity term):

$$\frac{\partial v}{\partial x} - \alpha v\frac{\partial v}{\partial y} - m'\frac{\partial}{\partial y}\int_{-\infty}^{y}\frac{\partial v}{\partial y'}\exp\left(-\frac{y-y'}{\tau}\right)dy' = 0 \tag{4.16}$$

where m' and τ are constants. Differentiating with respect to y one easily obtains the second-order differential equation

$$\left(\tau\frac{\partial}{\partial y}+1\right)\left(\frac{\partial v}{\partial x}-\alpha v\frac{\partial v}{\partial y}\right)=m'\tau\frac{\partial^2 v}{\partial y^2}. \tag{4.17}$$

Let us analyse the stationary solutions of this equation, in the form of a stationary shock wave (Khokhlov *et al.*, 1962; Pelinovsky *et al.*, 1984). Substituting $v(x,y)=v(\xi)$ where $\xi=y-bx$, b is constant, and integrating once, we have

$$\frac{\mathrm{d}v}{\mathrm{d}\xi}=\frac{\alpha}{2\tau}\frac{(v-v_1)(v_2-v)}{\alpha v+\mathcal{X}}, \tag{4.18}$$

where $\mathcal{X}=m'+b$ and v_1, v_2 are the equilibrium values such that

$$b=-(\alpha/2)(v_1+v_2).$$

Thus the shock wave velocity is equal, as before, to the average of the Riemann (simple) wave velocities in front of and behind it.

The integral of (4.18) can be written as

$$\alpha\xi=-2\tau\ln\left[(v-v_1)^{a_1}(v_2-v)^{a_2}\right]. \tag{4.19}$$

where

$$a_1=-\frac{\mathcal{X}+\alpha v_1}{v_2-v_1},\quad a_2=\frac{\mathcal{X}+\alpha v_2}{v_2-v_1}, \tag{4.20}$$

(Here we assume, for the sake of definiteness, that $v_2>v_1$ and then $v_1<v<v_2$.)

It is obvious from (4.18) or (4.19) that the case of $\mathrm{d}v/\mathrm{d}\xi>0$ (a compression front) is realized only if $\alpha v+\chi>0$ everywhere, or, on account of the monotonic nature of the solution, $\alpha v_1>-\mathcal{X}$, which gives the inequality

$$m>\frac{\alpha}{2}(v_2-v_1). \tag{4.21}$$

This is physically understandable: a continuous shock transition takes place only for rather strong relaxation, capable of competing with the nonlinearity.

If (4.21) is not fulfilled for $m'>0$, the solution to equation (4.18) acquires a singularity (the denominator in (4.18) vanishes at some $v=v_0$ lying between v_1 and v_2). This means that the solution becomes multivalued and, in fact, takes on discontinuous behaviour. In order to smooth it one has to allow for "ordinary" viscosity. It should be noted that this kind of situation is well known in the gas dynamics of a heat-conducting inviscid medium where an isothermal jump occurs (Landau & Lifshitz, 1986).

A more elaborate discussion of shock waves in a relaxing medium (for the case of $v_2 = -v_1$) is presented in the paper by Khokhlov and Soluyan (1964) and in the book by Rudenko and Soluyan (1977).

2.5 Propagation of intense acoustic noise†

Intense noise fields differ in their properties from low-amplitude statistical fields in that during the propagation process the wave profile evolution takes place and discontinuities appear, leading to energy redistribution over the wave spectrum.

Let us clarify the characteristic features of the process, taking plane noise waves as an example. Consider the propagation of a continuous noise wave, whose profile is described by a random function of time having a broadband spectrum with a maximum corresponding to a frequency ω_0.

When an intense wave propagates, its spectral components interact. As a result, the energy is distributed over the spectrum, propagating both into the higher– and lower–frequency ranges. The spectrum extension into the high-frequency range is limited by shock formation in the wave profile. The thickness of the resulting shock fronts defines the short-wave limit of the spectrum. Within the frequency range adjacent to this limit, a universal decay law of the intensity spectrum that is independent of the initial wave spectrum is set up, as will be shown later.

In the low-frequency range, the spectrum evolution after discontinuity formation is associated, in particular, with the fact that (due to the shock speed dependence on the amplitude) a gradual absorption of small shocks occurs, resulting in the growth of the characteristic scale of the process, and a corresponding energy transfer into the low-frequency region.

Let us focus our attention on some quantitative aspects of the problem. At $x = 0$ let a plane noise wave induced in a medium be specified by a random function $f(t)$. First one needs to clarify how the statistical features and energy of the wave change as a result of its nonlinear evolu-

† Here we confine ourselves to a description of a few physically illustrative models. The problems of statistical nonlinear acoustics are specially handled in a monograph by Gurbatov et al. (1991).

tion; the peculiarities of the wave spectrum evolution will be considered second.

Nonlinear evolution of the random profile wave, according to the conclusions formulated in Section 2, is governed by a Riemann solution (2.4a),

$$v(x,y) = f(y + \alpha v x). \tag{5.1}$$

Introduce the one-point velocity probability density function for the velocity v. By definition

$$w(v,x,y) = \langle \delta(v - v(x,y)) \rangle, \tag{5.2}$$

where δ is the Dirac delta function and the angle brackets denote an average over an ensemble of random realizations.

In the ensuing discussion we shall treat ergodic processes, for which the averaging over an ensemble is equivalent to that with respect to time for any chosen realization. Then the probability density function w will simply be proportional to the total relative period in which values of v are in the interval $(v, v + \Delta v)$, namely

$$w(v,x,y) = \lim_{\Delta v \to 0} \left(\frac{\sum_i \Delta y_i}{T \Delta v} \right) = \frac{1}{T} \sum_i \left| \frac{\partial y}{\partial v} \right|_{v=v_i}, \tag{5.3}$$

where Δy_i means a time during which v stays in the interval, and the sum is taken over all these intervals in the portion $|y| \le T$. The same result comes out directly from (5.2) if we replace the averaging by an integration with respect to time.

Consider first a Riemann (simple) wave (5.1) with no discontinuities. If we again write $\xi = y + \alpha v x$ then (see Section 2)

$$\frac{\partial v}{\partial y} = \frac{\partial v}{\partial \xi} (1 - \alpha x v_\xi)^{-1} \tag{5.4}$$

where the value $v_\xi = u$ is known from the boundary condition to be equal to the derivative $\partial v / \partial t$ at $x = 0$. This corresponds to a transition to the "Lagrangian" description (Gurbatov & Malakhov, 1984). Therefore

$$w(v) = \frac{1}{T} \sum_i \frac{(1 - \alpha x u_i)}{|u_i|} = w_0 - \sum_j \alpha \operatorname{sign} u_j, \tag{5.5}$$

where u_j is the value of u at a given v_j.

Hence, in a statistically homogeneous Riemann wave without discontinuities the probability distribution at any point is the same as that on the boundary. Indeed, in such a wave the positive and negative values of u are alternating and the last term in (5.5) vanishes, so that

$$w(u,x) = w_0(u). \tag{5.6}$$

The above result means that the interval contraction Δy with $u > 0$ is compensated for by an interval extension Δy with $u < 0$ at the same value of v (since $\Delta y \sim \alpha v$), so that $\Delta y_+ + \Delta y_-$ is constant. It is to be noted that if a determinate wave is given at the input, i.e. $v = v_0 \sin \omega\xi$, a well-known result is obtained:

$$w_0 = \frac{1}{\pi v_0 \cos \omega\xi} = \frac{1}{\pi\sqrt{1 - v^2/v_0^2}}. \tag{5.7}$$

Along with w_0, all statistical moments that are expressed by means of the one-point density function are conserved, in particular, the average values of any function of v, including the average energy, which is proportional to $\langle v^2 \rangle$. Naturally, in the region preceding the discontinuity formation the energy of a noise wave should be conserved. This is an obvious result of the area conservation in a Riemann wave (any function of v generates a Riemann wave as well). However, this result may not be applied to the multipoint distribution functions that define, in particular, the wave spectrum.

In the general case, a given v is represented by a continuous range of values of u so that the sums in (5.5) can be replaced by integrals. For the i-th interval $(v_i, v_i + \Delta v_i)$ we denote $w_{0i}(v_i) = \widetilde{w_0}(v,u)\mathrm{d}u$, where $\widetilde{w_0}$ is the velocity probability density on the axis of u; then

$$w_0(\tau) = \int_{-\infty}^{\infty} \widetilde{w_0}(v,u)\,\mathrm{d}u$$

and, in accordance with (5.5),

$$w(v,x,y) = \int_{-\infty}^{\infty} \widetilde{w_0}(v,y,u)(1 - \alpha u x)\,\mathrm{d}u. \tag{5.8}$$

Note that in a statistically homogeneous field, where the values u and $-u$ for a given v have equal probabilities, $\widetilde{w_0}$ is an even function of u and hence

$$\int_{-\infty}^{\infty} w_0 u\,\mathrm{d}u = 0,$$

i.e., $w = w_0$, which is in conformity with what was stated previously.

When discontinuities arise the situation is changed drastically since, firstly, some portions of a Riemann wave are "eaten up" and, secondly, the coalescence of discontinuities is possible. Here the discontinuity statistics are of interest. To begin with let us assess the average number of discontinuities that emerge at a specified distance x. As is clear from (2.6), the first condition for the discontinuities to arise is

$$1 - \alpha x u = 0. \tag{5.9}$$

Consequently, the discontinuities appear at distances $L_N = (\alpha u_{\max})^{-1}$. As long as the number of discontinuities is small and they do not merge (or only seldom) the average number of discontinuities at a given distance x and time interval T is equal to

$$n_T(x) = \lambda(1/\alpha x), \tag{5.10}$$

where λ is the average number of crossings of the level $u_x = 1/\alpha x$ by the derivative u from below.

Later on, as a result of coalescence, the stronger discontinuities absorb the weaker ones, and the characteristic field scale grows (Tatsumi & Kito, 1972). To estimate the number of discontinuities at this stage we resort to the following procedure (Gurbatov & Saichev, 1980, 1981). If n is the average number of discontinuities per unit time, then $\tau_s = n^{-1}$ will appear as an average time interval between them. The rate of decrease of n due to coalescence is proportional to n itself and to the ratio between the characteristic discontinuity convergence speed Δc_s and the characteristic distance between them,

$$\frac{dn}{dx} \simeq -n\frac{\Delta c_s}{\tau_s c_0^2} = -n^2\frac{\Delta c_s}{c_0^2}. \tag{5.11}$$

In order to estimate roughly the value of Δc_s one can proceed as follows. We shall suppose that the characteristic interval between the discontinuities τ_s is much longer than the initial correlation scale τ, i.e. the adjacent discontinuity velocities are uncorrelated. Then to evaluate Δc_s we can simply take a root-mean-square value of the discontinuity velocity (of its nonlinear portion δc_s, to be more exact). This portion can be estimated according to (2.7) if we consider a part of wave profile lying between two adjacent zeros t_1 and t_2 and having the form (3.7) (see Figure 2.7*b*). It is easy to see that the integral $\int_{t_1}^{t_2} v dt$, v being the particle velocity (which is obviously an invariant), is proportional to δc_s, namely

$$\varepsilon \int_{t_1}^{t_2} v(t) dt = \delta c_s (t_2 - t_1).$$

Then, finding the r.m.s. of δc_s we have

$$\langle \Delta c_s^2 \rangle \simeq \langle \delta c_s^2 \rangle = \frac{\varepsilon^2}{\tau_s^2} \int_0^{\tau_s} (\tau_s - \tau') B(\tau') d\tau', \tag{5.12}$$

where $B(\tau) = \langle v(t) v(t+\tau) \rangle$ is the correlation function of the input velocity. If, as has already been mentioned, the field correlation scale is small, i.e. $B(\tau)$ decays rapidly, then the integral in (5.12) will depend on the value of the initial energy spectrum density at zero frequency, $B(0) = D$, and thus

$$\Delta c_s^2 = \begin{cases} \frac{\varepsilon^2}{\tau_s^2} \int_0^\infty B(\tau')\, d\tau' = \frac{\varepsilon^2 D \tau_0}{\tau_s^2} = \frac{\varepsilon^2 \sigma^2 \tau_0}{\tau_s^2}, & D \neq 0, \\ \frac{\varepsilon^2}{\tau_s^2} \left(- \int_0^\infty \tau' B(\tau')\, d\tau' \right) & \\ = \frac{\varepsilon^2}{\tau^2} \int_{-\infty}^{\infty} \frac{g_0(\omega)}{\omega^2}\, d\omega = \frac{\varepsilon^2 \sigma^2 \tau_0^2}{\tau_s^2}, & D = 0. \end{cases} \tag{5.13}$$

Here τ_0 is a time scale, i.e. the characteristic field period, and σ_0^2 stands for the velocity variance (i.e. mean-square-wave amplitude) at $x = 0$.

Substituting these expressions into (5.11) and using $n = \tau_s^{-1}$ we readily find

$$\tau_s(x) = \begin{cases} \tau_0 \left(\frac{x}{x_N} \right)^{2/3}, & D \neq 0 \\ \tau_0 \left(\frac{x}{x_N} \right)^{1/2}, & D = 0. \end{cases} \tag{5.14}$$

Here $x_N \sim c_0^2 \tau_0 / \sigma$ is a characteristic length of the nonlinear processes. Therefore, the distance between the shocks is increased by their coalescence. It should be recalled for the sake of comparison that the duration of a single pulse grows as $x^{1/2}$ (i.e. slower than in the first formula of (5.14) where $D \neq 0$), while the periodic sawtooth wave period (this case conforms to $D = 0$) does not change at all.

Now, nothing impairs one from deducing the appropriate formulae for the wave energy density. In the domain of "developed" discontinuities the wave profiles over the regions between the discontinuities are close to linear and have about the same slopes, whereas the shock amplitudes decrease as x^{-1}, and the distances between them remain different. The order of the discontinuity amplitude at this stage is $\ell_0 c_0 / x$, in accordance with (2.17), while the wave intensity I is proportional to $(\ell_0 c_0 / x)^2$, $\ell_0 = c_0 \tau_0$ being a characteristic wavelength. However, over quite a long time the value of ℓ_0 is changed owing to the coalescence, as per (5.14), and thus

$$I = \begin{cases} \sigma^2 \left(\frac{x_N}{x} \right)^{2/3}, & D \neq 0 \\ \sigma^2 \frac{x_N}{x}, & D = 0. \end{cases} \tag{5.15}$$

Hence we see that the noise wave intensity falls off at a slower rate than the regular sawtooth one (for which $I \sim x^{-2}$), again because of the discontinuity motions leading to the growth of field variation scale, i.e. to energy transfer to low frequencies.

It can also be noted that the full saturation effect is lacking for a noise wave: in (5.15) the dependence on the initial amplitude is retained, which makes such a wave similar to a sequence of N-pulses, with the lengths changing due to the discontinuity coalescence.

Spectral characteristics of nonlinear noise

Of great interest is the analysis of the transformation of the field spectrum. As was stated at the beginning of this section, the energy redistribution over the spectrum takes place towards both the low- and high-frequency ranges, whereas at the discontinuous stage a universal law governing the high-frequency decay of the spectral intensity is established, which is determined by the shock fronts. Within the low-frequency range, however, the spectrum evolution continues after the discontinuity formation as well. Owing to their different speeds, a gradual absorption of the small discontinuities occurs, causing growth of the characteristic scale and of corresponding energy transfer into the low-frequency range.

As the appropriate theory is rather cumbersome, we shall offer some formulae without detailed derivation, referring the reader instead to the works by Rudenko and Chirkin (1974) and Saichev (1974). When calculating the spectrum at a given stage of the simple wave one can employ (2.25). To advance the analysis further we need to specify the statistical properties of the signal. As a rule, for the input signal (with $x = 0$) one uses Gaussian statistics. Setting the intensity spectrum from

$$S(\omega, x)\delta(\omega - \omega') = \langle C(\omega, x)C^*(\omega', x)\rangle, \tag{5.16}$$

where C is the complex signal spectrum, and taking into account the fact that for a Gaussian process $\langle \ell^z \rangle = \ell^{(1/2)\langle z^2 \rangle}$, one can obtain

$$S(\omega, x) = \frac{\mathrm{e}^{-\sigma^2\alpha^2\omega^2x^2}}{2\pi\omega^2\alpha^2x^2} \int_{-\infty}^{\infty} \left(\mathrm{e}^{B(\tau)\alpha^2\omega^2x^2} - 1\right) \mathrm{e}^{\mathrm{i}\omega\tau}\, \mathrm{d}\tau. \tag{5.17}$$

Here $B(\tau) = \langle v(t)v(t+\tau)\rangle$ is the correlation function of the random value of velocity $v(t)$ prescribed at $x = 0$, and again $\sigma^2 = B(0) = \langle v^2 \rangle$.

If we expand the exponential in the integrand as a power series in its argument, it is possible to distinguish between various orders of nonlinear interaction within the wave spectrum. In this case, however, we confine ourselves to small values of this argument, i.e. to the range of small distances x. Then (5.17) can be rewritten as

$$S(\omega, x) = S_0(\omega) + (\alpha\omega x)^2 \left[\frac{1}{4\pi}\int_{-\infty}^{\infty} B^2(\tau)\mathrm{e}^{\mathrm{i}\omega\tau}\,\mathrm{d}\tau - \sigma_0^2 S_0(\omega)\right], \quad (5.18)$$

where

$$S_0(\omega) = \frac{1}{2\pi}\int_{-\infty}^{\infty} B(\tau)\exp(\mathrm{i}\omega\tau)\,\mathrm{d}\tau$$

is the input power spectrum. First of all it is useful to remark that $S(0, x) \equiv S_0(0)$ is constant, i.e. the spectral density of the process does not vary at zero frequency. On the one hand this demonstrates the conservation of wave momentum (area), while on the other hand it results from the well-known Manley–Rowe relations which inhibit energy exchange with a zero frequency component in a nondissipative medium. As regards the right-hand side of (5.18), the first term in the square brackets reflects the energy supply to a given spectral component while the second is responsible for the energy decrease because of the generation of higher harmonics (qualitatively, this term corresponds to the Bessel–Fubini series at the initial stage).

Expression (5.18) can also be viewed as the low-frequency asymptotics of the process. Provided $\omega\tau_0 \ll 1$, where τ_0 is the characteristic time of the correlation function decay, the expression in the square brackets is finite as $\omega \to 0$ and, thus, the nonlinear correction to S_0 (the last term in (5.18)) proves to be proportional to ω^2. Therefore, if $S_0(\omega)$ decays faster than ω^2 as $\omega \to 0$, then this correction exceeds the first term as well, and one can consider $S \sim \omega^2$ as the universal low-frequency asymptotic behaviour of the spectrum, at least at the stage before discontinuity formation.

As to the high-frequency portion of the spectrum, (5.18) displays a trend (at small x) towards energy transfer into the higher harmonics due to the wave profile steepening in compression intervals. The build-up of the high-frequency components is observed before the discontinuities emerge in the wave profile. The formation of discontinuities makes a basic contribution to the creation of the high-frequency range of the spectrum. Indeed, for a discontinuous wave, the high-frequency spectrum is proportional to ω^{-1}, so that for the spectral intensity density we have (Kadomtsev & Petviashvili, 1973)

$$S(\omega) \sim \omega^{-2}. \quad (5.19)$$

This law, however, proves valid only for frequencies that are not too high, until the discontinuity width finiteness interferes. A more general result is then obtained, allowing for discontinuity broadening by dissipation, and one can use the formula (4.7) describing the profile of one wave. In this case, for the high-frequency signal asymptotics (cf. (4.9)):

$$C(\omega) = \frac{i\pi\Delta}{\omega_0 \mathrm{sh}\, \pi \frac{\omega\Delta}{2\omega_0}}. \tag{5.20}$$

It is evident from this that the spectrum defined in (5.19) appears only over a range of the wavelengths that are large compared with the discontinuity thickness, where

$$\frac{\pi}{2}\frac{\omega\Delta}{\omega_0} \ll 1,$$

while at higher frequencies, (5.20) implies that

$$S(\omega) \sim \exp\left(-\pi\omega\Delta/\omega_0\right), \tag{5.21}$$

i.e. at the high-frequency end the spectral intensity decays exponentially.

Experimental investigation of intense random waves is generally concerned with measuring the noise of aircraft and rocket engines. In particular, Morfey (1984) offered data on jet engine noise propagation in air. It was observed that the noise damping factor in a spectral band of 1 KHz and above is much less than that predicted by the linear theory. This effect can be explained by the energy transfer to high frequencies due to the discontinuity motions described above.

Laboratory experiments have involved noise propagation in a 29.3 m pipe filled with air subject to sound pressures up to 160 dB ($4.10^3 Pa$) (Pestorius & Blackstock, 1974). The waveforms in the vicinity of the source and 26 m away from it are shown in Figure 2.10. The discontinuity formation and increase of the distances between shocks are evident. One can also observe the signal spectrum extension into both the low- and high-frequency ranges.

Randomly modulated quasi-harmonic signal.

Let a modulated quasi-harmonic signal be given at $x = 0$ as

$$v(y) = v_0(t)\sin\left(\omega_0 t + \varphi(t)\right), \tag{5.22}$$

where $v_0(t)$ and $\varphi_0(t)$ are slowly varying random functions. The solution can again be written in the form of a simple wave:

$$v(y, x) = v_0(y)\sin\left(\omega_0 y + \alpha x v + \varphi(y)\right). \tag{5.23}$$

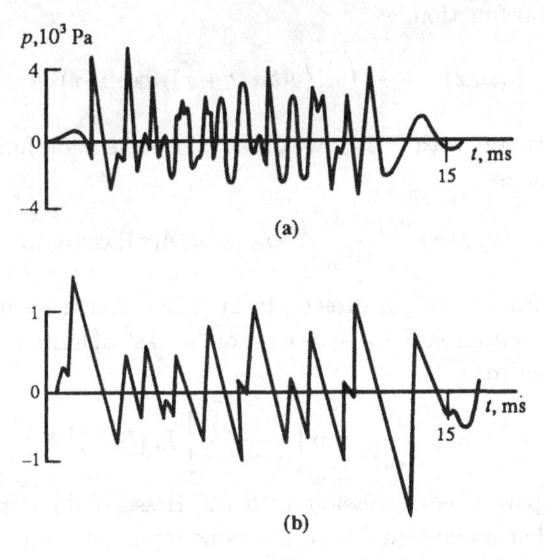

Figure 2.10. Nonlinear noise wave profile in air: (a) near a source, (b) at a distance of 26 m from the source.

Here we neglect nonlinear distortions of the slowly varying amplitude and phase; it is clear that they are small at least over the whole interval prior to discontinuity formation. This problem was considered by Rudenko and Chirkin (1974).

A "modulated" correlation function of the input noise is

$$B_0(\tau) = \sigma^2 A(\tau) \cos \omega_0 \tau,$$

where $A(\tau)$ is a dimensionless amplitude coefficient. Substituting it into (5.17) and making use of the Fourier transform of the function $\exp(z \cos\theta)$, we get

$$S(\omega, x) = \frac{\exp(-q^2)}{2\pi} \left[\int_{-\infty}^{\infty} \frac{I_0(q^2 A) - 1}{q^2} \cos \omega\tau \, d\tau \right.$$
$$\left. + \sum_{n=-\infty, n\neq 1}^{\infty} \int_{-\infty}^{\infty} \frac{I_0(q^2 A)}{q^2} \cos(\omega - n\omega_0)\tau \, d\tau \right], \quad (5.24)$$

where $q = \alpha\sigma\omega x$ and I_n denotes the modified Bessel function. Since the input signal is a quasi-harmonic one, the spectrum (5.24) is concentrated in narrow vicinities of the original signal harmonics. Because of this, we may assume $q_n = \alpha\sigma n\omega_0 x$.

Since the power spectrum is known to be expressed as an integral of the correlation function,

$$S(\omega, x) = \frac{1}{2\pi} \int_{-\infty}^{\infty} \langle v(t)v(t+\tau) \rangle \exp(i\omega\tau)\, d\tau \tag{5.25}$$

then, comparing it with (5.24), we obtain the correlation function of the n-th harmonic as

$$B_n(x, \tau) \sim \frac{\exp(-n^2 q_0^2)}{n^2 q_0^2} I_n \left(n^2 q_0^2 A(\tau)\right) \cos n\omega_0 \tau. \tag{5.26}$$

It is obvious from (5.26), or directly from (5.24), that the wave spectrum appears, in this approximation, as a set of "peaks" of finite width, having intensity equal to

$$I^{(N)} = \frac{4}{(n^2 q_0)^2} \exp\left[-\frac{(nq_0)^2}{2}\right] I_n \left((nq_0)^2 A\right). \tag{5.27}$$

If we compare this expression with the Bessel–Fubini spectrum for the initially harmonic signal with the same input intensity ($I_0^{(N)} = I_0^{(d)}$ where $I_0^{(d)} \sim v_0^2/2$ and $I_0^{(N)} \sim \sigma^2$) then, for example, for the Lorentz type initial signal

$$A(\tau) = \exp(-\tau/\tau_0) \tag{5.28}$$

we can write $I_n^{(N)}/I_n^{(d)} = n!$ for short distances ($nq_0 \ll 1$).

Therefore, harmonic generation in a randomly modulated random signal proves more efficient at the first stage than for a regular signal. This effect relates to the presence of separate (albeit rare) high peaks of the field which are especially pronounced in the higher harmonics.

These results have been, to a certain degree, experimentally verified by Pernet and Payne (1971) in work that was carried out prior to any theoretical developments. A wave with a carrier frequency within the 0.5–3.2 kHz interval and an average amplitude of 200 Pa (acoustic Mach number in air of 1.3×10^{-3}) propagated in a 75 m tube. The input signal was narrow-band noise with a linewidth of 6 per cent of the central frequency. The experiment showed the growth of the noise wave harmonic intensity as compared with the regular wave, the effect increasing with harmonic number: for $n = 2$ and 3 it conformed to the $n!$ law, whereas at large n the growth was slower (about 500 at the seventh harmonic). It might be connected with the formation and rapid dissipation of shocks (in air the discontinuity formation length is about 30 m, which is less than the pipe length).

With the aid of (5.26) one can also assess the correlation time τ_n and, thus, the particular width of each harmonic, $\Delta\omega_n \sim 1/\tau_n$. At short distances $\Delta\omega_n \sim n$, i.e. the bandwidth increases with the harmonic number. Hence we clearly see that these results are only valid for n not too large, so that $\Delta\omega_n \ll \omega$; it is appropriate to recall that at very high frequencies the nonviscous theory does not apply either.

Along with the change of the harmonic build-up rate in a randomly modulated wave, there is an additional broadening of the harmonic spectral components due to the interaction of a low-frequency wave resulting from "self-detection" with the basic frequency harmonics. After the discontinuity formation the occasional peaks decrease due to nonlinear absorption, and a quasi-periodic sawtooth wave with a harmonic composition close to the regular signal spectrum is generated. As it propagates further the low-frequency component continues to grow due to coalescence of discontinuities. The order of the low-frequency component amplitude can be evaluated as $v_n = \Omega\sigma/\omega_0$.

Eventually, all parts will overlap and form a continuous spectrum with the same universal asymptotics as for a broad-band signal treated above. Note that similar asymptotic spectra were investigated experimentally in a pipe by Bjørnø and Gurbatov (1985) (with influence of walls, however).

Notes on acoustic turbulence

Multiple attempts were made to describe quasi-equilibrium "acoustic turbulence" as a random ensemble of interacting waves having a self-similar spectrum with constant energy. Thus, Rudenko (1986) has considered the "forced" Burgers equation with a random right-hand part (as it may happen, say, in near-sonic flows). A stationary random wave process is possible in this case, sometimes with a singularity at zero frequency.

Along with the study of plane noise wave propagation, several attempts have been undertaken to describe self-similar spectra of acoustic turbulence, i.e. of a random ensemble of weakly nonlinear waves propagating in all possible directions. The wave excitation is supposed to occur within the low-frequency range, followed by energy transfer towards higher frequencies due to nonlinear interactions, without energy losses in the inertial frequency range. The condition of energy flow conservation over the spectrum gives rise to the following expression for the spectral energy density (Zakharov & Sagdeev, 1970):

$$s(\omega) \sim \omega^{-3/2}. \tag{5.29}$$

At the same time, due to the absence of dispersion the cumulative wave distortion takes place. This process is accompanied by the formation of the sawtooth wave profile characterized by the spectrum mentioned earlier (Kadomtsev & Petviashvili, 1973)

$$s(\omega) \sim \omega^{-2}. \tag{5.30}$$

A simultaneous allowance for both processes reveals (Naugol'nykh & Rybak, 1975) that within the inertial range of the acoustic turbulence spectrum one can distinguish two regions, one of which is determined by the spectrum of discontinuities and described by relation (5.30) while the second corresponds to the inertial interval in which (5.29) holds. The frequency boundary value ω is of the order shown:

$$\omega \sim \omega_0 / M. \tag{5.31}$$

Essentially, this is determined by a diffusion of the sawtooth wave fronts attributable to their interaction with a chaotic wave ensemble. Some features of three-dimensional acoustic turbulence with shocks has also been considered (Gurbatov et al., 1991).

2.6 Waves in media with "anomalous" nonlinearity

Now we shall discuss the formation and evolution of discontinuities in media with non-quadratic nonlinearity. Taking into account the fact that "anomalous" structural nonlinearity (see Chapter 1, Section 5) often proves much stronger than the "classical" quadratic elastic nonlinearity, we neglect the latter and allow only for an "anomalous" relation between the longitudinal stress and deformation $\varepsilon = \partial u / \partial x$ (here we refer again to one-dimensional waves). Then the following equation of motion will serve as the starting point (Nazarov & Ostrovsky, 1990; Ostrovsky, 1991)

$$\rho_0 \frac{\partial^2 u}{\partial t^2} = \frac{\partial \sigma}{\partial x} + \frac{\partial \sqcap}{\partial x}, \tag{6.1}$$

where ρ_0 is a nonperturbed density, $\sqcap = \sqcap_{xx}$ designates the longitudinal component of the viscous tensor, and $\sigma(\varepsilon)$ is prescribed by the structure of the medium. We do not specify the form of $\sqcap$ as yet (the classical models of viscosity may prove inadequate in such cases), and assume only that

$$\sqcap = O\left(\frac{\partial^2 u}{\partial x^2}\right).$$

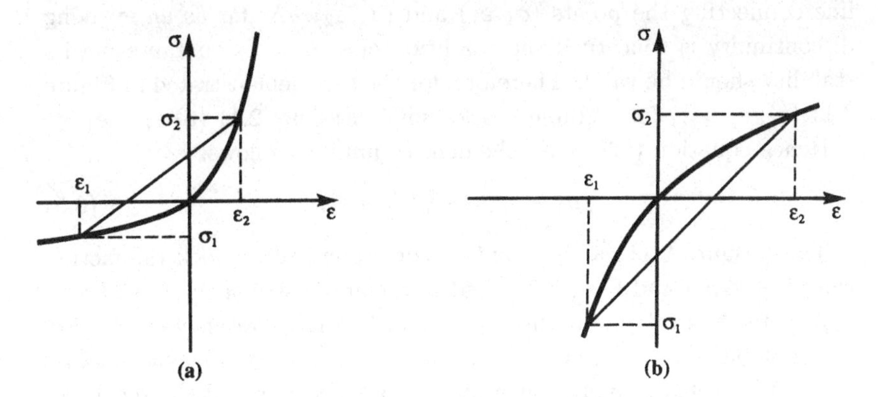

Figure 2.11. Two types of stress-strain relation for which shock is stable: (a) ($\varepsilon_2 > \varepsilon_1$) and (b) ($\varepsilon_2 < \varepsilon_1$).

Given the above assumptions, the following linear continuity equation for density will suffice:

$$\frac{\partial \rho'}{\partial t} + \rho_0 \frac{\partial^2 u}{\partial x \partial t} = 0. \tag{6.2}$$

In order to describe shock waves in such media we shall look for the solution of (6.1) in the form of a stationary travelling wave, i.e. we shall set u_i, σ functions of $\xi = x - c_s t$. Then integrating (6.1) over ξ gives us

$$\rho_0 c_s^2 (\varepsilon - \varepsilon_1) = (\sigma - \sigma_1) + \sqcap, \tag{6.3}$$

where $\varepsilon = \partial u / \partial \xi$ is the deformation, and index 1 is for constant values as $\xi \to \infty$ (ahead of the wave front). In particular, constant values of ε and σ as $\xi \to \infty$ (index 1) and $\xi \to -\infty$ (index 2) are related through a boundary condition giving the shock speed,

$$c_s = \left[\frac{\sigma_2 - \sigma_1}{\rho_0 (\varepsilon_2 - \varepsilon_1)} \right]^{1/2} \tag{6.4}$$

(as was said above, $\sqcap = 0$ if ε is constant).

This formula has a simple geometrical interpretation. In the quasi-static approximation σ is a function of ε (Figure 2.11). Obviously, slope of the tangent line to the curve $\sigma(\varepsilon)$ at each point is proportional to the local sound speed squared; the simple wave sound velocity is

$$c = \left(\frac{1}{\rho_0} \frac{\partial \sigma}{\partial \varepsilon} \right)^{1/2}.$$

According to (6.4), c_s^2 is proportional to the slope of the straight line connecting the points $(\sigma_1, \varepsilon_1)$ and $(\sigma_2, \varepsilon_2)$. As far as an evolving discontinuity is concerned the condition $c_1 < c_s < c_2$ guaranteeing its stability should be valid. Therefore, for the case demonstrated in Figure 2.11 (a), $\varepsilon_2 > \varepsilon_1$ for a stable shock, while in Figure 2.11 (b) $\varepsilon_2 < \varepsilon_1$.

Hence equation (6.2) gives the density jump as follows:

$$\rho_2 - \rho_1 = -\rho_0(\varepsilon_2 - \varepsilon_1). \tag{6.5}$$

Then Figure 2.11 (a) is found to correspond to a shock rarefaction wave ($\rho_2 < \rho_1$) and Figure 2.11 (b) to a compression shock wave ($\rho_2 > \rho_1$). Thus, hydrodynamic theorems on instability of rarefaction shocks† may not be adequate in the case under consideration, whereas entropy is of course certain to grow at a jump (note that losses at the shock are determined by the area enclosed between the curve $\sigma = \sigma(\varepsilon)$ and the line connecting the points 1 and 2).

In the case when the curve $\sigma = \sigma(\varepsilon)$ has a point of inflexion, both compression and rarefaction shock waves are possible.

It should be noted that the above treatment is analogous to that used for electromagnetic shock waves in a nonlinear magnetic or dielectric medium where the role of the curve $\sigma = \sigma(\varepsilon)$ is played by the relation between the magnetic flux and the current (Gaponov–Grekhov *et al.*, 1967).

In the following discussion some examples will be considered.

Modified Burgers equation

When analysing solid media with dislocations in Chapter 1 we saw that a cubic nonlinearity can be comparable with the quadratic one and even be far in excess of it.‡ Therefore, it seems helpful to begin with an idealized model, with $\sigma = E\varepsilon + \beta\varepsilon^3$ where E and β are constants (to be definite assume that $\beta > 0$) and $|\beta\varepsilon^2| \ll E$ (weak nonlinearity).

In (6.1), let

$$\sqcap = -q\frac{\partial^2 u}{\partial x^2} = -q\frac{\partial \varepsilon}{\partial x}$$

(a model with given viscosity). Then, moving to variables $y = t - x/c_0$,

† Incidentally, rarefaction shock waves can also be observed in ordinary "quadratic" elastic media, which was possibly (albeit indirectly, with respect to the second harmonic phase) confirmed experimentally for melted silica (Banas & Breazeale, 1975).

‡ And, of course, cubic terms play the main role for transverse nonlinear waves of different kinds (e.g. shear elastic waves) in isotropic media where due to the symmetry properties all quadratic terms vanish.

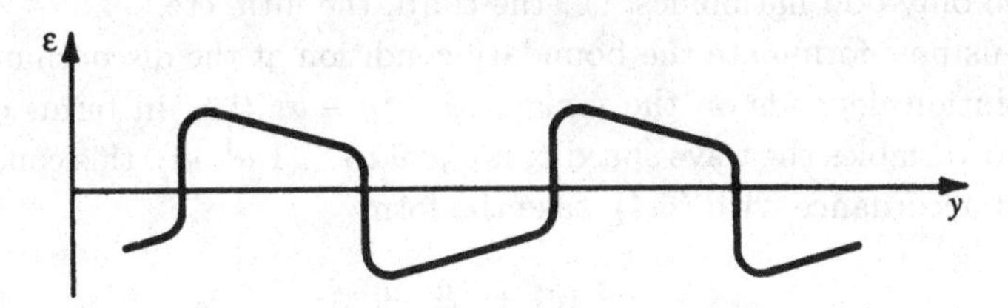

Figure 2.12. Typical profile of an initially harmonic wave propagating in a medium with cubic nonlinearity.

$x' = x$, ($c_0 = (E/\rho_0)^{1/2}$) and assuming the dependence on x to be slow, one can, as in Section 2, readily obtain as the evolution equation, the *modified* Burgers equation

$$\frac{\partial \varepsilon}{\partial x} - \gamma \varepsilon^2 \frac{\partial \varepsilon}{\partial y} = g \frac{\partial^2 \varepsilon}{\partial y^2}, \tag{6.6}$$

where $\gamma = 3\beta/c_0$ and $g = q/c_0^2$.

This equation has been introduced in several works and is given a detailed description, in particular, by Lee-Bapty and Crighton (1987). Let us begin with discussion of the non-viscous case, when $g = 0$ and the solution comes out in the form of a simple wave

$$\varepsilon = f(y + \gamma \varepsilon^2 x) \tag{6.7a}$$

or

$$y = -\gamma \varepsilon^2 x + F(\varepsilon). \tag{6.7b}$$

Here the evolution of the simple wave differs from that in a quadratic medium (Fig. 2.12). As earlier in this chapter, the condition for shock formation can be obtained by differentiating (6.7*b*) twice with respect to ε and assuming $\frac{\partial x}{\partial \varepsilon} = \frac{\partial y}{\partial \varepsilon} = 0$:

$$2\varepsilon_* \gamma x_* = F', \quad 2\gamma x = -F''(\varepsilon_*). \tag{6.8}$$

In particular, for a sinusoidal boundary condition, when

$$F = \frac{1}{\omega} \arcsin \varepsilon/\varepsilon_0,$$

we obtain (taking into account that according to (6.8) $F'' = -\varepsilon_*^{-1} F'(\varepsilon_*)$):

$$\varepsilon_* = \pm \frac{\varepsilon_0}{\sqrt{2}}, \quad x_* = \left(\gamma \sqrt{2} \omega \varepsilon_0^2\right)^{-1}. \tag{6.9}$$

Thus, here the discontinuity is formed simultaneously at two points, in the positive and the negative half-periods. In such a wave one can observe only odd harmonics, i.e. the third, the fifth, etc.

Let us now formulate the boundary condition at the discontinuity. If the solution depends on the variable $\xi = y - bx$ (i.e. in terms of the original variables the wave speed c_s is equal to $c_0(1+bc_0)$), this condition will, in accordance with (6.4), take the form

$$b = -\frac{\gamma}{3}\left(\varepsilon_1^2 + \varepsilon_1\varepsilon_2 + \varepsilon_2^2\right). \tag{6.10}$$

The above-mentioned evolution condition, claiming that the shock front velocity should satisfy the inequalities $c_1 < c_s < c_2$ where c_1, c_2 are the simple wave velocities behind and ahead of the discontinuity, gives, together with the relation (6.10),

$$\varepsilon_1^2 < \left(\frac{1}{3}\right)\left(\varepsilon_2^2 + \varepsilon_2^2 + \varepsilon_1\varepsilon_2\right) < \varepsilon_2^2. \tag{6.11}$$

As we have noted already, this condition will be violated after the point ε_1 coincides with the point at which a straight line coming out of point 2 touches the curve $\sigma(\varepsilon)$. At this condition, the left-hand inequality (6.11) becomes an equality, resulting in $\varepsilon_2 + 2\varepsilon_1 = 0$ (and $b = -\gamma\varepsilon_1^2$). This circumstance to a great extent governs the qualitative aspect of the periodic wave evolution. At some x the shock attains its maximum level when the shock wave velocity coincides with that of the simple wave in front of it. The relation $\varepsilon_2 = -2\varepsilon_1$ is preserved on the shock at all times, while ahead of the shock $c_1 = c_s$; thus the shock trajectory coincides with a characteristic in the region in front of it. Because of this the problem of the evolution of such "sonic" shocks needs special treatment (see Crighton & Scott, 1979; Lee-Bapty & Crighton, 1987).

As to the unipolar pulse with a leading discontinuity propagating into the nonperturbed medium, the shock amplitude can easily be shown to decay as $x^{-1/3}$ (i.e. slower than in the case of quadratic nonlinearity).

The stationary shock structure is deduced from (6.6) with $\varepsilon = \varepsilon(\eta)$, $\eta = y - bx$, yielding

$$g\frac{\partial\varepsilon}{\partial\eta} + (\varepsilon - \varepsilon_1)\left[b + \frac{\gamma}{3}\left(\varepsilon^2 + \varepsilon_1\varepsilon_2 + \varepsilon_1^2\right)\right] = 0, \tag{6.12}$$

where ε_1 is the equilibrium state ahead of the wave front. Let us write the explicit solution for the case $\varepsilon_1 = 0$ (Crighton 1986):

$$\varepsilon = \frac{\varepsilon_2 \exp(-p\xi)}{\left(1 + \exp(-2p\xi)\right)^{1/2}}, \quad \xi = x - c_s t, \tag{6.13}$$

where $p = -\gamma\varepsilon_2^2/3g$ (in this case, according to (6.10), $b = -\gamma\varepsilon_2^2/3$). Therefore, the shock front duration is of the order of $g/\gamma\varepsilon_2^2$.

In this, as well as in previous cases, the asymptotic behaviour of the solution as $\xi \to \pm\infty$ is exponential, but for the limiting sonic shock case mentioned, when $c_s = c_1$, things are different: as was shown in Crighton (1986), in this case one of the asymptotics of a stationary shock solution is of algebraic form. Indeed, if we assume in (6.12) that $\varepsilon(\eta \to \infty) = \varepsilon_1 + \varepsilon_1'$ where ε_1' is small, we obtain $g\partial\varepsilon'/\partial\eta = \gamma\varepsilon_1\varepsilon'^2$ or $\varepsilon' = g/\gamma\varepsilon_1\xi$.

Therefore, the features of the modified Burgers equation solution (this equation being nonintegrable) differ essentially from those of the standard "quadratic" Burgers equation.

Shocks in a bimodular medium

Another type of anomalous nonlinearity treated in Chapter 1 is that of a bimodular medium, where the elastic moduli for compression and tension are characterized by different constants (Nazarov & Ostrovsky, 1990):

$$\sigma = \begin{cases} E_+\varepsilon, & \varepsilon > 0 \\ E_-\varepsilon, & \varepsilon < 0. \end{cases} \tag{6.14}$$

Note that the standard quadratic nonlinearity can be neglected for waves that are not too strong, namely, for $|E_+ - E_-| \gg E_\pm\varepsilon_{\max}$ (provided the quadratic term in the expansion of σ in powers of ε relates to the linear one by a factor of order $\varepsilon_{\max}$). The discontinuity velocity, according to (6.4), equals

$$c_s = \frac{E_+\varepsilon_2 - E_-\varepsilon_1}{\rho_0(\varepsilon_2 - \varepsilon_1)}. \tag{6.15}$$

This formula holds true for any E_+ and E_-, but the case of a small nonlinearity is under consideration, as everywhere in these discussions, and so $|E_+ - E_-| \ll E_\pm$. For definiteness, let $E_+ > E_-$; then σ and ε grow while ρ decreases at the shock front.

Consider the evolution of an initially harmonic wave in such a medium. It is clear that in the absence of discontinuities each half period of the wave (either positive or negative) would propagate linearly without changing its shape at a velocity $c_\pm = (E_\pm/\rho_0)^{1/2}$. Due to the sound speed jump, however, discontinuities form immediately in the wave. Provided that reflections from the discontinuities can be neglected, the solution outside the discontinuities remains sinusoidal:

$$\varepsilon_\pm = \varepsilon_0 \sin\omega(t - x/c_\pm). \tag{6.16}$$

To embody the discontinuity evolution we resort to the method of characteristics. Write (6.16) in the form

$$\omega t - k_\pm x = \arcsin \varepsilon_\pm / \varepsilon_0, \tag{6.17}$$

where $k_\pm = \omega / c_\pm$. Subtracting these two equations in (6.17) yields

$$(k_- - k_+)x = \arcsin \frac{\varepsilon_+}{\varepsilon_0} - \arcsin \frac{\varepsilon_-}{\varepsilon_0}. \tag{6.18}$$

For weak nonlinearity, when E_+ and E_- are close to each other, the solution acquires a rather simple form. Set $E_\pm = \overline{E} \pm e$ where $\overline{E}$ is the mean value of E_+ and E_- and $e \ll \overline{E}$. Then

$$c_\pm = \bar{c}_\ell \left(1 \pm \frac{e}{2\overline{E}}\right),$$

where $\bar{c}_\ell = \sqrt{\overline{E}/\rho_0}$, and according to (6.16)

$$c_s = \bar{c}_\ell \left[1 + \frac{e}{\overline{E}}\left(\frac{\varepsilon_2 + \varepsilon_1}{\varepsilon_2 - \varepsilon_1}\right)\right]. \tag{6.19}$$

Return to equations (6.7) and (6.8) where, obviously,

$$k_\pm \approx \bar{k}\left(1 \mp \frac{e}{2\overline{E}}\right), \quad \bar{k} = \frac{\omega}{\bar{c}_\ell}.$$

The two equations (6.17) are clearly compatible provided $\varepsilon_- = \varepsilon_+$, $\omega t - \bar{k}x = 0$. This means that the jump can be brought to the point of zero phase and considered as symmetrical, i.e. $\varepsilon_2 = -\varepsilon_1 = \varepsilon_s$; here according to (6.19) $c_s = \bar{c}_\ell$, and symmetric deformation of the wave occurs as was observed in the "quadratic" case (note that the "equal area rule" may be proved valid in this case too) (see Fig. 2.13). Consequently (6.18) yields an explicit formula for the jump ε_s:

$$\varepsilon_s = \varepsilon_0 \sin \frac{(k_- - k_+)x}{2} \simeq \varepsilon_0 \sin \frac{\bar{k}e}{2\overline{E}}x. \tag{6.20}$$

Thus, after initially rising to reach ε_0, ε_s then vanishes at a finite time (in contrast to the "classical" quadratic case) and the wave energy is completely dissipated at finite distance $L_d = 2\overline{E}/\bar{k}e$. These results are, however, only valid in the first approximation with respect to e.

2.7 Sound excitation and amplification by moving sources

Sound generation by a moving distributed source whose velocity is near to that of sound in the medium can be so efficient that nonlinear effects manifest themselves in the wave being generated. Such a situation

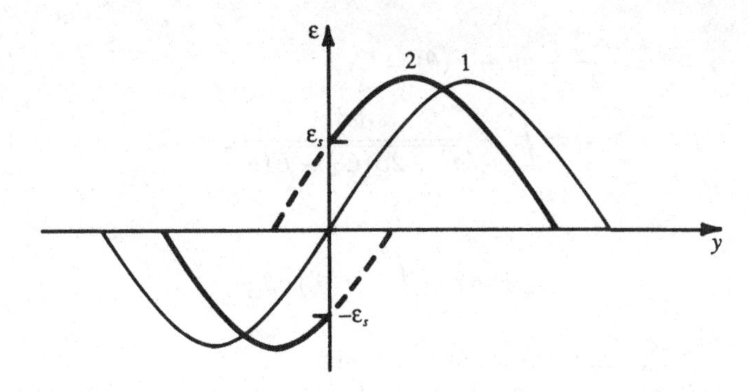

Figure 2.13. Evolution of an initially harmonic wave in a bimodular medium. Initial waveform (1); waveform at some distance x (2).

may, for example, arise in the case of the thermo-optical excitation of sound by a laser beam, or in stimulated Mandelstam–Brillouin scattering (Rudenko, 1974; Karabutov et al., 1976).

In simple cases a problem may be reduced to the solution of the forced Burgers equation

$$\frac{\partial v}{\partial x} = \alpha v \frac{\partial v}{\partial y} + \delta \frac{\partial^2 v}{\partial y^2} + f(\theta), \tag{7.1}$$

where $\theta = t - x/w = y + qx$ and $q = (w - c_0)/c_0^2$ is the parameter of desynchronism between the exciting force and the sound; it should be noted that $|w - c_0| \ll c_0$, otherwise the Burgers equation becomes inapplicable.

Consider first the case of large Reynolds numbers, when the viscous term can be neglected and the first-order equation

$$\frac{\partial v}{\partial x} = \alpha v \frac{\partial v}{\partial y} + f(y + qx) \tag{7.2}$$

must be solved. If we replace the variables x, y with x, θ, then (7.2) takes on the form

$$\frac{\partial v}{\partial x} = (\alpha v - q)\frac{\partial v}{\partial \theta} + f(\theta). \tag{7.3}$$

It can be solved by the method of characteristics. The characteristic system corresponding to (7.3) comes out as

$$\mathrm{d}x = -\frac{\mathrm{d}\theta}{\alpha v - q} = \frac{\mathrm{d}v}{f}, \tag{7.4}$$

with the following integrals:

$$\frac{\alpha v^2}{2} - qv + F(\theta) = C_1, \tag{7.5a}$$

$$x = \pm \int^{\theta} \frac{\mathrm{d}\theta'}{\sqrt{q^2 + 2\alpha(C_1 - F(\theta'))}} + C_2. \tag{7.5b}$$

Here

$$F(\theta) = \int_{-\infty}^{\theta} f(\theta')\,\mathrm{d}\theta',$$

and C_1 and C_2 are integration constants.

Let us discuss a localized source having the form of a single pulse ($f > 0$) such that

$$F_\infty = \int_{-\infty}^{\infty} f\,\mathrm{d}\theta$$

is finite. To begin with we shall handle stationary solutions depending only on θ, i.e. moving synchronously with the source. These solutions are expressed simply in formula (7.5a), which describes a wave in the form of a “step”.

Choosing the constant C_1 proves somewhat difficult under stationary conditions; it can be chosen based on the causality principle. For a subsonic source ($q < 0$) the field can only propagate ahead of it so that $v(\theta \to \infty) = 0$. Then it follows from (7.5a) that

$$\alpha v = \sqrt{q^2 + 2\alpha(F_\infty - F(\theta))} - |q|. \tag{7.6a}$$

In a supersonic situation ($q > 0$) the radiation is left behind so that $v(\theta \to -\infty) = 0$. Then

$$\alpha v = q - \sqrt{q^2 - \alpha F}. \tag{7.6b}$$

It is implied here that a smooth solution only exists subject to

$$q > \sqrt{2\alpha F_\infty} = q_0. \tag{7.7}$$

If $q > q_0$ the solution becomes multivalued, i.e. it acquires a singularity (of the discontinuity type) and (7.6b) becomes inadequate.

More physically realistic results have been obtained for nonstationary processes. Consider first the linear approximation, when the term $\alpha v v_\theta$ can be neglected in (7.3). If one handles a problem in a half-space $x > 0$ and sets a zero boundary condition $v(\theta, x = 0) = 0$, then the solution acquires the simple form

$$v = \frac{F(\theta) - F(y)}{q}. \qquad (7.8)$$

The solution (for $q \neq 0$) describes a positive pulse with diverging fronts. When $q < 0$ a "free" front goes ahead (with velocity c_0) while a "forced" front is moving behind (with velocity w). It is clear that allowance for even a weak nonlinearity leads to a discontinuity in the leading front. Therefore, in the case of a subsonic source a shock wave will always be generated ahead of it. If $q > 0$ the free solution is left behind and corresponds to a rarefaction wave. As regards the "forced" front moving ahead, it breaks only at sufficiently small q, as compared to the stationary case.

Finally, at $q = 0$ (exact synchronism) (7.8) yields

$$v = xf(\theta), \qquad (7.9)$$

i.e., the field grows secularly without limit in the linear approximation.

For the nonlinear stage one needs to use the full solution (7.5); besides, in the general case $C_1 = \psi(C_2)$ where ψ is an arbitrary function. Under the zero boundary condition we have

$$x = \mp \int_{\theta}^{G(\theta,v)} \left[(\alpha v - q)^2 + 2\alpha(F(\theta) - F(\theta'))\right]^{-1/2} d\theta', \qquad (7.10)$$

where

$$G(\theta, v) = \psi\left[\alpha\left(\frac{\alpha v^2}{2} - qv\right) + F(\theta)\right]$$

and ψ in this case is the function inverse to $F(\theta)$, i.e. $\theta = \psi(F)$.

Given $f(\theta) = A/\cosh^2\theta$, i.e. $F(\theta) = A\tanh\theta$, the integral in (7.10) can be reduced to elementary functions. The analysis shows that $q < q_0$ (in both subsonic and supersonic cases) leads to an overlap while $q > q_0$ implies a continuous solution. Provided $0 < q < q_0$ the field remains localized near the source whereas other values of q give rise to a stepline transition (cf. the linear solutions (7.6a), (7.6b) and (7.9)). The maximum value of αv is (Gusev & Rudenko, 1979)

$$\alpha v_{\max} = \begin{cases} \sqrt{q^2 + q_0^2} - |q|, & q < q_0 \\ q - \sqrt{q^2 - q_0^2}, & q \geq q_0. \end{cases} \qquad (7.11)$$

which corresponds to the solutions (7.6) for $q < 0$ and $q \geq q_0$, while the solution for $0 < q < q_0$ is the same as in the case of the "subsonic" formula (7.6*a*). Thus bifurcation occurs, followed by a jump of $v_{\max}$, setting up a regime unstable with respect to small field perturbations.

As follows from the estimates, the laser excitation of sound promotes quite realistically the manifestation of acoustic nonlinearity. To this end one should use absorbing media (organic solvents) with a light attenuation coefficient $\alpha_\ell \sim 10^2$–$10^3 \mathrm{cm}^{-1}$.

In an experiment on light pulse propagation in such a solvent, a pulse with a wavelength 1.06 μm had a power of $W_\ell = 28\,\mathrm{MW}$ (intensity I about 3 MW cm^{-2}) and duration t_0 slightly exceeding 10 ns. Then under the synchronous excitation, as was predicted, an acoustic pulse with a Mach number $M \approx 10^{-3}$ emerged in the medium (resulting in $P = 3\,\mathrm{MPa}$ for the pressure amplitude). Here the nonlinearity showed itself over a length L of order 10^{-1} cm.

Periodic source

Consider now the Burgers equation (7.1) in the general case of finite Reynolds numbers. Using a Hopf–Cole transform (4.1), we reduce it to the linear equation

$$\frac{\partial \zeta}{\partial x} = \delta \frac{\partial^2 \zeta}{\partial y^2} + \frac{\alpha \zeta}{2\delta} \int f\, dy. \tag{7.12}$$

Let us discuss here the case of a sinusoidal source travelling synchronously with the wave, namely $f = A \sin \Omega y$, so that (7.12) takes the form

$$\frac{\partial \zeta}{\partial x} = \delta \frac{\partial^2 \zeta}{\partial y^2} - B\zeta \cos \Omega y, \tag{7.13}$$

where $B = \alpha A / 2\delta\Omega$. We shall be seeking the time-periodic forced solutions in a half-space $x \geq 0$ satisfying a zero boundary condition $v(0, y) = 0$ or $\zeta(0, y) = 1$ (obviously, the constant value for ζ does not affect the result).

The solution to (7.13) can be obtained using separation of variables (Rudenko, 1974):

$$\zeta = \exp(-\lambda x) z(y), \tag{7.14}$$

and then we arrive at a Mathieu equation for z,

$$\frac{\partial^2 z}{\partial \xi^2} + (h - 2\mu \cos 2\xi) z = 0, \tag{7.15}$$

where $\xi = \Omega y/2$, $h = 4\lambda/\delta\Omega^2$, $\mu = 2B/\delta\Omega$. The periodic solutions to this equation (with period π with respect to ξ or $2\pi/\Omega$ with respect to

y) are the Mathieu functions $a_{2n}(\xi, \mu)$, h_{2n} (or λ_n) being the eigenvalues depending on μ. Then the complete solution for v appears as

$$v = \frac{\delta\Omega}{\alpha}\frac{\partial}{\partial\xi}\ln\sum_{n=0}^{\infty} a_{2n}\mathrm{e}^{-\lambda_{2n}x}a_{2n}(\xi, \mu). \tag{7.16}$$

The coefficients a_{2n}, subject to the above boundary condition, are equal to

$$a_{2n} = \frac{\int_0^\pi a_{2n}(\xi, \mu)\,\mathrm{d}\xi}{\int_0^\pi a_{2n}^2(\xi, \mu)\,\mathrm{d}\xi}. \tag{7.17}$$

It is noteworthy that, as is typical of any Sturm–Liouville problem, λ_{2n} increases with n, i.e. λ_0 is less than all other λ. Consequently, at large x the term with $\mathrm{e}^{-\lambda_0 x}$ dominates in the sum (7.16). Then (7.16) is reduced to the periodic stationary solution

$$v = \frac{\delta\Omega}{\alpha}\frac{\partial}{\partial\xi}\ln\left(a_0(\xi, \mu)\right). \tag{7.18}$$

The latter solution could have been found directly from (7.1), omitting $\partial v/\partial x$.

Therefore, the long-distance solution ceases to depend on x and enters the stationary periodic regime. The form of this solution depends on the value of μ, which has the sense of an equivalent source Reynolds number. At low μ the solution approaches the harmonic one, and with μ increasing it acquires steep fronts so that $\mu \to \infty$ results in discontinuities. In the latter case the solution is described by the elementary formula

$$v = \sqrt{\frac{A}{\alpha\Omega}}\,(\operatorname{sgn}\xi)\cos\xi, \quad \xi = \frac{\Omega y}{2} \in (-\pi, \pi). \tag{7.19}$$

The establishment of an acoustic wave profile occurs at distances $L \sim |\lambda_0 - \lambda_2|^{-1}$. For low μ this value is of the order of the wave attenuation length, while for higher μ it amounts to the order of the nonlinearity length.

On the basis of the nonlinear theory the intensity of the hypersound (at a frequency of the order of 10^{11} Hz) generated under the stimulated Mandelstam–Brillouin scattering was calculated. According to this, light of intensity 10^8 W cm^{-2}, in a quartz plate 0.05 cm thick, is capable of generating hypersound at maximum intensity of 300 W cm^{-2}, whereas the linear theory would predict a value 20 times higher, as was shown by Karabutov (1979).

Active medium

To conclude this section we shall briefly discuss sawtooth wave amplification in media with internal energy sources. Let us consider the generalized Burgers equation in the form

$$\frac{\partial v}{\partial x} = \alpha v \frac{\partial v}{\partial y} + \delta \frac{\partial^2 v}{\partial y^2} + \gamma v. \tag{7.20}$$

Equation (7.20) describes, for example, the thermo-optical generation of sound, not due to external source modulation, but because of light absorption coefficient modulation by the sound wave itself: changing the heat release in a medium, this modulation leads to a positive feedback and, eventually, to sound amplification (Kolomensky, 1989; Uvarov & Osipov, 1987).

At large Reynolds numbers, when the term with δ can be ignored, equation (7.20) lends itself to solution by the method of characteristics. The characteristic equations have the form

$$\mathrm{d}x = -\frac{\mathrm{d}y}{\alpha v} = \frac{\mathrm{d}v}{\gamma v},$$

and their integral is

$$v = \exp(\gamma x)\psi\left\{y - \frac{\alpha}{\gamma} v[1 - \exp(-\gamma x)]\right\}, \tag{7.21}$$

where ψ is an arbitrary function specified by a condition at $x = 0$. For a harmonic input signal $v(0, y) = v_0 \sin \omega t$, we have $\psi(\xi) = v_0 \sin \omega \xi$. Then one can easily obtain the distance at which a discontinuity occurs as

$$L = \frac{1}{\gamma} \ln\left(1 + \frac{\gamma}{\alpha v_0 \omega}\right)^{-1}. \tag{7.22}$$

Hence it is clear that when $\gamma > 0$ this distance proves less than that for $\gamma = 0$ (although when $\gamma < 0$, L may tend to infinity, which means that no discontinuity arises at all) due to loss.

The wave evolution at the discontinuous stage seems amenable to easy description as well; then, in analogy with (2.14), we obtain the equation for the jump amplitude

$$\frac{v_s}{v_0} \exp(-\gamma x) = \sin\left\{\frac{\alpha\omega}{\gamma} v_s [1 - \exp(-\gamma x)]\right\}. \tag{7.23}$$

It is obvious from the above that as $x \to \infty$ the jump amplitude tends to the finite limiting value

$$v_\infty = \frac{\pi\gamma}{\alpha\omega}. \tag{7.24}$$

Therefore, a sawtooth autowave is set up in which energy pumping is compensated by dissipation at a discontinuity. In acetone, at a temperature of 20° C an acoustic wave with frequency 200 MHz grows provided the light intensity I exceeds the threshold value $I = 3.3\,\mathrm{Wcm}^{-2}$, as suggested by Kolomensky (1989). Provided $I = 10\,\mathrm{Wcm}^{-2}$ the stationary wave will have a Mach number of 5×10^{-3}. The possibility exists for infrasound to be excited in the atmosphere by a similar mechanism due to solar radiation.

Instances of negative first (shear) viscosity may also be found. These questions have been tackled for some time in connection with models of two-dimensional turbulence; a Burgers equation with $\delta < 0$ has also been brought under consideration. Pelinovsky *et al.* (1984) showed that, given a certain class of initial conditions, this equation is characterized by an explosive-type instability. However, in the general case its solution can be shown to diverge instantaneously, such a model being inadequate for real situations. Meanwhile, the dependence of δ on v (nonlinear viscosity) can lead to physically reasonable stationary solutions, as suggested by Kogan *et al.* (1987).

2.8 Finite-amplitude standing waves

When waves are excited in a bounded system, or resonator, then provided the latter is of high quality, the oscillation amplitude in it appears many times higher than that of the exciting source, which favours the development of nonlinear effects. On the other hand, on reflection from the walls not only the amplitude but also the field phase changes (in fact, dispersion arises), which can dramatically alter the energy exchange conditions between harmonics, and thus the evolution of the wave profile. This point is discussed in more detail in Chapter 6; here we restrict ourselves to a consideration of the waves in a simple one-dimensional resonator made up by two rigid walls located at $x = 0$ and $x = \ell$. The main idea that allows us to simplify the solution of such problems is associated with the fact that in a "quadratic" approximation one can neglect the counterpropagating wave interaction since this is not resonant, and one need only take into account the accumulated nonlinear self-distortions of each travelling wave which is subject to successive reflections on walls with no phase change in pressure, which makes its evolution similar to that in an unbounded system. This is only valid for free oscillations; on their excitation by an oscillating boundary a wave interacts with the latter when travelling "to and fro" each time, and the energy transfer to the field compensates the losses (those at the discontinuities included).

This approach is in agreement with the results obtained using perturbation theory (Eichenvald, 1934; Andreyev, 1959). Let us examine the resonator free oscillations, described by the nonlinear wave equation (of (1.13), Chapter 1)

$$\frac{\partial^2 \varphi}{\partial t^2} - -c_0^2 \frac{\partial^2 \varphi}{\partial x^2} = (\gamma - 1) \frac{\partial \varphi}{\partial t} \frac{\partial^2 \varphi}{\partial x^2} + 2 \frac{\partial \varphi}{\partial x} \frac{\partial^2 \varphi}{\partial x \partial t}. \tag{8.1}$$

In the linear approximation the solution of this equation describes a set of modes in the form of standing waves. Consider one of them,

$$\varphi_1 = \varphi_0 \cos \omega t \cos(\omega x / c_0), \tag{8.2}$$

assuming that the initial condition is specified by (8.2) at $t = 0$. Here

$$\omega_n = \frac{n \pi c_0}{\ell} \quad (n = 1, 2, ...),$$

are the eigenfrequencies.

For the nonlinear problem, let us search for a correction to (8.2) by a perturbation technique, setting $\varphi = \varphi_1 + \varphi_2$, $\varphi_2 \ll \varphi_2$; then (8.1) yields a linearized equation for φ_2,

$$\frac{\partial^2 \varphi_2}{\partial t^2} - c_0^2 \frac{\partial^2 \varphi_2}{\partial x^2} = \frac{\omega^3}{c_0^2} \varphi_0^2 \sin 2\omega t \left(\frac{\gamma - 1}{2} \cos^2 \frac{\omega x}{c_0} - 2 \sin^2 \frac{\omega x}{c_0} \right), \tag{8.3}$$

subject to the initial condition $\varphi_2(0, x) = 0$ and boundary conditions

$$\frac{\partial \varphi_2}{\partial x} = 0 \quad \text{at} \quad x = 0, \ell.$$

The solution of (8.3) is easily obtainable and results in the following expression for the velocity correction v_2:

$$v_2 = \frac{\varepsilon v_0^2 \omega t}{c_0} \sin 2\omega t \sin \frac{2\omega x}{c_0}. \tag{8.4}$$

Therefore, in this approximation the second harmonic appears with its amplitude growing secularly with time. This corresponds to the initial stage of nonlinear wave distortions.

A more general approach has to do with the description of noninteracting travelling waves, each of which is described by the Burgers equation or at large *Re* corresponds to a simple wave which, generally, contains

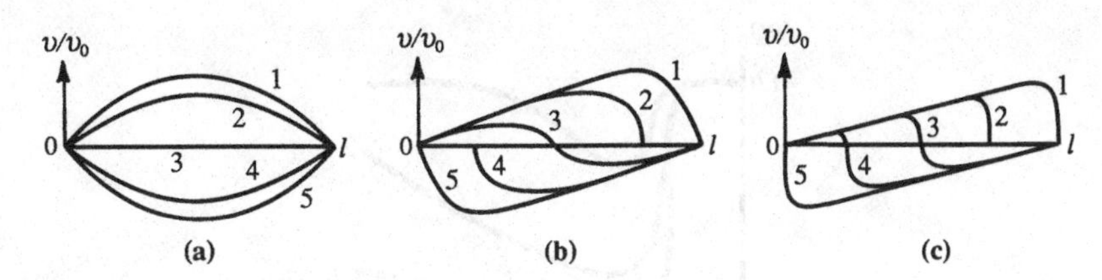

Figure 2.14. Standing waves of velocity v in a nonlinear resonator shown with time step $T/8$ at various values of σ: (a) $\sigma \ll 1$, (b) $\sigma = 1$, (c) $\sigma = \pi/2$.

discontinuities. In the latter case, under a sinusoidal initial condition it seems convenient to proceed from the solutions in characteristic form (Kaner *et al.*, 1977)

$$\begin{aligned} \frac{\omega x}{c_0} &= \arcsin\frac{v_1}{v_0} + \sigma\frac{v_1}{v_0} + \omega t, \\ \frac{\omega x}{c_0} &= \arcsin\frac{v_2}{v_0} + \sigma\frac{v_2}{v_0} + \omega t, \end{aligned} \tag{8.5}$$

where indices 1 and 2 refer to the waves travelling to the right and to the left; here $\sigma = \varepsilon v_0 \omega t / 2c_0$.

Equation (8.5) automatically satisfies the boundary conditions $v = 0$ at $x = 0,\ \ell$. Indeed, if one sets $v_2 = -v_1$, then at $x = 0,\ \ell$ the equations become identical (in view of the fact that $\omega_n = n\pi c_0/\ell$ is the system eigenfrequency, and $\arcsin v_2/v_0 = -\arcsin v_1/v_0 + 2n\pi$). Thus, the solution for the field inside the resonator is specified simply by the superposition $v = v_1 + v_2$; the oscillation profiles at any point are easily plotted. The result is depicted in Figure 2.14. The curves correspond to equal time intervals $\Delta T_i = T/8$ ($T = 2\pi/\omega$) at various values of the nonlinear scale σ. The nonlinear distortions accumulate with time; at $\sigma = 1$ (i.e. at $t = t_s = T/\varepsilon\pi M$, M being the Mach number) discontinuities emerge in each opposing wave, and when $\sigma \geq \pi/2$ the field in the resonator appears as a sum of sawtooth waves with discontinuity amplitudes $v_s = \pi v_0/(1 + \sigma)$.

Such a damped "standing" wave with a wide spectrum has a series of specific features. The particle velocity nodes are "tied" to the walls, as in the linear case, while the density nodes as well as the velocity and density peaks move across the resonator. Moreover, strictly speaking, an additional "running" node lying on a discontinuity is observed in the velocity distribution. As to the discontinuity itself, the ratio between

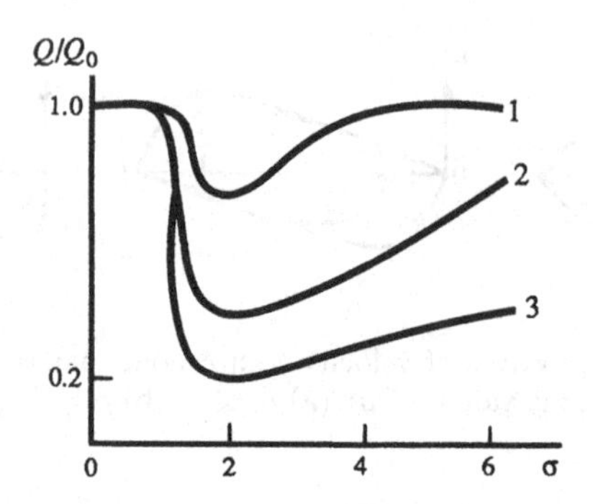

Figure 2.15. Time dependence of the nonlinear resonator quality factor Q at various values of a Reynolds number *Re*: curve 1, *Re* = 5; curve 2, *Re* = 10; curve 3, *Re* = 20.

its positive and negative parts changes with propagation so that the discontinuity arrives at another velocity node with its polarity changed.

Note that for the initial stage, calculating a second harmonic in each wave by the Bessel–Fubini formula for $\sigma \ll 1$, one obtains formula (8.4), as expected.

Allowing now for finiteness of the Reynolds number, an analogous description of oscillations in a resonator is achieved through a solution of the Burgers equation with respect to the two opposing waves (Kaner *et al.*, 1977). Here, at the final stage the sawtooth wave has again turned into a sinusoidal one at the fundamental frequency, as was the case for the travelling wave. Note that the resonator quality factor Q, defined as the ratio of the total oscillation energy to the energy dissipated over one period, is not constant. Initially Q is close to its value in the linear approximation, while in the course of the nonlinear evolution, and in particular after the discontinuity formation, Q drops abruptly due to nonlinear dissipation; at the final stage, however, when nonlinearity is practically insignificant, Q returns again to its "linear" value (Figure 2.15, Kaner *et al.*, 1977).

Let us discuss now the problem of forced oscillations of a gas column in the resonator (a pipe) of length ℓ. The oscillations will be assumed to be excited by a piston mounted at one end of the pipe performing harmonic oscillations with a velocity

$$v = v_0 \sin \omega t, \tag{8.6}$$

the other end of the pipe being plugged (Gor'kov, 1963). The nonlinear oscillations are described by (8.1). In the linear approximation we have

$$v = -\frac{v_0 \omega_0}{2\pi \Delta\omega} \cos kx, \tag{8.7}$$

where ω_0 is the resonance frequency and $\Delta\omega = \omega - \omega_0$ is the frequency shift from resonance. In the second approximation the solution depends on the intensity of the oscillations, this depending in turn on the "detuning" $\Delta\omega$ (Goldberg, 1987). If its value is less than the critical one,

$$\frac{\Delta\omega_k}{\omega_0} = \frac{2}{\pi}\left(\frac{2\varepsilon v_0}{\omega \ell}\right)^{1/2},$$

the solution contains shocks which, upon reflection from the resonator ends, travel across the pipe twice per oscillation period.

The discontinuous solution satisfying equation (8.1) and the condition of mass conservation in a tube lead to the following expression for the shock amplitude:

$$p_s = 4\rho_0 c_0^2 \sqrt{\frac{v_0}{\omega \ell(\gamma+1)}} \sqrt{1 - \frac{\pi^2 \ell \omega}{4(\gamma+1)v_0} \frac{\Delta\omega}{\omega_0}}. \tag{8.8}$$

At $\Delta\omega = 0$ (resonance case) we obtain

$$p_s = 4\rho_0 c_0^2 \sqrt{\frac{v_0}{\omega_0 \ell(\gamma+1)}}. \tag{8.9}$$

It is noteworthy that one can deduce this formula from simple energy considerations (Ostrovsky, 1974). It will be assumed that two sawtooth waves travelling in opposite directions are generated by a resonator. Then the field damping per single "journey" to and fro across the resonator is small. Therefore (2.16) yields

$$1 - \frac{v_s}{v_{s0}} \simeq \frac{2\varepsilon k \ell v_{s0}}{c_0} = g(v_{s0}), \tag{8.10}$$

where $k = \omega/c_0$ and $\omega = \omega_0 = n\pi c_0/\ell$, and v_{s0} is the discontinuity amplitude at $x = 0$.

Under the boundary condition (8.6), the discontinuity amplitude v_{s0} can be extracted from the condition of energy balance for the first field harmonic, the amplitude v_{10} of which in a sawtooth wave is uniquely related to v_{s0}: $v_{10} = 2v_{s0}/\pi$. The average energy flux from the source at frequency ω is equal to $\rho_0 \ell c_0 v_{10} x_{10} \omega$, x_{10} being the displacement amplitude of the piston. This flow is consumed by the field damping at discontinuities according to (8.10) as well as by the variations of the field mean energy w in time ($w = \rho_0 v_s^2/3\ell$). As a result one can readily write

$$\frac{2\ell}{c_0}\dot{v}_1 + g(v_1)v_1 = \frac{3}{\pi}x_{10}\omega. \tag{8.11}$$

This equation yields a solution in the form

$$\frac{v_s}{c_0} = \frac{2\pi v_1}{c_0} = \sqrt{a_s}\frac{v_{s0} + \sqrt{a_s}\tanh((t-t_0)/\tau)}{\sqrt{a_s} + v_{s0}\tanh((t-t_0)/\tau)}. \tag{8.12}$$

Here

$$a_s = \frac{\pi^3 x_0}{2\ell(\gamma+1)}, \quad \tau = 2\omega^{-1}\sqrt{2\ell/(\gamma+1)x_{10}}.$$

From this it is evident that the time needed to establish the n-th mode of oscillation amounts to $\tau \sim (n\sqrt{x_0})^{-1}$ and the discontinuity amplitude in the stationary regime is $v_{s0} \sim \sqrt{x_{10}}$ as suggested by (8.9).

A more elaborate analysis of the nonlinear oscillation excitation in a resonator (pipe) of a finite length has been performed by Seymour and Mortell (1973). They carried out the calculations, relying on the characteristic equations of type (8.5) neglecting the opposite wave interaction. In the general case the calculation was only performed numerically, but at large Re its results agree with (8.12). Here the effect of sound radiation losses due to the opening in the pipe wall was also accounted for. The decrease in discontinuity magnitude was found to be directly proportional to the radiation loss. This is confirmed qualitatively in the experiments conducted by Sturtevant described in the paper of Seymour and Mortell (1973). Air oscillations were excited in a pipe of length $\ell = 132.5$ inches and inner diameter 3.0 inches. One pipe end contained a piston that performed harmonic oscillations. The other end was closed by a rigid cover in one experimental series whereas in other tests a cover with an opening was used, sized so as to vary the loss factor. It follows from the experimental data that at sufficiently large amplitudes of piston displacement ($x_0/\ell = 0.02$) and moderate losses, the wave amplitude is inversely proportional to radiation losses as mentioned above. With decrease in x_0/ℓ to 0.01, deviations from the linear law are observed. The resonance curve of the system was measured in detail.

2.9 Energy and momentum in a plane sound wave

When considering wave fields, their equations are often formulated as conservation laws (transport equations), which allows one to assess the wave energy and momentum as well as their fluxes. Frequently the equations of transport are developed from the linearized equations of the field. However, the values obtained in this manner that are quadratic with respect to the wave amplitude do not necessarily coincide with the actual energy and momentum associated with the wave.

From the physical point of view, the wave energy is naturally interpreted as the difference between the total energy of a system "medium plus wave", and that of the unperturbed medium, i.e. the energy expended on the wave field excitation.

General expressions for the instantaneous values of the plane wave energy density E and its flux I are shown, for example, in the book by Landau and Lifshitz (1986). Allowing for the linear and quadratic terms,

$$E = \frac{\rho_0 v^2}{2} + \frac{c_0 \rho'^2}{2\rho_0} + w_0 \rho'; \quad I = p'v + w_0 \rho' v + w_0 \rho_0 v, \tag{9.1}$$

where $w_0 = c_0^2/(\gamma - 1)$ is the enthalpy of the unperturbed medium per unit volume.

Analogously, for the density of the x-component of the momentum j and its flux $\sqcap$ we have

$$j = \rho_0 v + \rho' v, \quad \sqcap = p' + \rho_0 v^2. \tag{9.2}$$

As a rule, the time-averages of the above values appear to be the most interesting. In linear problems, when all values are assumed harmonic, one should have omitted, by averaging, all the linear terms in (9.1) and (9.2). In reality, however, as will be shown here, the averages of these terms have, like other terms, a quadratic order of amplitude, and cannot be dropped (Andreyev, 1955).

Accordingly, consider first the general procedure of averaging for any perturbation f corresponding to a one-dimensional simple wave (Ostrovsky, 1968):

$$f = F(\xi) = F\left(t - \frac{x}{u}\right), \quad u = c_0 + \varepsilon v. \tag{9.3}$$

We estimate the average value of f over time. Recall that

$$\mathrm{d}t = \mathrm{d}\xi \left(1 - \frac{x}{u^2}\frac{\mathrm{d}u}{\mathrm{d}\xi}\right) \simeq \left(1 - \frac{x\varepsilon}{c_0^2}\frac{\mathrm{d}v}{\mathrm{d}\xi}\right). \tag{9.4}$$

By definition, at fixed x

$$\int_{t_1}^{t_2} f\,\mathrm{d}t = \int_{\xi_1}^{\xi_2} \mathrm{d}\xi \left(F - \frac{\varepsilon x}{c_0^2} F \frac{\mathrm{d}F}{\mathrm{d}\xi}\right)$$

$$= \int_{\xi_1}^{\xi_2} F \, \mathrm{d}\xi - \frac{\varepsilon x}{c_0^2} \left(F^2(\xi_2) - F^2(\xi_1) \right). \tag{9.5}$$

Meanwhile, the time average is known to have the form

$$\bar{f} = \lim_{T \to \infty} \frac{1}{T} \int_t^{t+T} f \, \mathrm{d}t. \tag{9.6}$$

Since, within this integration interval

$$\xi_2 - \xi_1 = t_2 - t_1 + \frac{\varepsilon x}{c_0^2} (f_2 - f_1) \simeq T$$

and the last term in (9.5) remains finite as $T \to \infty$, we have

$$\bar{f}^t = \bar{f}^\xi = \bar{f}, \tag{9.7}$$

i.e. the averaging can be accomplished with respect to the explicit argument ξ. It is worth mentioning here that the average of the function (9.3) with respect to x differs from that over ξ.

Therefore, $\bar{f}$ depends directly on the function F, which is determined by a boundary condition. For the problem of an oscillating piston with coordinate specified as $x = x_p(t)$ and velocity given by $v = v_p(t) = \dot{x}_p$, the boundary condition for (9.3) is

$$F\left(t - \frac{x_p(t)}{c_0 + \alpha f} \right) = v_p(t). \tag{9.8}$$

Provided $x_p(0) = 0$, then $|x_p| < t|v_p|_{\max}$ always, while in a sound wave $|v_p|_{\max} \ll c_0$; hence the second term in the argument of F in (9.8) is always small compared with the first term. Then one can readily approximate the explicit form of F from (9.8) and write down (9.3) as

$$v = \dot{x}_p(\xi) + \frac{1}{c_0} x_p(\xi) \ddot{x}_p(\xi). \tag{9.9}$$

Now using (9.7) one can easily find the averages of v, and other physical quantities associated with v, in a simple wave, through

$$\rho' = \frac{\rho_0}{c_0} \left(v + \frac{2 - \varepsilon}{2c_0} v^2 \right), \quad p' = \rho_0 c_0 \left(v + \frac{\varepsilon}{2c_0} v^2 \right). \tag{9.10}$$

Given that the average piston speed is zero, the results will have the form

$$\bar{v} = -\frac{\overline{v_p^2}}{c_0}, \quad \overline{\rho'} = -\frac{\varepsilon}{2\rho_0} \overline{\rho'^2}, \quad \overline{p'} = \left(\frac{\varepsilon - 2}{2\rho_0 c_0^2} \right) \overline{p'^2}. \tag{9.11}$$

These results hold for a nondissipative medium up to the discontinuity formation point.

Substituting the expressions (9.10) into (9.1) and (9.2) one can obtain the average wave energy and momentum, together with their fluxes.

If we insert the relations (9.10) for ρ' and v into (9.1) we obtain the universal relation, which is independent of the boundary condition,

$$I = c_0 E + \frac{\varepsilon w_0 c_0}{2\rho_0}\rho'^2. \tag{9.12}$$

This formula is naturally valid for the averages with respect to both t and x. Averaging (9.1) and taking into account (9.11) we have

$$\overline{E} = \rho_0 \overline{v^2} - \frac{\varepsilon w_0}{2\rho_0}\overline{\rho'^2}, \quad I = c_0 \rho_0 \overline{v^2}. \tag{9.13}$$

In particular, for an ideal gas $\overline{E} = \rho_0 \overline{v^2}(3\gamma - 5)/4(\gamma - 1)$ so that for monatomic gas $\overline{E} = 0$ while for polyatomic gas $\overline{E} < 0$, i.e. the energy of a fixed layer of the medium drops as a result of the wave excitation. The fact is that when the wave excitation starts, its leading part entrains mass. Indeed, the perturbation of mass m' over an interval between the piston and an arbitrary point x_0 amounts to $x_0\overline{\rho'}$ and according to (9.11) this value is negative; consequently, in the wave head $\overline{m'} > 0$, which follows from mass conservation and can be shown directly as well.

In the same way, using (9.2) for the momentum density j and the momentum flux $\sqcap$ associated with the wave and substituting (9.10), we obtain the universal relation

$$\sqcap = c_0 j + \frac{\varepsilon \rho_0}{2} v^2. \tag{9.14}$$

Concerning the piston oscillating with zero average velocity, one can write

$$\overline{j} = 0, \quad \overline{\sqcap} = \frac{\varepsilon \rho_0}{2}\overline{v^2}. \tag{9.15}$$

Note that the pressure of such a wave against an immobile wall is equal to $2\overline{\sqcap}$ since the nonlinear interaction of the opposing waves (incident and reflected) has its average effect only at the third order in amplitude.

All these expressions essentially rely only on the condition of no average mass flow in a wave.

Other results are obtained, for example, for a stationary sound beam of finite cross-section. The expressions are formulated in the book by Landau and Lifshitz (1986) for this model. Here, as one might expect, the average pressure is equalized over the cross-section to match that in the beam environment, i.e. $\overline{p}' = 0$. Then using (9.10) one can easily formulate

$$\overline{v} = -\frac{\varepsilon}{2c_0}\overline{v^2}, \quad \overline{\rho'} = \frac{\rho_0}{c_0^2}(1-\varepsilon)\overline{v^2}. \tag{9.16}$$

This yields energy and momentum expressions different from those given above. One needs, however, to bear in mind that, as far as real beams are concerned, even minor dissipation can lead to the appearance of acoustic flows much stronger than $\overline{v}$ in (9.16).

It is instructive to discuss dissipative effects on the average values in a plane wave. The physical difference from the ideal liquid case shows itself clearly for the case of low Reynolds numbers. The solution to the problem of wave excitation by a sinusoidally oscillating piston is, in this case, readily obtained by the perturbation method as was used in Section 1 of this chapter. It has the form $v = v_1 + v_2$, $p' = p'_1 + p'_2$, etc., where index 1 stands for linear approximation while 2 stands for the small nonlinear correction, for which the average values are easily calculated as

$$\overline{v} = -\frac{\overline{v^2}}{c_0}, \quad \overline{\rho'} = -\frac{\varepsilon}{\rho_0}\overline{\rho'^2}, \quad \overline{p'} = -\frac{\overline{p_1'^2}}{\rho_0 c_0^2}. \tag{9.17}$$

A comparison with (9.11) shows that the averaged results at low and high *Re* are entirely different for all values except $\overline{v}$, which is determined by the mass flow value as has already been mentioned. Correspondingly instead of (9.13) one has

$$\overline{E} = \left(\frac{\gamma-2}{\gamma-1}\right)\rho_0\overline{v}^2, \quad \overline{I} = c_0\rho_0\overline{v}^2, \quad \overline{j} = 0, \quad \overline{\Pi} = 0. \tag{9.18}$$

Recently a more elaborate analysis for the problem of average values in a plane wave, allowing for reflections from discontinuities, has been performed (Makarov & Khamzina, 1986). This yielded the same formulae as (9.17), (9.18) for the average values at the sawtooth wave stage.

References

Andreyev, N.N. (1955). On some quantities of the second order in acoustics, *Akhust. Zh.* **1**, 3–11.

Andreyev, N.N. (1959). On standing waves of finite amplitude. Investigations in theoretical and experimental physics, in *Collection in Memory of G. S. Landsberg*, (Izd. AN SSSR., Moscow), 53–55.

Bacon, D.R. (1990). Nonlinear acoustics in ultrasound calibration and standards, in *Frontiers of Nonlinear Acoustics; Proc. of 12th ISNA*, ed.

M.F. Hamilton & D.T. Blackstock (Elsevier Science Publish. Ltd., London).

Bainas, J. & Breazeale, M. (1975). Nonlinear distortion of ultrasonic wave in solids: approach using a stable backwards sawtooth, *JASA* **57,** 745–746.

Bjørnø, L.& Gurbatov, S. (1985). Evolution of universal high-frequency asymptotic forms of the spectrum in the propagation of high-intensity acoustic noise, *Sov. Phys. Acoustics* **31,** 179–182.

Blackstock, D.T. (1962). Propagation of plane sound waves of finite amplitude in nondissipative fluids, *JASA* **34,** 9–30.

Blackstock, D.T. (1964). Thermoviscous attenuation of plane, periodic finite-amplitude sound waves, *JASA* **36,** 534–542.

Blackstock, D.T. (1966). Connection between the Fay and Fubini solutions for plane sound waves of finite amplitude, *JASA* **39,** 1019–1026.

Burov, V.A. & Krasil'nikov, V.A. (1958). Direct observation of the intensive ultrasonic wave shape distortion in a liquid, *Dokl. Acad. Nauk. SSSR* **118,** 920–923.

Cole, J.D. (1951). On a quasi-linear parabolic equation occurring in aerodynamics, *Quart. Appl. Math.* **9,** 225–236.

Courant, G. & Friedrichs, K. (1948). *Supersonic flow and shock waves* (Interscience, New York, London).

Crighton, D.G. & Scott, J.F. (1979). Asymptotic solution of model equations in nonlinear acoustics, *Phil. Trans. R. Soc. Lond.* **292,** 101–134.

Crighton, D.G. (1986). The Taylor internal structure of weak shock waves, *J. Fluid Mech.* **17 (3),** 625–672.

Dean, W. (1962). Interaction between sound waves, *JASA* **34,** 1039–1044.

Eichenvald, A.A. (1934). Acoustic waves of large amplitude, *Usp. Fiz. Nauk.* **14,** 553–555.

Fay, R.D. (1931). Plane sound waves of finite amplitude, *JASA* **34,** 222–241.

Fridman, V.E. (1976). An interaction of pulses with discontinuous fronts, *Akust. Zh.* **22,** 780–782.

Fridman, V.E. (1980). On the nonlinear acoustics of explosive waves, in *Nonlinear Acoustics (collection of papers),* ed. V. Zverev & L. Ostrovsky, (IAP, Gorky, 68-97).

Fubini-Giron (1935). Pressione di radiatione acustica e onde di grande ampiezza, *Alta Frequenza* **4,** 530–581.

Gaponov-Grekhov, A.V., Ostrovsky, L.A. & Freidman, T.I. (1967). Electromagnetic shock waves, *Izv. Vuzov. Radiofizika.* **10,** 1376–1413.

Gol'dberg, Z.A. (1957). On the propagation of plane waves of finite amplitude, *Akust. Zh.* **3,** 322–326.

Gol'dberg, Z.A. (1987). Parametric instability of nonlinear plane standing waves in a liquid, *Problems of nonlinear acoustics, Proc. XI ISNA Novosibirsk* **Part 1,** 154–158.

Gor'kov, L.P. (1963). Nonlinear acoustic oscillations of a gas column in a closed pipe, *Inzh. Zh.* **III,** 246–250.

Gurbatov, S.N. & Malakhov, A.N. (1984). Random waves in nonlinear acoustics, Preprint, *IRE* **11** (383), 48.

Gurbatov, S.N., Malakhov, A.N. & Saichev, A.I. (1991). *Nonlinear Random Waves in Nondispersive Media* (Manchester University Press, Manchester).

Gurbatov, S.N. & Saichev, A.I. (1981). Degeneracy of one-dimensional acoustic turbulence at large Reynolds numbers, *JETP* **80,** 689–703.

Gusev, V.E. & Rudenko, O.V. (1979). Nonstationary quasi-one-dimensional acoustic flows in unbounded spaces allowing for hydrodynamic nonlinearity, *Akust. Zh.* **25,** 875.

Hiedelmann, E.A. & Zankel, K.L. (1958). Simple demonstration of the presence of the second harmonic in progressive ultrasonic waves of finite amplitude, *JASA* **30,** 582–583.

Hopf, E. (1950). The partial differential equation $u_t + uu_n = \mu u_{nn}$, *Comm. Pure Appl. Math.* **3,** 201–230.

Inoue, Y. & Yano, T. (1993). Weakly nonlinear wave radiated by pulsation of a cylinder, *JASA* **93,** 132–141.

Kadomtsev, B.B. & Petviashvili, V.I. (1973). Sound turbulence, *Dokl. Acad. Nauk SSSR* **208,** 794–796.

Kaner, V.V., Rudenko, O.V. & Khokhlov, R.V. (1977). On the theory of nonlinear oscillations in acoustic resonators, *Akust. Zh.* **23,** 756–765.

Karabutov, A.A. (1979). *Excitation of Nonlinear Waves by Distributed External Sources in Nondispersive Media*, Kand. dis. MGU.

Karabutov, A.A., Lapshin, E.A. & Rudenko, O.V. (1976). On the interaction of light radiation with sound under the condition of acoustic nonlinearity, *JETP (Zh ETF)* **71,** 111–121.

Karpman, V.I. (1973). *Nonlinear Waves in Dispersive Media* (Nauka, Moscow).

Khokhlov, R.V. & Soluyan, S.I. (1964). Propagation of acoustic wave of moderate amplitude through absorbing and relaxing media, *Acustica* **14,** 242–247.

Khokhlov, R.W., Naugol'nykh, K.A. & Soluyan, S.I. (1964). Waves of moderate amplitude in absorbing media, *Acustica* **14,** 248–253.

Kogan, E.Ya., Malevich, N.E. & Oraevsky, A.N. (1987). Structure of nonlinear acoustic waves in non-equilibrium vibrationally excited gases, *Pisma ZhTF* **13,** 836–839.

Kolomensky, A.A. (1989). Acoustic autowave in light-absorbing media, *Akust. Zh.* **35,** 370–372.

Kuznetsov, V.P. (1970). Equations of nonlinear acoustics, *Akust. Zh.* **16,** 548–553.

Landau, L.D. & Lifshitz, E.M. (1986). *Hydrodynamics* (Moscow, Nauka).

Lee-Bapty, I.P. & Crighton, D.G. (1987). Nonlinear wave motion governed by the modified Burgers equation, *Phil. Trans. R. Soc., Lond.* **A 323,** 173–209.

Lighthill, M.T. (1956). Viscosity effects in sound waves of finite amplitude, in *Surveys in Mechanics,* ed. G.K. Batchelor & R.M. Davies (Cambridge University Press, Cambridge).

Makarov, S.N. & Khamsina, B.C. (1989). Determination of average parameters of sound field in Earnshaw problem on the base of a nonlinear theory, *Akust. Zh.* **35,** 308–312.

Mendousse, J.S. (1953). Nonlinear dissipative distortion of progressive sound waves at moderate amplitude, *JASA* **25,** 51–54.

Mikhailov, I.G. & Shutilov, V.A. (1960). Finite amplitude ultrasonic wave shape distortion in various liquids, *Akust. Zh.* **6,** 340–346.

Morfey, G.L. (1984). Aperiodic Signal propagation of finite amplitude. Some practical applications, *Proc. X ISNA, Kobe, Japan* , 199.

Nakamura, A. (1984). Recent developments of the research for N-waves, *Proc. Tenth Congress of Non-linear Acoustics,* Kobe, 31–36.

Naugol'nykh, K.A. (1958). Finite amplitude sound wave absorption, *Akust. Zh.* **4,** 115–24.

Naugol'nykh, K.A. & Rybak, S.A. (1975). Sound turbulence spectrum, *ZhETF.* **68,** 78–83.

Naugol'nykh, K.A. & Romanenko, E.V. (1958). On the problem of finite-amplitude wave propagation in liquids, *Akust. Zh.* **6,** 200–2.

Ostrovsky, L.A. (1968). Second order values in the progressive sound wave, *Akust. Zh.* **14,** 140–2.

Ostrovsky, L.A. (1974). Discontinuous oscillations in an acoustic resonator, *Akust. Zh.* **20,** 140–2.

Pelinovsky, E.M. (1976). The spectral analysis of simple waves, *Izv. Vuzov Radiotechnica* **19 (3),** 373–83.

Pelinovsky, E.M., Fridman, V.E. & Engelbrecht, J. (1984). *Nonlinear evolution equations* (Valgus, Tallinn).

Pernet, A.F. & Payne, R.C. (1971). Nonlinear propagation of sound in air, *J. Sound & Vib.* **17,** 383.

Pestorius, F.M. & Blackstock, D.T. (1974). Finite-amplitude wave effects, in *Fluids,* ed. (IPC Science and Technology Press, London).

Pierce, A.D. (1993). Nonlinear acoustics research topics stimulated by the sonic boom problem, *Proc. XIII ISNA, World Scientific, Singapore* , 7–20.

Rudenko, O.V. (1974). Powerful hypersound generation possibility using laser radiation, *Pis'ma ZhETF* **20,** 445–8.

Rudenko, O.V. (1986). Interaction of intensive noise waves, *Usp. Fiz. Nauk.* **149,** 413–47.

Rudenko, O.V. & Chirkin, A.S. (1974). Theory of nonlinear interaction of monochromatic and noise waves in weakly dispersive media, *ZhETF.* **67,** 1903–11.

Rudenko, O.V. & Soluyan, S.I. (1977). *Theoretical Foundations of Nonlinear Acoustics* (Plenum Press, New York).

Saichev, A.I. (1974). Spectra of some random waves propagating in nonlinear media, *Izv. Vuzov. Radiofizika* **18,** 1025–33.

Seymour, B.R., & Mortell, M.P. (1973). Resonant acoustic oscillations with damping: small rate theory, *J. Fluid. Mech.* **58,** 352–73.

Soluyan, S.I. & Khokhlov, R.V. (1961). Finite amplitude acoustic wave propagation in a dissipative medium, *Vest. MGU. Ser 3 Fizika i Astronomia* **3,** 52–61.

Tatsumi, T. & Kida, S. (1972). Statistical mechanics of the Burgers model of turbulence, *J. Fluid Mech.* **55,** 659–75.

Thuras, A.L., Jenkins & O'Neill (1934). Extraneous frequencies generated in air carrying intense sound waves, *JASA* **6,** 173–80.

Uvzarov, A.V. & Osipov, A.I. (1987). Propagation of nonlinear hydrodynamic perturbations in oscillatory-nonequilibrium gas, *Khim. Fizika* **6,** 385–9.

Vinogradova, M.B., Rudenko, O.V. & Suhorukov, A.P. (1979). *The Theory of Waves* (Nauka, Moscow).

Werth, G.C. (1953). Attenuation of repeated shock waves in tubes, *JASA* **25,** 821.

Whitham, G. (1974). *Linear and Nonlinear Waves* (Wiley, New York).

Zakharov, V.E. & Sagdeev, R.Z. (1970). On the acoustic turbulence spectrum, *Dokl. Acad. Nauk SSSR* **192,** 297–300.

Zarembo, L.K. & Krasil'nikov, V.A. (1956). On the absorption of finite amplitude ultrasonic waves in liquids, *Dokl. Acad. Nauk SSSR* **109,** 731–4.

3

Nonlinear geometrical acoustics

3.1 Rays and ray tubes

Having studied the basic laws of plane wave propagation let us proceed to the discussion of waves with a more complicated spatial structure. First of all we analyse a wide class of waves whose deviations from the plane ones develop sufficiently smoothly, on scales far in excess of a characteristic wavelength. In linear theory this approximation corresponds to geometrical acoustics, where the wave geometry is described by a system of rays such that propagation takes place independently along each ray tube. Finite-amplitude waves can have analogous geometric properties, in which case we speak about nonlinear geometrical acoustics (NGA). In this case, one has to analyse a sometimes complicated "interplay" between nonlinear effects on the one hand, and the effects of wave divergence, focusing, refraction, etc., on the other hand. The following circumstance seems to be noteworthy. The methods of linear geometrical acoustics and linear geometrical optics (dealing with the propagation of short electromagnetic waves) are in general similar, being most frequently based on the consideration of harmonic or quasi-harmonic processes. Nonlinear geometrical optics and acoustics, in contrast to their linear counterparts, have pursued diverse paths of development: while the former is mainly concerned with quasi-harmonic waves, the latter essentially treats the wave profile distortions, which, even in one dimension, do not lend themselves easily to description, as is evident from the previous chapter.

The first contribution touching upon NGA (in the broad sense of the word) seems to be that of Landau (1945), who considered spherical and cylindrical waves at a long distance from an explosion site, taking into account the simultaneous action of nonlinearity and spherical divergence. A more detailed analysis of finite (albeit small) amplitude spherical waves was suggested in the 1950s and 1960s (see, e.g., Christianovich, 1956, or Whitham, 1974). A discussion of rays and ray tubes in this connection was given by Whitham (1953), Keller (1954), and Gubkin (1958); note that in the latter paper the term "nonlinear geometrical acoustics" was used, perhaps for the first time. However, the applications of this approximation to particular problems came under consideration much later, with a few exceptions.

In this chapter, we shall discuss the basic concepts of NGA and offer some examples of physical or applied interest.

The majority of linear geometrical acoustics (and optics) schemes are developed for quasi-harmonic wave fields of the type $A(\mathbf{r})e^{i\psi}$, where A is a slowly varying amplitude and ψ a phase (eikonal). In this case $\omega = -\partial\psi/\partial t$ and $\mathbf{k} = \nabla\psi$ are the slowly changing frequency and wave vector, respectively, related through the same dispersion law as that used in a harmonic plane wave; thus, for a static acoustic medium, $\omega^2 = k^2c^2$. Being formulated for a phase function, this dispersion law appears as a Hamilton–Jacobi equation, so that the appropriate canonical equations describe the rays. This theory can be found in all texts where the problem is considered. In this book, however, we employ another approach (different in form alone, the physical sense being the same), which is convenient for describing non-harmonic waves and permits one to carry out a natural extension to nonlinear problems. This method envisages the separation, in the general case, of three stages. The first stage is devoted to kinematic wave front and ray tube analysis, to be conducted according to Whitham (1974). The second stage deals with the study of wave evolution along the ray tube, and the third offers a solution, should it be necessary, to the self-consistent nonlinear problem of the mutual influence of the ray structure and the wave amplitude properties.

The ray geometry

Consider an arbitrary wave front, i.e. constant phase surface. The position of the front is time-dependent, and its equation can be written as $t = \alpha(\mathbf{r})$.

From the condition $d(\alpha(\mathbf{r}) - t) = 0$ we obtain the normal front speed at any point of the front, $c = |\nabla\alpha|^{-1}$. Normal lines with respect to the front, being tangent to the vector $\mathbf{v}$ at each point, we shall call rays. Take an arbitrary ray tube with cross-sectional area variation given by the function Q, such that the cross-sectional area of the elementary tube is $ds = Q ds_0$ where ds_0 is the initial area (at $\alpha = \alpha_0$); without loss of generality one can assume the value of ds_0 is the same for all tubes. Taking Q as a coordinate function one can readily observe that the following equality is satisfied:

$$\operatorname{div}\left(\frac{\mathbf{l}_0}{Q}\right) = 0, \tag{1.1}$$

where $\mathbf{l}_0 = \nabla\alpha/|\nabla\alpha|$ is a unit vector oriented along the ray. Indeed, choosing an elementary volume Σ as a tube segment (Figure 3.1 (a)) we can write

$$\int_\Sigma \operatorname{div}\left(\frac{\mathbf{l}_0}{Q}\right) dV = \oint \frac{\mathbf{l}_0\mathbf{n}}{Q} ds,$$

where $\mathbf{n}$ denotes a unit vector normal to the tube surface. The right-hand side of this equality is zero: at lateral tube walls $\mathbf{l}_0 \perp \mathbf{n}$ and for the others by definition

$$\frac{ds_2}{Q_2} - \frac{ds_1}{Q_1} = 0.$$

Therefore, the left-hand side also equals zero which, due to the arbitrary choice of volume, results in (1.1). Thus finally we have

$$\operatorname{div}\left(\frac{c\nabla\alpha}{Q}\right) = 0. \tag{1.2}$$

If we add here a relation between c and Q, a closed set of equations will result, that defines the relation $\alpha(\mathbf{r})$ and, hence, the front shape and position. For linear media $c = c_0$ where $c_0(\mathbf{r})$ is the unperturbed sound velocity, and from (1.2) we obtain

$$(\nabla\alpha)^2 = \frac{1}{c_0^2}, \quad \operatorname{div}\left(\frac{c_0\nabla\alpha}{Q}\right) = 0. \tag{1.3}$$

The first equation of this pair is the well-known eikonal equation, while the second defines the wave amplitude, since for a linear nondispersive medium it follows from the law of energy conservation that $Q = A^{-2}$, where A is a wave amplitude. These results prove quite consistent with those of common geometrical acoustics.

More evident results can be obtained for two-dimensional waves, when the rays and successive front positions form an orthogonal reference frame α, β (Figure 3.1 (b)). As the function α defines the front movement, the distance covered by the front within time $\delta\alpha$ will amount to $c\delta\alpha$. The ray tube width is $Q\delta\beta$. Let us consider a curvilinear quadrilateral PSRG shown in Figure 3.1 (b). Its sides are not parallel, due to

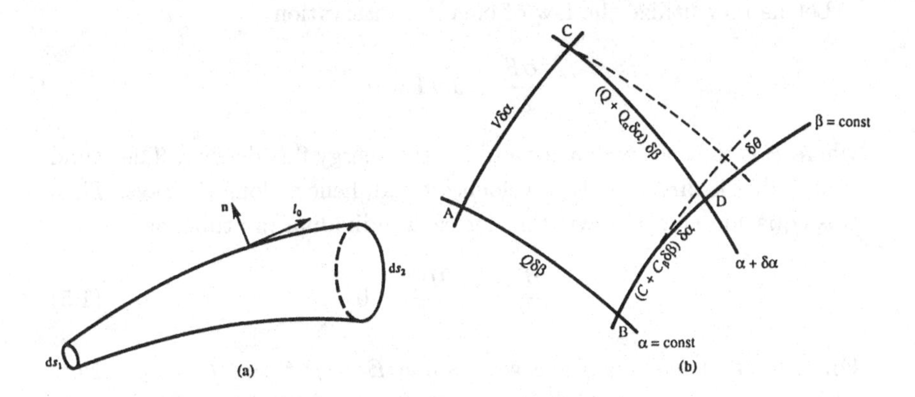

Figure 3.1. (a) a ray tube; (b) shock front positions for the moments α(AB) and $(\alpha + \delta\theta)$(CD). AC and BD are the rays.

bending of the ray and front by an angle δQ. The figure shows that the respective variations satisfy the relations

$$\delta(Q\delta\beta) = c\delta\alpha\delta\theta, \quad \delta(Q\delta\alpha) = -Q\delta\beta\delta\theta,$$

which yield two kinematic equations (Whitham, 1974)

$$\frac{\partial\theta}{\partial\beta} = \frac{1}{c}\frac{\partial Q}{\partial\alpha}, \quad \frac{\partial\theta}{\partial\alpha} = -\frac{1}{Q}\frac{\partial c}{\partial\beta}. \tag{1.4}$$

It is clear from the above that for a linear homogeneous medium, when $v = c_0$, θ does not depend on α, which implies that the rays are straight. In this case the first equation of (1.4) embodies the linear dependence of Q on α. For a nonlinear medium, however, the velocity v depends on the field intensity, the system (1.4) is not closed and one has to add the solutions determining the wave profile evolution along the rays.

Waves in ray tubes

It seems useful to recall that for a plane linear sound wave propagating in the direction $\mathbf{l}$ all physical perturbations can be written as functions of a phase variable $\psi = \mathbf{ql} - t$ where $\mathbf{q} = \mathbf{l}_0/c_0$ and $\mathbf{l}_0$ is a unit vector in the propagation direction such that the particle velocity $\mathbf{v}$ is directed along $\mathbf{l}_0$. In a geometro-acoustical situation for "almost plane" waves, everything said above concerning a plane wave remains "locally" valid while "globally" the direction of $\mathbf{l}$ may vary from one ray to another (and, of course, along the curved rays) and the field amplitude may change smoothly along the ray.

Let us now utilize the law of energy conservation

$$\frac{\partial E}{\partial t} + \operatorname{div} \mathbf{I} = 0,$$

where E is the energy density and $\mathbf{I}$ is the energy flux density. The sound energy flux is directed along velocity $\mathbf{v}$ and, hence, along the rays. Then this equation can be rewritten, for each individual ray tube, as

$$\frac{\partial EQ}{\partial t} + \frac{\partial IQ}{\partial \ell} = 0. \tag{1.5}$$

For a linear travelling wave we assume $E = \rho v^2$ and $I = cE$. Here and subsequently the subscript "0" of the sound velocity symbol c_0 and density ρ_0 are omitted. Thus from (1.5) we have

$$v_t + cv_\ell + \frac{1}{2}\frac{(c\rho Q)_\ell}{\rho Q} v = 0. \tag{1.6}$$

Here the medium parameters are considered to be constant in time, so that for a given ray system the quantities c, ρ, Q can be supposed to depend on ℓ alone, the last term on the right-hand side of (1.6) being small compared with the other two. Replacing the variables ℓ, t by ℓ, and $\tau = t - \int d\ell/c$ yields the ordinary differential equation

$$cv_\ell + \frac{(c\rho Q)_\ell}{2\rho Q} v = 0 \tag{1.7}$$

(here both terms are of the same order). Provided c, ρ and Q are assumed to be known functions of ℓ (which is true once the rays have been constructed) the solution of (1.7) has the form

$$v = (c\rho Q)^{-1/2} F(\tau),$$

where $F(\tau)$ is an arbitrary function of τ. For a harmonic wave F is just a sinusoid. From this it is evident that both the wave amplitude and its spatial scale change along the rays, in proportion to $(c\rho Q)^{-1/2}$ and c, respectively.

3.2 Nonlinear wave evolution along the rays

We now ask, how are we to revise these results to allow for nonlinearity? Recall that for a plane wave a term $\varepsilon v v_\tau$, and another term $\delta v_{\tau\tau}$ to take viscosity into account, should be added to an equation of type (1.7) (see Chapter 1). As these terms have already proved locally small in a plane wave, their variations due to the curved geometry appear to be of the next order of smallness, and can be neglected in the first approximation. For the same reason, the x- and t-derivatives can be replaced with those with respect to τ for small terms, as the direct dependence on ℓ is a slow one. Therefore (1.7) can be extended to

$$c v_\ell - \frac{\varepsilon}{c} v v_\tau - c\delta v_{\tau\tau} + \frac{(c\rho Q)_\ell}{2\rho Q} v = 0. \tag{2.1}$$

This equation extends the Burgers equation to the case of an inhomogeneous medium. Let us consider, for the present, an inviscid medium, with $\delta = 0$. Then having changed the variables to

$$u = v\sqrt{c\rho Q}, \quad X = \int_0^\ell \frac{d\ell'}{c^2\sqrt{c\rho Q}}, \tag{2.2}$$

we obtain

$$u_X - \varepsilon u u_\tau = 0, \tag{2.3}$$

which has the same form as that for a homogeneous medium. Because of this, in the "reduced" variables (2.2) the acoustic field varies along a tube just as a simple wave,

$$u = F(\tau + \varepsilon u X), \tag{2.4}$$

while the effects of the medium inhomogeneity and ray distortion are reduced, in this approximation, to scale transformations of the variables. However, as will be shown later, these transformations can lead to non-trivial physical results.

Naturally, one can offer a more rigorous and consistent derivation of equations of the type (2.1), based on the use of perturbation theory schemes which, in addition, provide the possibility of estimating the error in the solutions obtained (see, for example, Ostrovsky & Pelinovsky, 1974).

The above equations can be extended to the case of a moving medium. Let us begin again with a linear plane wave in a homogeneous moving medium. The wave velocity $\mathbf{V}$ can be written as $c_0\mathbf{l}_0 + \mathbf{u}$ where $\mathbf{u}$ is the medium velocity, $c_0\mathbf{l}_0$ the velocity of sound with respect to the medium, and $\mathbf{l}_0$ characterizes the wave propagation direction with respect to the motion of the medium. It is essential here that the particle velocity $\mathbf{v}$ in a wave is directed along $\mathbf{l}_0$, as before. Indeed, if we write the linearized Euler equation

$$\mathbf{v}_t + (\mathbf{u} \cdot \nabla)\mathbf{v} + \frac{\nabla p'}{\rho_0} = 0,$$

and substitute $\mathbf{v}$ and ρ' into it as functions of the variable $\eta = \ell/V - t$, where ℓ is a coordinate along the propagation direction (in a fixed reference system), we get

$$\mathbf{v}_\eta \left(\frac{u_\ell}{V} - 1\right) + \frac{\mathbf{l}_0 p'_\eta}{V\rho_0} = 0,$$

where $\mathbf{l}_0$ is a unit vector in the direction $\mathbf{l}$, and u_ℓ is the projection of the medium velocity in this direction. Hence we find

$$\mathbf{v} = \frac{\mathbf{l}_0 p'}{\rho_0(V - u_\ell)} = 0.$$

Thus $\mathbf{v}$ is indeed directed along $\mathbf{l}_0$.

For non-plane waves in a moving medium one can again introduce rays as a tangent family with respect to the vector $\mathbf{V}$ (or $\mathbf{v}$, which is the same), and using the energy conservation law, describe wave propagation along such a tube. Beforehand, however, one has to adequately formulate expressions for the wave energy in a moving medium (Ryzhov & Shefter, 1962). It appears that the following quadratic quantities satisfy the energy conservation law in a moving medium:

$$E = \frac{V}{c}\rho v^2, \quad \mathbf{I} = \mathbf{V}E. \tag{2.5}$$

In a stationary moving medium (with time-independent parameters) the wave energy is conserved at a fixed point, and equation (1.5) yields the variation of $\mathbf{v}$ along the tube. Indeed, substituting equations (2.5) and passing again to variables ℓ, $\tau = t - \int \mathrm{d}\ell/V$ we have $v_\tau - vq_\ell/q = 0$, or

$$v = q^{-1}F\left(t - \int \frac{\mathrm{d}\ell'}{V(\ell')}\right), \tag{2.6}$$

where now $q = (c_0/Q\rho V^2)^{1/2}$.

As before, to extend these results to the nonlinear case, it suffices to add the term $\varepsilon v v_\tau/V$ to (2.6), obtaining

$$v_\ell + \frac{\varepsilon v v_\tau}{V} - \frac{q_\ell}{q}v = 0. \tag{2.7}$$

Hence, a Riemann (simple) wave equation comes out again, provided the following substitution is performed:

$$u = v\sqrt{V\rho Q}, \quad X = \int_0^{\ell} \frac{d\ell'}{V^2\sqrt{c\rho Q}}. \tag{2.8}$$

A more intricate case is that of nonstationary media (for instance, an inhomogeneous medium travelling as a whole) (Ostrovsky, 1963). Here the wave energy is not conserved.

As was remarked at the end of Chapter 2, the expressions for the wave energy and its flux suggested here do not necessarily correspond to the natural physical definition of these quantities as a difference between a medium with or without a wave. However, for our purposes here it is in fact sufficient to write any pair of quantities that are quadratic with respect to the wave amplitude and related through the law of conservation.

3.3 Spherical waves

In order to proceed to applications of the theory it is instructive to start with relatively simple problems where the ray trajectories are known beforehand, thanks to the symmetry of the problem. Here, in particular, we refer to the case of a spherically symmetrical wave (converging or diverging) in a homogeneous medium. For such a wave the coordinate ℓ along the ray coincides with the radius r, and evidently $Q \sim r^2$. Then the formulae describing the relation between the reduced and "primary" variables yield

$$u = vr\sqrt{c\rho}, \quad x = \pm\frac{|\ln(r/r_0)|}{c^2\sqrt{c\rho}},$$

where r_0 is the initial radius, taken far enough from the centre of symmetry to be much larger than the characteristic wavelength: only then can the tube width Q variation be viewed as sufficiently slow. The plus sign here refers to a diverging wave while the minus refers to a converging wave. Therefore, the formula (2.4) takes the form

$$v = \frac{1}{r}F\left(y + \frac{\varepsilon vr}{c^2}\left|\ln\frac{r}{r_0}\right|\right), \quad y = t - \frac{r}{c}. \tag{3.1}$$

Analysing this, one can easily find the discontinuity formation coordinate by making use of the formulae deduced in Section 2.3 for a plane wave. For a wave sinusoidal at $r = r_0$, we have from (3.1) (Naugol'nykh,

1971)

$$r_* = r_0 \exp\left(\pm\frac{c_0^2}{\varepsilon\omega v_0 r_0}\right) = r_0 \exp\left(\pm\frac{L_N}{r_0}\right), \quad L_N = \frac{c_0^2}{\varepsilon\omega v_0}, \tag{3.2}$$

where ω is the wave frequency, L_N is the discontinuity formation distance in a plane wave with the same initial amplitude.

Hence it is obvious that in a diverging wave a discontinuity always arises after a longer distance than in a plane wave ($r_* - r_0 > L_N$) whereas in a converging wave this distance is less ($r_0 - r_* < L_N$). These results can be interpreted in terms of wave intensity decreasing due to spreading in the former case and growth in the latter.

Let us also formulate an asymptotic expression for the spherical wave amplitude at the discontinuous stage where an initially sinusoidal field turns into a sawtooth wave. Formula (2.16) of Chapter 2 suggests that the evolution of such a wave can be described by the equation (Naugol'nykh, 1959)

$$\frac{vr}{v_0 r_0} = \left(1 + \sigma_0 \left|\ln\frac{r}{r_0}\right|\right)^{-1} (-\omega y + \pi), \tag{3.3}$$

$$-\pi \leq \omega y \leq \pi$$

$$\sigma_0 = \varepsilon\frac{v_0}{c} k r_0, k = \frac{\omega}{c},$$

i.e. as $r \to \infty$ the diverging wave "forgets" its initial amplitude v_0.

It should be noted that while the amplitude of a converging linear wave increases as $1/r$, the nonlinear wave demonstrates a more complicated pattern. Discontinuity formation results in strong absorption, leading to an amplitude drop in a converging sawtooth wave in spite of the energy concentration due to the wave convergence (Naugol'nykh *et al.*, 1963). Thus it turns out that $v_s \to 0$ as $r \to 0$. These effects are of importance in consideration of focusing systems (Naugol'nykh & Romanenko, 1959).

In a similar way, we can obtain the evolution equation for the cylindrical wave, which has amplitude v_0 at distance r_0 (Naugol'nykh, 1971)

$$v = \frac{1}{\sqrt{r}} F\left(y + \frac{2\varepsilon v r_0}{c^2}\left|\sqrt{\frac{r}{r_0}} - 1\right|\right), \tag{3.4}$$

the discontinuity formation distance for divergent (+) and convergent (−) initially harmonic waves of frequency ω:

$$r_* = r_0 \left(1 \pm \frac{1}{2\sigma_0}\right)^2 \tag{3.5}$$

$$\sigma_0 = \varepsilon \frac{v_0}{c} k r_0, k = \frac{\omega}{c},$$

and the sawtooth cylindrical wave equation:

$$\frac{v}{v_0}\sqrt{\frac{r}{r_0}} = \left(1 + 2\sigma_0\left|\sqrt{\frac{r}{r_0}} - 1\right|\right)^{-1}(-\omega y + \pi). \tag{3.6}$$

The equations of the evolution of the harmonic wave in its initial stage can be obtained from the expansion of equation (2.18) of Chapter 2, (3.1) and (3.4) at $F(y) = \sin \omega y$: for plane waves,

$$\frac{v}{v_0} = \sin \omega y + \frac{r}{2L_N} \sin 2\omega y, \tag{3.7}$$

for cylindrical waves,

$$\frac{v}{v_0}\sqrt{\frac{r}{r_0}} \sin \omega y + \sigma_0 \left|\sqrt{\frac{r}{r_0}} - 1\right| \sin 2\omega y, \tag{3.8}$$

for spherical waves,

$$\frac{v}{v_0}\frac{r}{r_0} = \sin \omega y + \frac{\sigma_0}{2}\left|\ln \frac{r}{r_0} - 1\right| \sin 2\omega y. \tag{3.9}$$

The problem of spherical pulse propagation is of considerable practical value, owing to the need to study explosion waves. One needs to keep in mind that even in a linear medium such a pulse cannot be unipolar, and pressure and velocity must change their signs in such a way that

$$\int_0^\infty p' \, dt = 0.$$

This observation was made by Landau back in 1945, and can be attributed to the fact that in a spherical pulse of finite duration the potential $\varphi = \frac{1}{2} f(t - r/c)$ should become zero both in front of and behind the pulse, and $p' \sim \partial\varphi/\partial t$. With respect to a nonlinear wave this means that, strictly speaking, not fewer than two discontinuities must arise in a pulse so that an "N-wave" occurs (Figure 3.2). Far from the centre, however, where the wave is locally close to a plane one, the front discontinuity propagation can be handled independently. If, for example, a positive portion of the pulse has a triangular shape at $r = r_0$, then, making use of the formulae in Chapter 2 describing plane wave behaviour at long distances, one can obtain the corresponding "spherical" formulae for the pulse amplitude v_0 and length ℓ (Landau, 1945; Landau & Lifshitz, 1986)

$$v_s = v_0 r_0 \Big/ r\sqrt{1 + \frac{\varepsilon v_0 r_0}{c\ell_0} \ln \frac{r}{r_0}}, \quad \ell = \ell_0 \sqrt{1 + \frac{\varepsilon v_0 r_0}{c\ell_0} \ln \frac{r}{r_0}}. \tag{3.10}$$

Figure 3.2. Profile of an N-wave.

On transition to the linear wave, however, these simple formulae have only a limited applicability. This can be attributed primarily to the need to allow for finite dissipation, within the framework of a generalized Burgers equation (2.1). Such an equation cannot be reduced to one with constant coefficients, and only a few approximate solutions of it are known. In (2.1) the pulse shock front is considered to be close to a stationary wave, and has the same structure as a plane wave (since the front thickness $\ell_s \sim \nu/v_s$ is small compared with its radius of curvature). Evidently, however, this only holds true as long as the acoustic Reynolds number $Re \sim \ell/\ell_s$ remains large enough. For a plane wave in the form of a solitary unipolar pulse this requirement is always initially met (see Chapter 2).

Things are different in a spherical wave. Let us take as an example a pulse wave. According to (3.4), the acoustic Reynolds number,

$$\mathrm{Re} = v_s \ell/\nu \sim \frac{1}{r},$$

where ν is the kinematic viscosity, decreases and becomes of the order of unity at a distance r_1 defined by the condition $\ell_s \sim \ell$, i.e. when

$$\ell_0 \sqrt{1 + \sigma_0 \ln \frac{r_1}{r_0}} = \frac{\nu}{v_s} = \frac{\nu}{v_0 r_0} r_1 \sqrt{1 + \sigma_0 \ln \frac{r_1}{r_0}},$$

or

$$\frac{r_1}{r_0} = \frac{v_0 \ell_0}{\nu} = Re. \tag{3.11}$$

For $r > r_1$, the front duration becomes comparable with the total pulse duration and its front cannot be considered as quasi-stationary. Later on Re becomes small, and the wave is transformed into a linear one. Here one speaks of a "shock-to-acoustic wave transition".

This transition, however, can be triggered by another mechanism as well (Naugol'nykh, 1972). The fact is that spherical divergence brings about such a rapid amplitude decrease that dissipation has no time to broaden the shock front to its stationary width (quasi-stationary width, to be more exact) and, even though $Re \gg 1$ for the pulse as a whole, the local Reynolds number at the wave front becomes small, while it is always of the order of unity for the stationary shock wave.

In order to assess the distance at which this transition takes place we resort to the generalized Burgers equation (2.1). With $Q \sim r^2$ it is clear that the shock wave front can remain close to a stationary wave front only if the last term, which is responsible for the spherical divergence, remains small compared with the others. Thus the above formulae become inapplicable from a radius r_2 at which the nonlinear term $\varepsilon v v_\tau / c$ will become of the order of the last term $cv/2r$, i.e.

$$\frac{v_s}{c\tau_s} = \frac{v_s^2}{\nu} \sim \frac{c}{2r}.$$

Substituting the solution (3.10) here yields

$$\xi(1 + \sigma_0 \ln \xi) = T. \tag{3.12}$$

Here $\xi = r_2/r_0$, $T = v_0^2 r_0 / c\nu = M\, Re(r_0)$, where M is the initial Mach number, and $Re(r_0)$ is the Reynolds number with respect to the scale r_0. If $T < 1$ this equation has no solution in the region $\xi > 1$, i.e. where $r_2 > r_0$. This means that from the very beginning the front behaviour is practically linear. If, however, $T > 1$, (3.10) yields a finite value of r_2 corresponding to the shock-to-acoustic wave transition, such that for $r > r_2$ the front width grows according to the law of linear diffusion, i.e. as $\sqrt{r}$. The value of r_2 can be less than that of r_1, and then the trailing edge of the pulse will be described by (3.1) as before. However, as the term $\varepsilon v v_\tau / c$ is less in this region than that at the front (the trailing edge is much longer than the front), nonlinearity has no time to manifest itself behind the front either, where the basic terms are cv_ℓ and $cv/2r$, so that the total wave amplitude decays as r^{-1} and, in fact, the wave becomes completely linear even at $r = r_2$.

Consider an example: let an explosion pulse have amplitude equal to $p_s = 5 \times 10^7$ Pa and duration $\tau = 100\,\mu$s at radius $r_0 = 1$ m in water (which corresponds to an explosive charge of about 1 kg). Then $v_s = 3.3\,\mathrm{m\,s^{-1}}$, and $M = 2.2 \times 10^{-3}$. In this case $Re = 2 \times 10^5$, $T = 4.4 \times 10^2$ and $\sigma_0 = 6 \times 10^{-2}$. Eventually we find that $r_1 = Re\, r_0 = 2 \times 10^5$ m, whereas $r_2 = 2 r_0 T = 9 \times 10^2$ m, i.e. the linear stage at the front is reached in a shorter time than that needed for Re to become of the order of unity over the whole pulse. Reducing the wave amplitude by two orders we obtain $T = 4.4 \times 10^{-2}$, in which case the wave can be considered as linear from the very beginning (although $Re = 2 \times 10^3 \gg 1$).

It should be noted that recently Hammerton and Crighton (1989) performed a detailed analysis of the evolution of an initially harmonic spherical wave within the framework of generalized Burgers equations of type (2.1). Using a matching procedure they found, in particular, the harmonic wave amplitude at the "old-age", linear stage of the process.

The nonlinear acoustics approximation turns out to be rather effective for the description of diverging spherical waves even of a comparatively large amplitude.

As an example, let us consider the compressional wave, radiated by a sphere, expanding with constant velocity U. The movement started at $t = 0$. The exact self-similar solution of this problem was obtained numerically by G.I. Taylor (Taylor, 1946).

To obtain the solution in the nonlinear acoustics approximation (Naugol'nykh, 1964) we begin with the equation (2.1). Neglecting the dissipative term and making the substitution $v = \partial(u/r)/\partial r$ in this equation, after one integration we obtain, keeping only the main terms:

$$-\frac{\partial u}{\partial r} = \frac{1}{r}\frac{\alpha}{2c_0}\left(\frac{\partial u}{\partial y}\right)^2 + f(r), \tag{3.13}$$

where $y = t - r/c_0$, and $f(r)$ is an arbitrary function.

We shall use an approximate boundary condition coinciding with the one for the solution obtained in the linear approximation, which can be written as follows:

$$u(r, y) = (U^3 y^2)/(1 - M^2), r = R, \tag{3.14}$$

where $M = U/c_0$ is the Mach number.

By the method of characteristics one can get the solution of equation (3.13), which gives the velocity distribution $v(r)$ in the compressional wave radiated by an expanding sphere:

$$\frac{1}{\xi} - 1 = \left[-1 + \left(1 + \frac{v}{c_0}\frac{1-M^2}{M^3}\right)^{1/2}\right] \times \frac{(\gamma+1)M^3}{1-M} \ln\left[\frac{M}{1-M}\left(-1 + \left(1 + \frac{v}{c_0}\frac{1-M^2}{M^3}\right)^{1/2}\right) + 1\right]. \tag{3.15}$$

Here $\xi = r/c_0$ is a dimensionless coordinate. According to (3.15) the fluid velocity v is a multivalued function of the coordinate in a vicinity of front, which implies the existence of a shock wave. Its position is obtained by the standard procedure, applying the rule of "area conservation".

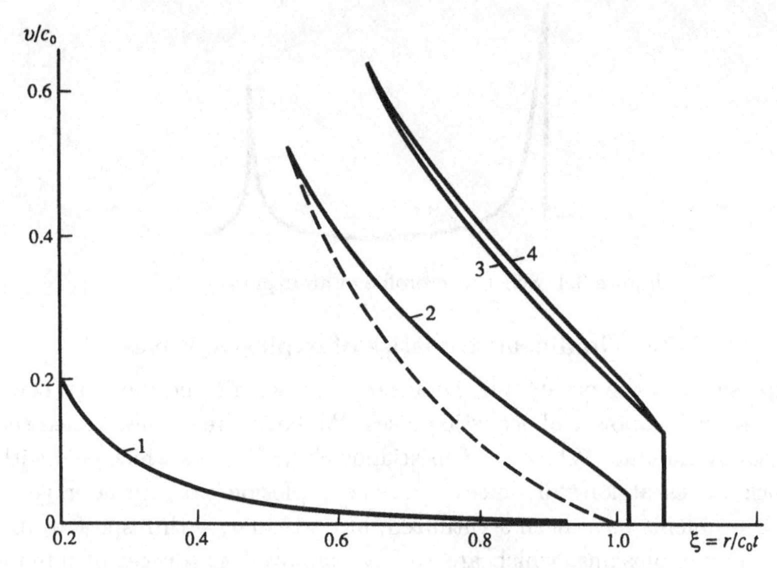

Figure 3.3. The fluid velocity distribution v/c_0, where c_0 is the velocity of sound in the compressional wave generated by an expanding sphere, for different speeds U of expansion. Direct numerical solution (Taylor, 1946): curve 1, $M = U/c_0 = 0.203$; curve 2, $M = 0.523$; curve 4, $M = 0.638$. Nonlinear acoustics approximation (Naugol'nykh, 1964): curve 1, $M = 0.203$; curve 2, $M = 0.523$; curve 3, $M = 0.638$. The linear acoustics approximation is shown by line 1 for $M = 0.203$ and by the dashed line for $M = 0.523$. The horizontal axis is the dimensionless distance from the sphere.

The solution (3.15) is presented in the Figure 3.3 in comparison with the exact self-similar solution and the solution obtained in the linear acoustic approximation. The fluid velocity distribution in the compressional wave is presented for different expansion speeds $U = Mc_0$. Curves 1, 2 and 4 correspond to the exact numerical solution by Taylor for $M = 0.203$, 0.323 and 0.638 respectively. On the same plot the results of the analytical solution (3.15) of the problem in the nonlinear acoustics approximation are given. For $M = 0.203$ and 0.323 they practically coincide with the exact solutions, while for $M = 0.638$ this solution is presented by the curve 3. The corresponding data in the linear approximation for $M = 0.203$ coincide with the exact solution presented by the curve 1, while for $M = 0.323$ the linear solution is given by the dashed line. The linear solution turns out to be quite inappropriate for $M = 0.638$, and it is not presented on the plot.

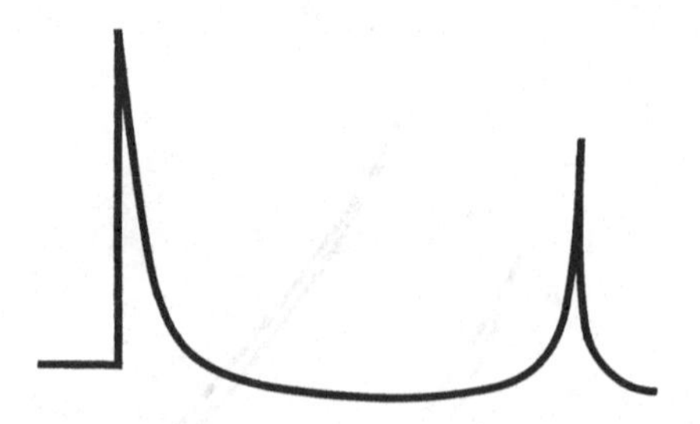

Figure 3.4. Schematic profile of an explosive signal.

3.4 Nonlinear acoustics of explosive waves

Explosions in the ocean and the atmosphere seem to be the most common source of powerful acoustic pulses. We have already mentioned the works by Landau, Whitham, Christianovich and others which deal with shock waves at long distances from the explosion site. In later years such problems have been scrutinized, in particular, with respect to underwater explosions, which are widely employed as sources of intense broad-band signals. Such signals are used to solve problems of bottom profiling, in sea geology, earthquake and underwater volcano investigations, underwater communication systems, navigation, etc.

A qualitative description of an underwater explosion is as follows. A spherical detonation shock wave enters water escaping from an explosive material. It is followed by a weaker alternating pressure wave associated with pulsations of the gas bubble created by the detonation products (Figure 3.4). Here our main concern will be with the leading pulse, which has a discontinuous front and a tail close to exponential: $p = p_s \exp(-t/\tau)$. Quite a long time ago empirical formulae were obtained defining pulse parameters dependent on the distance r and the charge mass q (Cole, 1948):

$$p_m = G_1(q^{1/3}/r)^{1.13}, \quad \tau = G_2 q^{1/3}(q^{1/3}/r)^{-0.22}. \tag{4.1}$$

Coefficients G_1 and G_2 are deduced from averaging the numerous experimental data, and differ from one author to another; on average we can take $G_1 = 11 \times 10^7$, $G_2 = 90$ (q being expressed in kilograms, p_s in Pascals, and τ in microseconds).

Here a relevant question can be asked: how do formulae (4.1), also valid at distances such that the nonlinearity in a wave is sufficiently small, match those of nonlinear acoustics (and could not the former be replaced by the latter, which are more rigorous)? For an exponential initial pulse these formulae acquire the forms

$$\bar{p}_m = 2\left(1+\sqrt{1+2\sigma_0 \ln\frac{r}{r_0}}\right)^{-1},$$

$$\frac{\tau}{\tau_0} = e^{-1}\left[1+(e-1)\left(1+2\sigma_0 \ln\frac{r}{r_0}\right)^{1/2}\right], \tag{4.2}$$

where

$$\bar{p}_m = \frac{p_m r}{p_{m0} r_0}, \quad \sigma_0 = \varepsilon v_0 r_0 / c_0^2 \tau, \quad \tau_0 = \tau(r_0), \quad p_{m0} = p_m(r_0).$$

(The pulse, of course, loses its exponential shape, and τ has to be understood here as its width at a level e^{-1} below the peak value.)

Comparing (4.1) and (4.2) one can see the difficulties of matching them: the former equations involve only an arbitrary parameter q while the latter equations have two arbitrary parameters v_0 and τ_0 which, therefore, should be connected with those of an explosion. More important is the fact that the wave decay laws are different: $r^{-1.13}$ and $r^{-1}(\ln r)^{-1/2}$, respectively. Nevertheless, if we compare the solutions (4.1) and (4.2) at a certain radius r_0, the values of v_s prove to agree quite closely (the difference not exceeding 10 per cent) over a rather wide range of distances (of several orders) (Figure 3.5); see Fridman (1980). This makes one think that in reality the law $r^{-1.13}$ is simply an approximation to the logarithmic one (it was only natural to make use of a power law in the empirical formula). More significant, however, is the difference in pulse durations, since the actual broadening of a pulse proved to develop faster than the theoretical one (Figure 3.6). There are various explanations for this difference: some authors attribute it to the influence (in salty water) of a molecular relaxation process (Fridman, 1980) while others ascribe it to the incorrect representation of the spherical wave as a locally plane one even if comparatively far from an explosion site (Petukhov, 1983).

A description of the leading pulse using the NGA formulae seems to be of value for several reasons. Thus, it allows us to evaluate the distances r_1, r_2 beyond which a wave front can be considered as linear. If the empirical formulae (4.1) are substituted into those corresponding to the condition $Re = 1$, the following condition results:

$$r_1 = 10^8 q^{0.37}\,(\mathrm{m}),$$

while the condition of front linearity (3.2) yields (for $\sigma_0 \ll 1$, $T \gg 1$)

$$r_2 = 2.7 \times 10^4 q^{0.6}\,(\mathrm{m}). \tag{4.3}$$

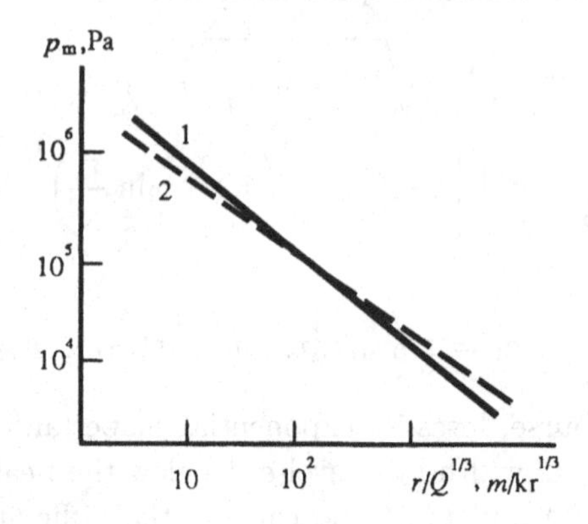

Figure 3.5. Magnitude of the leading explosive wave versus distance: 1 – experiment, 2 – theoretical dependence according to (4.1).

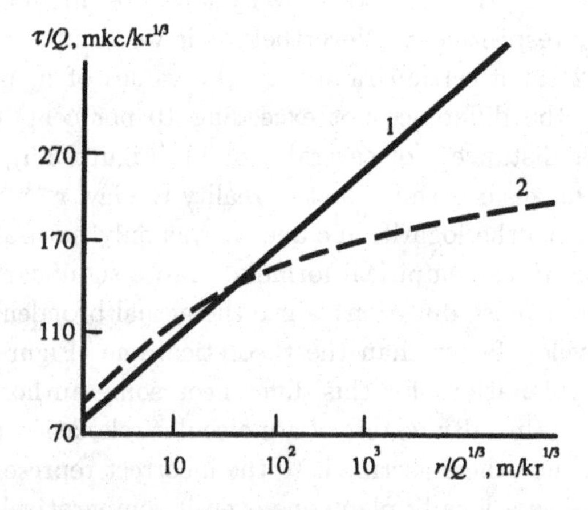

Figure 3.6. Duration of the leading explosive pulse versus distance. 1 – experiment, 2 – theoretical dependence according to (4.2).

For real sea water, however, these formulae hold only at high enough frequencies (when the pulse duration is less than 10 μs). For lower frequencies the sound damping cannot be described by a constant viscosity coefficient since the linear decrement is proportional to $\omega^{3/2}$ rather than to ω^2. In this case, for instance, instead of (4.3) we have

$$r_2 = 1.1 \times 10^3 q^{0.47}. \tag{4.4}$$

The NGA formulae can also be adequate for such large distances from the explosion site that the ocean must be treated as inhomogeneous. Thus Ostrovsky *et al.* (1979) studied the influence of the free sea surface in the simplest case when surface waves and gas bubbles are absent. Then the main effects are attributed to the pressure sign change in the reflected wave, i.e. when the compression and rarefaction portions of the pulse replace each other. Eventually the shock front begins to broaden while the trailing edge becomes sharper and, provided the nonlinearity is strong enough, a new discontinuity emerges in the reflected wave. Without resorting here to the appropriate formulae, we offer just a single estimate. Let an explosive charge of 10^{-4} kg of trinitrotoluene (TNT) detonate at a depth of 10 m. Then according to formulae (4.1) the shock wave amplitude near the surface above the explosion site will be 1 atmosphere. In this wave the discontinuity broadens gradually, while a new one is formed at different distances depending on the reflection angle; e.g. in a ray reflected at 60° to the vertical this distance amounts to 80 m horizontally and 27 m vertically downwards from the explosion centre.

The formulae obtained for the spherical explosion wave parameters in a homogeneous medium can also be employed as the starting formulae when calculating the wave propagation in a smoothly inhomogeneous sea.

3.5 One-dimensional propagation in inhomogeneous media. Waves in the atmosphere

Here we address another class of one-dimensional problems where the question of rays does not arise. Let the medium properties change only in the direction x (a stratified medium) and let a plane acoustic wave propagate exactly in this direction. In this case one can directly use formulae (2.2), (2.4). Here the following important circumstance should be immediately emphasized. If as $x \to \infty$ we simultaneously have $X \to \infty$, while the quantity u remains finite together with v, then, as in a homogeneous medium, a discontinuity always emerges and the wave dissipates completely. For an inhomogeneous medium, however, it is possible that the integrand for X decreases rapidly with x, or u decreases so that the term εuX in (2.4) remains finite as $x \to \infty$. Given these conditions, there may be no time for a discontinuity to arise, and, even if it does, a periodic wave has no sawtooth asymptotics and the pulse remains of finite duration. Evidently, this "freezing" of nonlinear distortions results either from an abrupt decrease of nonlinearity with distance (e.g. due to damping) or from the wave velocity growth so that nonlinear distortions have no time to accumulate.

Let us consider as an illustrative example an initially harmonic wave propagating in an isothermal atmosphere (Ostrovsky, 1963; Romanova, 1970). This atmosphere is known to have c = constant and $\rho = \rho_0 \exp(-x/H)$ where $H = c^2/\gamma g$ is an effective atmosphere height (for the Earth $H \approx 8\,\mathrm{km}$, for the solar chromosphere $H \approx 100\,\mathrm{km}$). Substituting these into (2.2) with Q constant, one readily obtains

$$u = v\sqrt{c\rho}\,\mathrm{e}^{x/2H}, \quad X = \frac{2M}{c^2\sqrt{c\rho}}\left(\mathrm{e}^{x/2H} - 1\right). \tag{5.1}$$

For $H > 0$ these formulae correspond to sound propagating upwards (towards the density decrease); in order to describe the downward propagation one only has to change the sign of H.

Now it is easy to give a complete description of the process. If at $x = 0$ a sinusoid with amplitude v_0 and frequency ω is given, a discontinuity will appear at $X_* = (\varepsilon\omega u_0)^{-1}$ or, in terms of primary variables, at

$$x_* = 2\left[H \ln\left(1 \pm \frac{L_N}{2H}\right)\right]^{-1}, \tag{5.2}$$

where $L_N = c^2/\varepsilon\omega v_0$ is the distance of discontinuity formation in a homogeneous medium with the same initial parameters. "+" and "−" correspond to upward and downward propagation respectively. It is apparent from this that during the upward propagation we always have $x_* < L_N$, i.e. a discontinuity develops more rapidly than in a homogeneous medium. This should be attributed to the velocity amplitude growth as $\rho^{-1/2} \sim \mathrm{e}^{x/2H}$. In contrast, the amplitude decreases in the downward travel so that $x_* > L_N$. If, thereby, we have $L_N > 2H$ the above-mentioned "freezing" of the wave profile is realized and a discontinuity does not appear at all.

Should a discontinuity emerge, its further evolution will also be determined (for variables u and X) by the formulae in Chapter 2, from which one can easily return to the original variables. Upon doing this the asymptotic behaviour of a periodic wave travelling upwards is given by $v_s \to \pi c^2/2\omega H$, i.e. the velocity amplitude tends to a constant value: the growth of v due to density decrease is compensated by the damping at the discontinuity. It stands to reason that the wave energy flux tends to zero, as the pressure amplitude drops rapidly. For a unipolar triangular pulse

$$v_s \to c\sqrt{\frac{2v_0}{\varepsilon\omega H}}\mathrm{e}^{x/4H},$$

i.e. the growth of v_s is still retained.

In the case of downward travel, if a discontinuity has enough time to emerge, the value of u in it remains finite and hence $v_s \sim \mathrm{e}^{-x/2H}$ as in a linear wave, and the discontinuity does not vanish asymptotically.

These straightforward formulae may have rather interesting geophysical and astrophysical applications. Most important here seems to be the possibility of very rapid nonlinear distortions in the upward propagation. As a result of them, even moderate perturbations of the Earth's surface can be transformed into discontinuous waves undergoing heavy damping in the upper atmosphere. For example, for strong tsunami waves in the ocean, vertical water surface displacements ξ_0 can be of the order of 1 m, over an area of, say, 100 km in diameter, within a time T of the order of several seconds. Making use of (5.2) for $T = 10\,\mathrm{s}$, $\omega = 3 \times 10^{-1}\,\mathrm{rad\,s^{-1}}$, $v_0 = \omega\xi_0 = 0.3\,\mathrm{m\,s^{-1}}$, $c = 330\,\mathrm{m\,s^{-1}}$, $\varepsilon = 1.2$, $H = 8\,\mathrm{km}$, one obtains the distance of discontinuity formation as $X_* = 8.2H = 66\,\mathrm{km}$. Though the isothermal approximation is not strictly observed up to such altitudes, the sound velocity variation is still moderate, so that this estimate seems to reflect reality. At this altitude the particle displacement is already equal to $\mathrm{e}^{x/2H}\xi_0 = 60\xi_0 = 60\,\mathrm{m}$ and further on, at the sawtooth stage, it attains a limiting value of about 200 m. More realistically, however, one may observe the effects of earthquakes and atmospheric explosions producing rather strong displacements of the upper layers. Such displacements, which may occur within the ionospheric F-layer, were registered by means of radar sounding (Alperovich *et al.*, 1979; Kosin & Saifutdinov, 1988).

Another interesting example is the solar chromosphere. Nowadays astrophysicists have quite elaborate information about its structure at their disposal. It is a fact that the chromospheric temperature first drops in a pronounced way with altitude (approximately from 5500° K down to 4700° K at a height of several hundreds of kilometres), to rise subsequently, achieving a value of about 6500° K at approximately 1700 km above the convective zone level, i.e. at the upper boundary of the chromosphere. Since there are no direct heat sources in the chromosphere, there must be some external heating factors capable of maintaining the observed temperature rise. A possible qualitative explanation, suggested long ago, is as follows. The energy of convective motions prevailing in subphotospheric layers is transmitted upwards in the form of sonic, magnetosonic and other waves. When propagating upwards they are transformed into shock waves whose dissipation leads to the gas heating. Energy estimates confirm such a possibility. To accomplish a quantitative analysis of this process, the methods of nonlinear acoustics can be employed.

Let us first estimate the characteristic distances at which nonlinearity shows itself. Assuming the chromosphere to be isothermal we set $g = 2.7 \times 10^4\,\mathrm{cm\,s^{-2}}$, $c = 10\,\mathrm{km\,s^{-1}}$, $\varepsilon = 1.2$, $H = 200\,\mathrm{km}$. One also needs to know the characteristic amplitudes and frequencies of perturbations coming out of the convective zone into the chromosphere. Details of these parameters are lacking; probably the perturbations occupy a wide frequency range, with a spectral maximum corresponding to periods within the range 30–60 seconds. For the oscillation amplitude one can assume $v_0 = 0.4$–$2\,\mathrm{km\,s^{-1}}$, i.e. the nonlinearity is small, at least as regards the initial amplitude values. Approximating (for rough estimates) this convective noise by a sinusoid with a frequency $\omega = 0.15\,\mathrm{s^{-1}}$ ($2\pi/\omega = 40\,\mathrm{s}$) and amplitude $v_0 = 1\,\mathrm{km\,s^{-1}}$, and substituting into (5.2) we see that $x_* = 400\,\mathrm{km}$, i.e. discontinuities appear at a rather short range. After travelling another 400 km the wave will be near to a sawtooth, with its amplitude taking an asymptotic value of $v_s = 5\,\mathrm{km\,s^{-1}}$. Thus, nonlinearity in the chromosphere is rather pronounced. However, to consider the possibility of chromospheric heating one should treat a self-consistent problem; the waves conveying the energy from the convective region propagate into an inhomogeneous medium whose parameters are, in their turn, defined by the wave energy dissipation for $x > x_*$, after the shock front formation.

Such a problem, also of general physical interest, was considered by Kaplan et al. (1972), taking into account different mechanisms of heat transfer by radiation.

Computed results for waves of different frequencies are depicted in Figure 3.7. The computations demonstrate that if one chooses the wave parameters near to $\omega = 0.15\,\mathrm{s^{-1}}$, $v_m = 2.5 \times 10^5\,\mathrm{cm\,s^{-1}}$, $\rho_0 = 5 \times 10^{-7}\,\mathrm{g\,cm^{-3}}$, $T_0 = 5500°$ K (ρ_0 and T_0 are the values of ρ and T at $x = 0$), fair quantitative agreement takes place between the predicted data and the observed profile, and, thereby, the temperature inversion effect is explained (i.e. its initial drop before the discontinuity appears, to be followed by a pronounced rise). Incidentally, the density distribution remains close to exponential, while the velocity amplitude in a discontinuity tends to a constant value, as was the case with an isother-

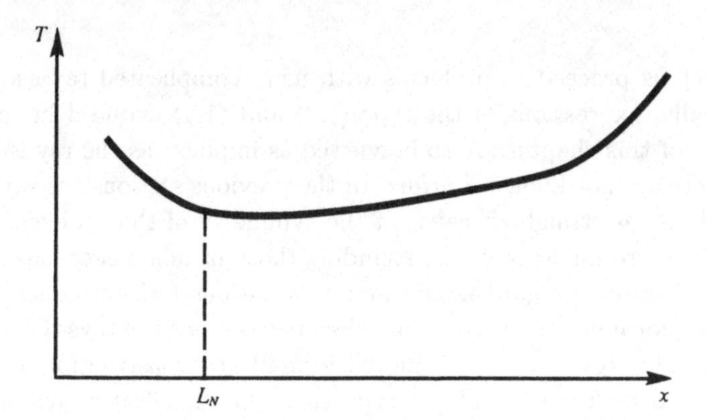

Figure 3.7. Temperature profile in the Sun's chromosphere; L_N is the shock formation distance.

mal atmosphere. Of course there are still many obscure questions in this situation. First, a model of heat transfer in the chromosphere should be defined more precisely. Second, acoustic waves constitute but one particular type of perturbation radiated to the chromosphere from below; in addition, one should consider magnetoacoustic waves, Alfven waves, etc., as other contributors to the upward energy transfer. An analysis of the nonlinear distortions of magnetosound in an exponential atmosphere was carried out by Ostrovsky and Rubakha (1972), who demonstrated that in a strong magnetic field h (for $c \ll h/\sqrt{4\pi\rho}$), the discontinuities in slow magnetosonic waves emerge at a still higher rate than in nonmagnetic sound. In fast magnetosonic waves, however, which are moving upwards, a discontinuity may not appear at all, because of an unbounded wave acceleration (the velocity tends to infinity as $\rho \to 0$, and the upward propagation time within this model remains finite as $x \to \infty$). These waves seem to be capable of entering the solar corona, the heating of which also needs to be explained: the temperature rise inside the corona is far in excess of that in the chromosphere. An adequate model of corona wave heating has not been developed as yet; recently the Joule mechanism for corona heating has been discussed as the most likely one.

3.6 Refraction of nonlinear waves in a stratified medium

Now let us proceed to problems with more complicated ray geometry. Generally, expressions of the type (1.2) and (1.4) deduced in the first section of this chapter are to be viewed as implicit, as the ray tube parameters are not known *a priori*. In the previous sections the rays were considered as straight because of the symmetry of the problem. However, in more intricate cases, including those of inhomogeneous media, one can often disregard nonlinear ray variations (self-refraction) while allowing for nonlinear wave profile distortions along the rays (Ostrovsky, 1976). This result can be achieved formally by constructing a system leading to an equation of the type (2.3), but qualitative justification of it is quite obvious. Essentially, relative displacements of the points in the wave profile due to nonlinear distortions attain values of the order of a wavelength λ at distance $L_N \sim \lambda M^{-1}$ (M being the acoustic Mach number). The same holds true for lateral front distortions due to self-refraction. However, the wave scale in the longitudinal direction is itself of the order of λ, so that these deformations alter the wave profile dramatically, whereas the transverse variation scale (within the frame of NGA) is far greater than λ and shifts by a value of the order of λ do not significantly influence the ray pattern, on the whole.

This does not necessarily mean that self-refraction effects are altogether unimportant; in fact, the above reasoning is not at all universal. These effects, along with the restrictions of a "linear-ray" approximation, will be considered below, while at this point of our discussion we assume that this approximation holds so that the ray pattern can be recovered from the linear approximation. Let us now turn our attention to several physical examples.

We begin with problems conserving plane symmetry (Ostrovsky *et al.*, 1976). At a horizontal plane $z = 0$ all the rays are directed at the same angle θ_0 with respect to the z axis. Such a geometry is realized, for instance, for the oblique incidence of a plane wave (in the xz-plane) from a homogeneous medium onto a stratified layer such that in the linear approximation the whole wave pattern is moving along the x-axis with a phase velocity $v_{ph} = c_0/\sin\theta_0$. The same holds true if at $z = 0$ there exists a travelling source created, for example, in the atmosphere or the ocean by a travelling seismic wave. In this case the rays are curved but remain parallel everywhere, so that the horizontal distance dx between two rays remains constant (Figure 3.8). Therefore, the ray

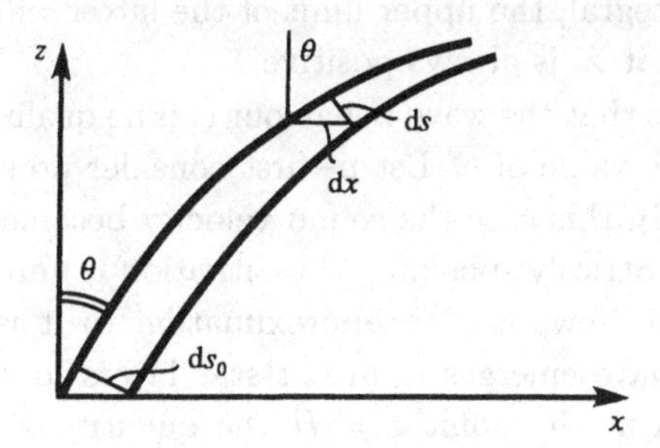

Figure 3.8. Rays in the oblique wave propagation in a stratified medium.

tube width ds equals d$x \cos\theta$ and, hence,

$$Q = \frac{\mathrm{d}s}{\mathrm{d}s_0} = \frac{\cos\theta}{\cos\theta_0},$$

where ds_0 is the tube width at the level $z = 0$, while θ is the local angle of the ray with respect to the vertical direction. The angle θ is defined by Snell's law: $c_0 \sin\theta = c(z)\sin\theta_0$. Therefore, the tube width parameter is equal to

$$Q(z) = (\cos\theta_0)^{-1}\sqrt{1 - \left(\frac{c}{c_0}\sin\theta_0\right)^2}. \tag{6.1}$$

This allows us to reduce the problem to one of integrating with respect to the vertical coordinate z. Because $\mathrm{d}\ell = \mathrm{d}z/\cos\theta$ the reduced variables (2.2) can be easily written as

$$u = v\left(c\rho\frac{\sqrt{1-\alpha^2/n^2}}{\sqrt{1-\alpha^2}}\right)^{1/2}, \tag{6.2}$$

$$X = \int_0^z \frac{\sqrt{1-\alpha^2}}{\rho^{1/2}c^{5/2}(1-\alpha^2/n^2)^{3/4}}\mathrm{d}z. \tag{6.3}$$

Here we write $\alpha = \sin\theta_0$; $n = c_0/c$ is the refractive index of the medium.

Suppose, for example, that the function $c(z)$ may be approximated by a power law:

$$c = c_0(1 \pm z/H)^b, \quad b, H = \text{const}.$$

Then the integral (6.3) is readily reduced to

$$X = \pm\int_1^a \frac{H\sqrt{1-\alpha^2}\,\mathrm{d}a}{b\rho^{1/2}c_0^{5/2}a^{(z-2/b)}(1-\alpha^2a^2)^{3/4}}, \tag{6.4}$$

where $a = c/c_0$ (Ostrovsky *et al.*, 1976). Here, depending on the sign preceding the integral, the upper limit of the latter will be more or less than unity, so that X is always positive.

It is easy to see that the wave behaviour can be qualitatively different, depending on the value of b. Let us first consider propagation towards a decrease of c. In this case the sound velocity becomes zero at a finite distance $z = H$. Strictly speaking, this situation is unrealistic, but from the astrophysical viewpoint an approximation to it is possible when, for example, a wave emerges from hot star layers to enter cold layers. If $b \geq 1/3$, then at the point $z = H$ the quantity X diverges, which means, as has been mentioned already, an unbounded nonlinear wave distortion and a complete dissipation of energy at a discontinuity (the velocity amplitude at this point tends to infinity, while the pressure tends to zero, such that the energy in a harmonic wave, as can be readily demonstrated, decays as $c^{6-2/b}$). If, however, $b < 1/3$, then X remains finite at $z = H$, so that a discontinuity would either not emerge at all or not have time to dissipate completely.

Let us proceed to the case when c grows with z. Here is a somewhat special case of normal propagation ($\alpha = 0$). If in this case $b < 1/3$, then X is subject to unbounded increase with z so that the wave always becomes discontinuous, and its energy dissipates. If $b > 1/3$, X remains finite and the wave energy does not vanish, because of the wave "acceleration" and velocity amplitude decay in it.

At oblique incidence ($\alpha \neq 0$) under the circumstances described, a turning point of the rays, i.e. a caustic, always emerges; this takes place at an altitude where $Q = 1/\alpha$. The wave behaviour near a caustic cannot be described within the framework of NGA; this problem will be tackled in the next chapter. Let us, however, highlight an important fact: when analysing the integral (6.4) near a caustic, one can easily see that X remains finite up to the caustic itself. Because of this we may expect that, despite the unlimited growth of the wave amplitude (in this approximation) ($v \sim (1 - a^2\alpha^2)^{-1/4}$), nonlinear distortions remain finite, which will allow us to offer a simplified study of the nonlinear wave behaviour in the caustic vicinity.

In certain instances the integral (6.4) can be calculated analytically. As a result one can easily find the height z_* of discontinuity formation in an initially sinusoidal wave

$$v(z = 0) = v_0 \sin\left[\omega\left(t - \frac{\alpha x}{c_0}\right)\right].$$

Thus for $b = 2/9 < 1/3$ we obtain

$$\frac{z_*}{H} = \alpha^{-9/2}\left\{1 - \left[(1-\alpha^2)^{1/4} \mp \frac{\alpha^2 L_N}{2H(1-\alpha^2)^{1/4}}\right]^{1/4}\right\}^{9/4} - 1. \quad (6.5)$$

Here, as above, $L_N = c_0/\varepsilon\omega v_0$ is the distance of discontinuity formation in a homogeneous medium (the $-$ sign here corresponds to the $+$ sign in (6.4) and vice versa). When c is decreasing, a discontinuity arises only if $b < 0.55$. If c is increasing, however, a discontinuity may well not appear before the caustic, provided that the incidence angle exceeds a certain critical value θ_s defined by the formula

$$\theta_s = \arcsin\frac{2\sqrt{2}}{y}\left(\sqrt{1+\frac{y^2}{4}} - 1\right)^{1/2}, \quad y = \frac{x_*}{H}. \quad (6.6)$$

For a case when relative variations of c are always small, a linear approximation of $c(z)$ is of interest, i.e. $b = 1$, provided that $z \ll H$. Here substituting $a = 1 \pm bz/H$ into (6.4) we find the same result, to the first approximation, as was observed for a homogeneous medium, while to the next approximation there is a minor correction to it. An exception is the case of grazing angles, when α is close to 1, and even small deviations of a from unity are of importance. Assuming in (6.4) that $a = 1 + \mu$ and $\alpha = 1 - s$, where μ, s are small, we easily obtain

$$X = B\left[s^{1/4} - (s-\mu)^{1/4}\right], \quad \mu < s,$$

where B is a constant factor, namely

$$B = \frac{2^{7/4} H\sqrt{s}}{\rho_0^{1/2} c_0^{5/2}},$$

and μ is evidently equal to z/H, so that according to (6.2) $u = v_0\sqrt{\rho c_0} \times (1-\mu/s)^{1/2}$. Equating X to $L_N/\sqrt{\rho c_0^5}$ we find a value of $\mu_* = z_*/H$, which corresponds to discontinuity formation in the initially harmonic wave,

$$\frac{z_*}{H} = s - \left(s^{1/4} - \frac{c_0^2}{\varepsilon\omega v_0 2^{7/4} H\sqrt{s}}\right)^4, \quad (6.7)$$

from which it follows that a discontinuity is formed prior to a caustic if the difference $\beta = \pi/2 - \theta_0 = \sqrt{2s}$ is the value

$$\beta_{cr} = (2\varepsilon kMH)^{-1/3}, \quad (6.8)$$

where, customarily, $M = v_0/c_0$. This is possible (when $\beta \ll 1$) only for sufficiently short and intense waves for which $kMH \gg 1$. For an ocean layer, for example, with a variation in c of 10 $\mathrm{m\,s^{-1}}$ per km (i.e. for $c_0 = 1.5 \times 10^3\,\mathrm{ms^{-1}}$ we have $H \simeq 1.5 \times 10^5\,\mathrm{m}$) one can observe from (6.8) that in a wave with frequency 50 kHz and pressure amplitude 6.2 Pa ($M \simeq 1.4 \times 10^{-4}$), the critical angle amounts to 3×10^{-2} rad, or about 1.7°.

A monopole in a linearly stratified medium

For a point source in a plane-layer medium the ray pattern is also known (see, for example, the book by Brekhovskikh, 1973). We shall impose the boundary condition $a = a_1$ on a sphere of radius R_0 surrounding the source such that R_0 must be large compared with a wavelength while being small compared with the characteristic inhomogeneity scale, so as to make the wave parameters similar over the whole sphere. If we use a linear stratification with respect to sound velocity, the following expression results for the ray tube width:

$$Q = \frac{H^2}{\alpha^4 R_0^2}\left(1 \mp \sqrt{\frac{1-\alpha^2 a^2}{1-a^2}}\right), \tag{6.9}$$

where the same notation is employed as previously: $\alpha = \sin\theta_0$, $a = c/c_0$ where c_0 corresponds to the point of source localization (or to radius R_0, which is the same) while θ_0 takes on arbitrary values. The minus sign in (6.9) corresponds to ray travel up to the turning point whereas the plus sign applies beyond this point. Here again Q can be expressed using a vertical coordinate z, but this relation depends, of course, on the direction (on the value of θ_0). For a vertical ray ($\theta_0 = 0$) we obtain, for example:

$$Q(z) = \frac{z^2}{R^2}\left(1 \pm \frac{z}{H}\right),$$

i.e. when moving towards larger c the ray tube section enlarges faster than in a homogeneous medium while in the opposite direction this process is slower. Note also that at a turning point, where $a = \alpha^{-1}$, we have $Q = H^2/\alpha^4 R_0^2$. Substituting (6.9) into the expression for the reduced coordinate yields

$$X = \alpha^2 R_0 \int_{a_1}^{a} \frac{a^{-5/2}(1-\alpha^2 a^2)^{-1/2}}{1 \mp \left[(1-\alpha^2 a^2)/(1-\alpha^2)\right]^{1/2}}\, da, \tag{6.10}$$

where $a_1 = a(R_0) = 1 \pm R_0/H\sqrt{1-\alpha^2}$.

Now the points of discontinuity formation in various directions, in an

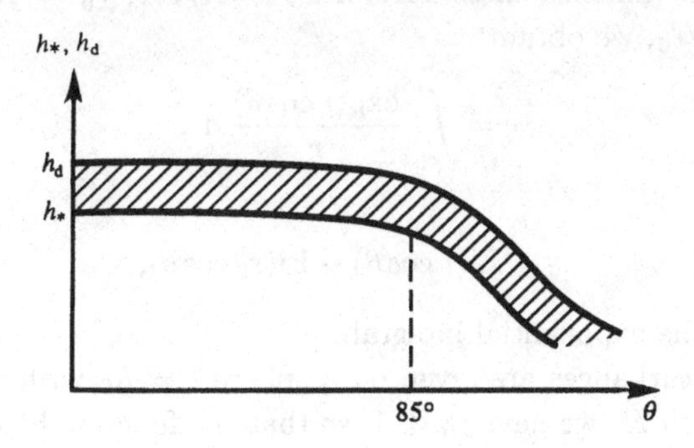

Figure 3.9. Height of shock formation h_s and the layer of dissipation h_d as a function of the angle of incidence.

initially harmonic or pulsed wave, can readily be found. In the general case the integral (6.10) can be calculated numerically. Such a solution was obtained by Ostrovsky *et al.* (1976) for the case $R_0/x_* = 0.05$, $R_0/H = 0.01$ (as regards the ocean, for example, realistic values seem to be $H = 10^4$ m, $x_* = 2 \times 10^3$ m, the latter for a spherical pulse having length $\ell = 2$ m and Mach number $M_s \sim 10^{-4}$).

A point source in an exponential atmosphere

In the conclusion of this section let us consider an oscillating point-source in an isothermal atmosphere (Ostrovsky & Fridman, 1985). Apart from physical attraction this problem is of certain practical significance, for instance with regard to acoustic sounding of the atmosphere by explosive pulses.

In an isothermal gas the sound velocity is constant and hence the "linear" rays remain straight, whereas nonlinear effects depend heavily on direction due to the rapid change of density with altitude. For this case $\rho = \rho_0 \exp(-2z/H)$.

The solution to this problem can, as previously, be written in terms of the reduced variables so as to follow the ray coordinate radially from the source. Given the angle θ of the ray slope with respect to the vertical direction, and the source altitude z_0, we can interpret ρ as

$$\rho(\ell) = \rho_0 \mathrm{e}^{-(z_0 + l\cos\theta)/H}. \tag{6.11}$$

Then the reduced coordinate can be represented in the form

$$X = c^{-5/2} \int_{\ell_0}^{\ell} \frac{\mathrm{d}\ell}{\sqrt{\rho Q}} = \frac{\ell_0}{c^2\sqrt{\rho_0 Q_0 c}} \int_{\ell_0}^{\ell} \exp\left(\frac{z + \ell\cos\theta}{2H}\right) \frac{\mathrm{d}\ell}{\ell}. \tag{6.12}$$

Proceeding to dimensionless variables $r = \ell/2H$, $r_0 = R_0/2H$, $\eta = Xc^2\sqrt{c\rho_0 Q_0}/\ell_0$, we obtain

$$\eta = \int_{r_0}^{r} \frac{\exp(r\cos\theta)}{r}\,\mathrm{d}r,$$

or

$$\eta = \mathrm{Ei}(r\cos\theta) - \mathrm{Ei}(r_0\cos\theta), \tag{6.13}$$

where Ei is the exponential integral.

As the disturbances are given on a sphere $\ell = R_0$ with radius small compared with H, we have $r_0 \ll 1$, so that the function $\mathrm{Ei}(r_0\cos\theta)$ can be expanded in a series where only the first two terms are significant,

$$\eta = \mathrm{Ei}(r\cos\theta) - G - \ln(r_0\cos\theta) \tag{6.14}$$

where $G = 0.5772$ is the Euler constant.

Now one can effortlessly find, for example, the distance of discontinuity formation in an initially harmonic wave with amplitude v_0 and frequency ω. This distance r_*, or the corresponding altitude $h_* = r_*\cos\theta$, is defined by the equality

$$\mathrm{Ei}(h_*) = R + G + \ln r_0 + \ln(\cos\theta). \tag{6.15}$$

Here $R = (x_*/R_0)\mathrm{e}^{-z_0/2H}$ and again $L_N = c^2/\omega\varepsilon v_0$ is the distance of discontinuity formation in a plane wave with amplitude v_0.

If $R \ll 1$, h_* is also small and (6.15) yields $r_* \simeq r_0$. This result is obvious: for small R the discontinuity forms at distances for which the atmosphere can be treated as homogeneous.

Still more interesting is the case of $R \gg 1$. Given this condition the discontinuity formation altitude depends on the angle θ, but this dependence is weak everywhere with the exception of grazing angles approaching $\pi/2$: for all other directions the term $\ln\cos\theta$ is small compared with R.

Figure 3.9 depicts the dependence $h_*(\theta)$ obtained from a numerical calculation at $r_0 = 10^{-2}$, $R = 100$. A discontinuity is observed to arise at practically the same altitude up to angles of about 85°. If, for example, a source with frequency 300 Hz located near the Earth's surface generates a wave with Mach number 10^{-4} at a distance $\ell \simeq 100\,\mathrm{m}$ ($R = 100$) the discontinuity formation altitude will be 62 km.

After shock formation the wave energy dissipates, so that the "dissipation layer" thickness ℓ_d estimate seems important. As the upper boundary of this layer we take the altitude h_d at which a discontinuity

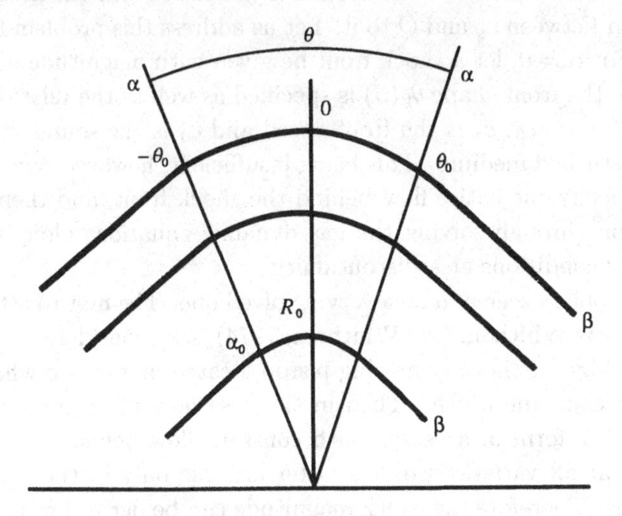

Figure 3.10. Self–refraction of a converging shock front.

evolves so as to attain a wave maximum with about one third of its initial energy being dissipated. It can easily be found that h_d satisfies an equation analogous to (6.15),

$$\mathrm{Ei}(h_\mathrm{d}) = \frac{\pi}{2}R + G + \ln r_0 + \ln(\cos\theta). \tag{6.16}$$

Obviously h_d also depends only weakly on θ. Figure 3.9 also depicts the upper boundary of the layer. With the parameters cited above, $h_\mathrm{d} \simeq 67\,\mathrm{km}$, i.e. the "dissipation layer" thickness equals only 5 km.

Therefore, over a very large sector the wave energy is conserved up to almost the same height, then eventually dissipates quite quickly. This phenomenon seems important when analysing the conditions of acoustic remote probing of the atmosphere by intense sound.

3.7 Self-refraction of shock waves

In this section we at last put aside the linear-ray approximation and embark on considering self-refraction effects, i.e. ray direction variations due to nonlinearity. We shall confine ourselves here to two-dimensional problems and use the ray equations in the form (1.4),

$$c_s\frac{\partial\theta}{\partial\beta} = \frac{\partial Q}{\partial\alpha}, \quad Q\frac{\partial\theta}{\partial\alpha} = -\frac{\partial c_s}{\partial\beta}. \tag{7.1}$$

As was mentioned above, this system is not closed and one needs to add a relation between c_s and Q to it. Let us address this problem for shock waves. For $\alpha = 0$, let a shock front be given with magnitude depending on β, i.e. the front shape $\theta_0(\beta)$ is specified as well as the relation $M_s(\beta)$ where $M = c_s/c_0$, c_s is the front speed and c_0 is the sound velocity in the undisturbed medium. This being insufficient, however, one generally has to specify the entire flow behind the shock front, and then analyse its motion through solving the gas dynamic equations along with the boundary conditions at a discontinuity.

This problem seems to be a very involved one. The first to attempt its solution was Whitham (see Whitham, 1974), who considered the shocks generated by a uniformly moving piston located in a region where a ray tube has constant width. Then in the case of a plane front the wave acquires the form of a "step" with constant flow behind a discontinuity, so that all variations of the latter are due only to the tube width variations. Therefore the shock magnitude can be derived from the conditions of mass and momentum conservation, so that the discontinuity amplitude and velocity will be functions of the local tube cross-section Q alone. Relying upon an appropriate solution for a channel of variable cross-section, Whitham formulated such a relation for a shock wave of arbitrary amplitude in an ideal gas. For weak shock waves, however, this relation is obvious: as in the linear approximation, for two-dimensional waves we have $v_s \sim Q^{-1/2}$, i.e.

$$\frac{Q}{Q_0} = \left(\frac{M_0}{M_s}\right)^2, \tag{7.2}$$

where M_0, Q_0 are arbitrary initial values.

From the NGA viewpoint we are more interested in short waves than in "infinite steps"; e.g. a wave in the form of a triangular pulse is implied, with length ℓ small compared with a characteristic ray curvature radius. In this case the discontinuity value is subject to the simultaneous influence of two factors: variation of Q, and nonlinear evolution of the simple wave behind the discontinuity. To complete the system (7.1) in the case of small nonlinearity one can use (Fridman, 1982) an equation of the type (2.10) for the wave evolution along the ray, employing the reduced variables (2.2), which have the following form for a homogeneous medium:

$$u = v\sqrt{c\rho Q}, \quad X = c^{-5/2}\rho^{-1/2}\int_0^\ell \frac{d\ell'}{\sqrt{Q(\ell')}}.$$

Then, in particular, the equation for the triangular pulse amplitude, with front velocity $c + \varepsilon v_s/2$, has the form

$$\frac{\partial}{\partial \alpha}\left(\frac{1}{QM_s^2}\right) = \frac{\varepsilon c}{2\sqrt{Q}s(\beta)}, \tag{7.3}$$

where $s = \ell M_s/2$ and $M_s = v_s/c_0$, ℓ being the pulse length and s its area. As the results in Chapter 2 indicate, the value of s is conserved along each tube, i.e. it does not depend on α, while dependence on β is possible provided initially it is specified as varying over the front. This dependence is entirely defined by the initial conditions.

Therefore, in the general case we obtain a set of three equations, given in (7.1) and (7.3), for the variables θ, Q and M_s. The two aforementioned factors, however, such as tube inhomogeneity and nonlinear dissipation, may show up to different extents. In particular, for a long enough pulse (large ℓ) the right-hand side of (7.3) is small and one can write $QM_s^2 = Q_0 M_{s0}^2$, which coincides with the quasi-stationary case (7.2). Substituting $c_s = c(1 + \varepsilon M_s/2)$ and assuming, as above, that $Q_0 = 1$, we derive the following equations from (7.1):

$$c_0 \frac{\partial \theta}{\partial \beta} = -\frac{2M_{s0}^2}{M_s^3}\frac{\partial M_s}{\partial \alpha}, \quad \frac{1}{c_0}\frac{\partial \theta}{\partial \alpha} = -\frac{\varepsilon M_s^2}{2M_{s0}^2}\frac{\partial M_s}{\partial \beta}. \tag{7.4}$$

This set reminds us of the equations of one-dimensional gas dynamics, being quasilinear and second-order. For it one can easily formulate two families of characteristics along which weak disturbances propagate against the background of constant values of M_s and θ:

$$\frac{\mathrm{d}\beta}{\mathrm{d}\alpha} = \pm c_0 \frac{\sqrt{\varepsilon M_s}}{2}\left(\frac{M_s}{M_{s0}}\right)^2 = \pm D. \tag{7.5}$$

As the value of D is real, our set is a hyperbolic one. (It seems noteworthy that $D/c_0 \sim \sqrt{M_s}$.) Certainly, for any fixed level of M_s one could set $M_s = M_{s0}$; however, it is convenient to retain the difference between these quantities, since (7.5) also describes the propagation of a simple wave with variable M_s. In order to find such a wave one can suppose $\theta = \theta(M_s)$, and then obtain from (7.4)

$$\frac{\mathrm{d}\theta}{\mathrm{d}M_s} = \pm\sqrt{\frac{\varepsilon}{M_s}}, \quad \theta - \theta_0 = \pm 2\left(\sqrt{\varepsilon M_s} - \sqrt{\varepsilon M_{s0}}\right). \tag{7.6}$$

The corresponding solution for M_s is framed as

$$M_s = F(\beta - D(M_s)\alpha). \tag{7.7}$$

Note that (7.4) can now be rewritten in terms of Riemann invariants $\mathcal{I}_\pm = \theta \pm 2\sqrt{\varepsilon M_s}$:

$$\left(\frac{\partial}{\partial\alpha} \pm D\frac{\partial}{\partial\beta}\right)\mathcal{I}_\pm = 0. \tag{7.8}$$

Let us finally speculate about possible discontinuities in θ and M_s propagating along the front of a "primary" shock wave. It is clear that propagation of a simple wave (7.7) can result in its breaking, i.e. in abrupt variations of θ and M_s at a certain point. Could a secondary shock wave appear in the form of a jump of θ and M_s, i.e. could the front kink have different values of M_s on the two sides (a "shock-shock" according to Whitham's definition)? If the existence of such a secondary shock is postulated, then integrating (7.4) with respect to β, we shall have in its vicinity the following boundary conditions:

$$c(\theta_2 - \theta_1) = -\Gamma M_{s0}^2\left(\frac{1}{M_{s2}^2} - \frac{1}{M_{s1}^2}\right), \quad \Gamma(\theta_2 - \theta_1) = \frac{\varepsilon c_0}{6M_{s0}^2}\left(M_{s2}^3 - M_{s1}^3\right) \tag{7.9}$$

where indices 1 and 2 refer to quantities to the right and left of a front kink, and Γ is the velocity of travel of the kink along the front ($\Gamma = \mathrm{d}\beta_s/\mathrm{d}\alpha$ where β_s is the kink coordinate). For low M_s the value of Γ, as may be expected, is small compared with c_0 (of the order of $c_0\sqrt{M_s}$) and therefore $\theta_2 - \theta_1 \sim \sqrt{M_s}$, i.e. the kink – a "shock-shock" – is characterized by a small angle.

Eliminating Γ from the above expression we obtain a relation between the jumps of θ and M_s – an analogue of the Hugoniot adiabat in gas dynamics ((7.6) can then be viewed as an analogue of the Poisson adiabat).

However, the real existence of such secondary shocks remains questionable until a possible "viscosity" mechanism is clarified supporting their structure, just as ordinary viscosity is necessary for gas dynamic shock waves to exist. It seems highly probable that such a "viscosity" can be connected with diffraction effects leading to energy radiation from the kink vicinity, where the geometrical approach is inapplicable. In this case a stationary kink can be set up, in principle, with a sharp but continuous variation of θ and M along the front.† Its characteristic width δ_s can be estimated by equating the distance L_d at which the "secondary" shock broadens to a width δ_s due to diffraction ($L_\mathrm{d} \sim \delta_\mathrm{s}^2/\delta_\mathrm{p}$ where δ_p is

† Note that similar conjectures hold true for solitons in dispersive media; in the latter case the Burgers equation may be derived for the soliton front parameters taking radiation into account (Shrira, 1980).

the thickness of the primary shock front) to the corresponding nonlinear length for a wave of type (7.7) ($L_N \sim c_0\delta_s/(c_2 - c_1)$). Hence we obtain

$$\delta_s \sim \delta_p \frac{2M_{s0}^2}{\sqrt{\varepsilon}(M_{s2}^{5/2} - M_{s1}^{5/2})} \sim \frac{\delta_p}{\sqrt{\varepsilon M_s}}. \tag{7.10}$$

(The second expression is for the case when $M_{s2} - M_{s1} \sim M_s \sim M_{s0}$.) Therefore, it can be seen that δ_s is far in excess of δ_p, as could be anticipated in the case of small nonlinearity ($M_s \ll 1$).

Using the above approach one can solve some problems of a "gas dynamics" type describing the front evolution of a weak shock wave. Let us consider two examples.

Stability of a cylindrical shock front.

One of the obvious solutions to (7.4) is cylindrically symmetric:

$$\theta = \frac{\beta}{c_0\alpha_0}, \quad Q = \frac{\alpha}{\alpha_0}, \quad M_s = M_{s0}\sqrt{\frac{\alpha_0}{\alpha}}. \tag{7.11}$$

Let us discuss the behaviour of small disturbances against the background of such a wave. Let

$$\theta = \frac{\beta}{c_0\alpha_0} + \theta'(\alpha, \beta), \quad M_s = M_{s0}\sqrt{\frac{\alpha_0}{\alpha}}(1 + s(\alpha, \beta)), \tag{7.12}$$

where θ' and s are small. Then linearizing equation (7.4) yields

$$\frac{\partial \theta'}{\partial \beta} = -\frac{2}{\alpha_0 c_0}\frac{\partial}{\partial \alpha}(\alpha s), \quad \frac{\partial \theta'}{\partial \alpha} = -\frac{c_0\varepsilon M_{s0}}{2\alpha}\frac{\partial}{\partial \beta}(\alpha s).$$

Denoting $\alpha s = \xi$ and eliminating θ' from these equations we obtain

$$\frac{\partial^2 \xi}{\partial \alpha^2} = K\alpha^{-5/2}\frac{\partial^2 \xi}{\partial \beta^2}, \quad K = \frac{1}{4}c_0^2\alpha_0^{5/2}\varepsilon M_{s0}. \tag{7.13}$$

The solution to this equation can be attempted in the form of "angular modes" for which $\xi \sim e^{im\beta}$ where m is an integer. Then the dependence of ξ on α is given by way of cylindrical functions. The resulting expression for the modulation factor is

$$s \sim \alpha^{-1/2} Z_{-2}(4mK^{1/2}\alpha^{-1/4})e^{im\beta}, \tag{7.14}$$

which gives $\theta' = (2i/\alpha_0 c_0 m)(\alpha s)_\alpha$.

In the general case this solution includes cylindrical functions Z of the first and second kinds; it is determined through specifying an initial distribution (at $\alpha = \alpha_0$) of s and θ' over β. In particular, one can observe the asymptotic disturbance at large α for a diverging wave, and as $\alpha \to 0$ for a converging one. As $\alpha \to \infty$ the argument of the function Z in (7.14) tends to zero and then it can easily be seen that s and θ' have an α-independent limit. If $\alpha \to 0$ then $s, \theta' \sim \alpha^{-3/8}$. This implies that in a converging wave disturbances grow at a higher rate than the initial symmetric shock front amplitude. Therefore, a converging shock wave appears unstable with respect to small curvatures of its front (this conclusion had been arrived at earlier by Whitham (1974) for strong shock waves). Consequently, the many efforts to describe converging shock wave behaviour near the centre of symmetry become somewhat senseless, i.e. this symmetry must be violated.

Let us mention also the case of delta function disturbances localized on a cylindrical front. Their behaviour is different from that of the angular modes considered above with finite m, since they correspond to the asymptotic value $m \to \infty$. Such disturbances propagate just along the characteristics of (7.13), i.e. along $\mathrm{d}\beta = \sqrt{K}\alpha^{-5/4}\,\mathrm{d}\alpha$ or

$$\beta - \beta_0 = \pm 4\sqrt{K}(\alpha_0^{-1/4} - \alpha^{-1/4}). \tag{7.15}$$

Therefore, in particular, it follows that in a diverging wave ($\alpha \to \infty$) localized disturbances only have time to separate at a finite angle. From the expression for K it can readily be seen that the asymptotic value of this angle $\Delta\theta \approx \Delta\beta/c_0\alpha_0$ is equal to $\sqrt{\varepsilon M_s/2}$, i.e. within the framework of nonlinear acoustics ($M_s \ll 1$), disturbances diverge weakly.

Transformation of a cylindrical shock wave into a plane wave

Let the shock wave front at the initial instant $\alpha = \alpha_0$ be specified as cylindrical, with a curvature radius R_0, within the angular range $|\theta| < \theta_0$, and as a plane outside this range (Figure 3.10), that is

$$\theta(\alpha = \alpha_0) = \begin{cases} \beta/R_0 & |\beta| < \beta_0 = R_0, \\ -\theta_0 & \beta < -\beta_0, \\ \theta_0 & \beta > \beta_0. \end{cases} \tag{7.16}$$

The initial value of the amplitude $M(\alpha_0) = M_0$ is assumed constant along the front.

Let us consider first a converging wave (Whitham (1974) addressed this problem qualitatively). We are, therefore, interested in the domain $\alpha < \alpha_0$ and as in (7.11) it is convenient to set $\alpha_0 = R_0/c_0$, so that $\alpha = 0$ corresponds to the circular wave focusing point in a linear problem. The solution behaviour can easily be judged by characteristics on the $\alpha\beta$-plane (Figure 3.11).

In the domain (I), corresponding to $|\theta| > \theta_0$, these characteristics are

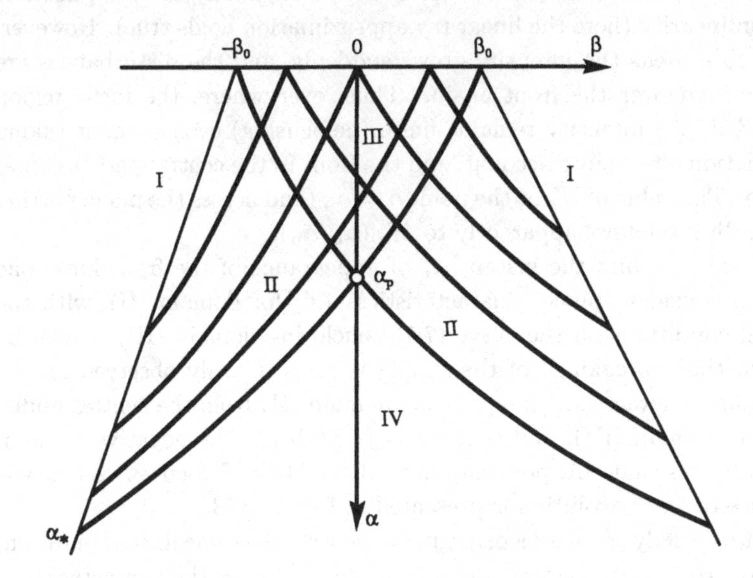

Figure 3.11. Characteristics of the system of the equations (7.4).

straight, and hence the boundary characteristics will also be straight. Inside this angle an interaction domain (III) is observed where two families of curvilinear characteristics are present, corresponding to a circular wave; these diverge later to form the simple wave domains (II) where one of the Riemann invariants is constant. Finally, starting at $\alpha = \alpha_*$ the characteristics begin to intersect, forming the above-mentioned kink. But still earlier, at $\alpha = \alpha_\mathrm{p} < \alpha_*$, the interaction region vanishes, to leave two diverging simple waves separated by domain (IV), i.e. by a new plane front portion.

Now the "critical" solution points α_p and α_* can easily be found. As domain (III) is bounded by the characteristics (7.15), which originate from the points $\pm\beta_0$ to intersect at the point α_p for $\theta = 0$, it is an easy matter to find

$$\alpha_\mathrm{p} = \alpha_0 \left(1 + \frac{\theta_0 c_0 \alpha^{5/4}}{4\sqrt{K}}\right)^{-4} = \alpha_0 \left(1 + \frac{\theta_0}{2\sqrt{\varepsilon M_\mathrm{s}}}\right)^{-4}. \tag{7.17}$$

Therefore, it is clear that at very small θ_0 ($\theta_0 \ll \sqrt{\varepsilon M_{\mathrm{s}0}}$), a disturbance is rather quickly distributed over the characteristics (α_p is close to α_0) leaving a small plane zone. If, however, $\theta_0 \gg \sqrt{\varepsilon M_{\mathrm{s}0}}$, then α_p is close to a focusing point ($\alpha_\mathrm{p} \ll \alpha_0$) and down to the smallest vicinity of this point the front convergence proceeds in the same way as in the absence of nonlinearity (here the linear ray approximation holds true). However, near to a focus the intensity grows suddenly, and the disturbances are distributed over the front again. Thus, everywhere, the focus region included, the intensity remains finite (defocusing) even without taking diffraction effects into account, and the front in the central part becomes plane. The value of M_s at the point $\alpha = \alpha_p$ (and across the plane portion generally) amounts apparently to $M_{s0}(\alpha_0/\alpha_p)$.

In order to find the instant α_* of appearance of the front kink, one should consider "plane" characteristics (7.5) for domain (II), with the initial conditions on the curve (7.15) enclosing domain (III). It can be shown that "breaking" of the simple wave is initially observed on the boundary characteristics separating domain (II) from the central undisturbed domain (IV), and that $\alpha_* = (-3/5)\alpha_p$. The negative value of α_* indicates that this point lies behind the "linear" focus. A schematic picture of wave evolution is presented in Figure 3.11.

Let us briefly consider a diverging wave with the same initial condition, (7.16). Here disturbances are also distributed over the characteristics, but much more slowly because of the drop in wave amplitude. For the plane portion formation coordinate we obtain

$$\alpha_p = \alpha_0 \left(1 - \frac{\theta_0}{2\sqrt{\varepsilon M_{s0}}}\right)^{-4}. \qquad (7.18)$$

This means that the plane portion only emerges for rather strong and weakly diverging waves; if, however, $\theta_0 > 2\sqrt{\varepsilon M_{s0}}$, then the characteristics enclosing the interaction domain (III) do not intersect at all, and the central region remains cylindrical. Within the domain (II) of simple waves the characteristics diverge, and the front kinks do not develop.

Whitham (1974) addressed some other problems as well: shock wave diffraction at a rounded wall, a wedge or a sphere using the known gas dynamics solutions and compared some results with experimental data. For weak shock waves, however, the appropriate solutions do not seem entirely adequate, and are therefore not tackled here.

Short pulses

The previous results, based on the assumption of a local relation between M_s and Q, only hold for pulses with a large enough area. As can be readily observed from (7.3), this implies that a condition of the type

$$\frac{\partial M_s}{\partial \alpha} \gg \varepsilon c_0 M_s^2 \frac{\sqrt{Q}}{s(\beta)}$$

should be satisfied. Let us return now to the general case, and analyse the solution of (7.1) and (7.3) in complete form. For the case of small nonlinearity which is important to us, this system is somewhat simplified, and for a triangular pulse acquires the form

$$c_0 \frac{\partial \theta}{\partial \beta} = \frac{\partial \theta}{\partial \alpha}, \quad \frac{\partial Q}{\partial \alpha} = -\frac{\varepsilon c_0}{2Q} \frac{\partial M_s}{\partial \beta}, \quad \frac{\partial}{\partial \alpha}\left(\frac{1}{QM_s^2}\right) = \frac{\varepsilon c_0}{2\sqrt{Q}s(\beta)}. \tag{7.19}$$

This system does not lend itself to an elaborate investigation. However, a few nontrivial solutions can still be found.

Let us begin with the problem of the stability of a plane wave in the form of a decaying triangular pulse. We shall search for a solution to (7.19) in the form

$$\theta = \theta', \quad Q = 1 + Q', \quad M_s = M_0(\alpha)(1 + g(\alpha, \beta)), \quad s = s_0(1 + s(\beta)),$$

where θ', Q', g, s are small values and $M_0(\alpha)$ corresponds to a plane undisturbed pulse. This permits us to linearize the equations (7.19) with respect to the unperturbed solution (2.4). Taking into account that

$$\frac{\partial M_0}{\partial \alpha} = -\varepsilon c_0 \frac{M_0^3}{4s_0},$$

we obtain after elementary transformations the equations for Q' and g:

$$\frac{\partial^2 Q'}{\partial \alpha^2} = -\frac{c_0 \varepsilon}{2} M_0 \frac{\partial^2 g}{\partial \beta^2},$$
$$2\frac{\partial g}{\partial \alpha} + \frac{\partial Q'}{\partial \alpha} = \frac{2}{M_0}\frac{\partial M_0}{\partial \alpha}\left(2g + \frac{Q'}{2} - 2s\right). \tag{7.20}$$

It will be assumed that all disturbances are proportional to $e^{i\mathcal{X}\beta}$ where $\mathcal{X}$ is real, retaining previously introduced notation for the corresponding amplitudes. The amplitude M_0 will be given in the form $M_0 = A/\sqrt{\alpha}$ (A constant), corresponding to triangular pulse asymptotics. Then we obtain

$$\frac{\partial^3 Q'}{\partial \alpha^3} + \frac{3}{2\alpha}\frac{\partial^2 Q'}{\partial \alpha^2} + \frac{P}{2\sqrt{\alpha}}\frac{\partial Q'}{\partial \alpha} + \frac{PQ'}{4\alpha\sqrt{\alpha}} = \frac{Ps}{\alpha\sqrt{\alpha}}, \tag{7.21}$$

where

$$P = c_0 \varepsilon \mathcal{X}^2 A/2.$$

First, it should be noted that if we assume $M_0(\alpha)$ to be constant in (7.20), which in (7.21) implies that all the terms of degree higher than

$1/\sqrt{\alpha}$ are neglected and $\sqrt{\alpha}$ is "frozen", then (7.21) will yield

$$Q' \sim \mathrm{e}^{\mathrm{i}\mu\alpha}, \quad \mu = \left(\frac{P}{2\sqrt{\alpha}}\right)^{1/2} = \frac{c\mathcal{X}}{2}\sqrt{\varepsilon M_0},$$

which corresponds to the case considered above in (7.5), when nonlinear decay can be neglected compared with ray tube variations. Coming back to the general case we define the asymptotic solution as $Q \to \infty$. The solution of the third-order equation (7.21) involves three independent functions. One of the functions can be promptly derived by assuming that the first two terms in (7.21) are small, which is confirmed by the result obtained. It can easily be found that

$$Q' = 4s + \nu\alpha^{-1/2}, \quad \theta' \sim \alpha^{-3/2}, \quad q \sim \alpha^{-2}, \tag{7.22}$$

where ν is constant. In this solution the pulse amplitude disturbance disappears at a higher rate than the rate at which disturbance of the tube area tends to constancy at infinity.

The two remaining solutions have an oscillating character and can be found using the WKB technique. Let us set

$$Q = 4q + a\exp\left(\mathrm{i}\int f\,\mathrm{d}\alpha\right),$$

where a and f are slowly changing real functions. Then, keeping only the first and second terms in (7.21) at the lowest approximation (the result obtained confirms the smallness of the remaining term) we get $f = \pm(p/2\sqrt{\alpha})^{1/2}$, as in the above case of M_0 constant. At the next approximation terms with Q' and $\partial f/\partial\alpha$ in the two terms mentioned have to be accounted for, while for the remaining ones the zero order will suffice. As a result we obtain

$$a \sim \alpha^{-1/8}, \quad Q' = 4s + \nu_{\pm}\alpha^{-1/8}\exp\left(\pm\frac{4}{3}\sqrt{\frac{p}{2}}\mathrm{i}\alpha^{3/4}\right), \tag{7.23}$$

where $\nu_{\pm}$ are constants. Thus the amplitude of Q' decays as $\alpha^{-9/8}$, while that of g decays as $\alpha^{-13/8}$, and again disturbances of Q are observed to live longer than others. As disturbances in (7.23) decay at a slower rate than those in (7.22), this solution has to be treated as the asymptotic limit. Fridman (1982) conducted a numerical investigation of linearized equations equivalent to (7.21) (with $s = 0$) and confirmed that the solution is oscillatory with a decaying frequency, and that disturbances of Q decay extremely slowly.

As regards the nonlinear solutions of (7.19), it was shown (Fridman,

1982) that one can obtain different versions (eight altogether) where variables are separated such that some of them are nontrivial, one corresponding to the initially circular front, another to the initially plane front with a linear variation of shock amplitude along the front. Both are valid for finite intervals of β.

In concluding this section, we note that validity requirements for a "discontinuous" approximation when describing the shock waves impose certain restrictions on the front curvature, otherwise the shock wave duration will have no time to adjust to its stationary value (in analogy with the spherical shock wave transformation to an acoustic wave considered earlier). It is noteworthy that this is again equivalent to requiring that in a generalized Burgers equation (2.1) the last term, associated with the ray tube width variation, would at the shock front be small compared with the remaining terms, whose balance determines the stationary shock wave structure. When comparing, for example, with the nonlinear term $(\varepsilon/c_0)vv_\tau$ and recalling that $Q_\ell/Q \sim 1/R$ where R is the front curvature radius, we obtain the quasi-stationary condition in the form $\delta_p \ll RM_s$, where δ_p is the shock front width. Were this condition not fulfilled, the shock wave would transform into an acoustic wave.

References

Al'perovich, L.S., Gokhberg, M.B., Sorokin, V.M. & Fedorovich, G.V. (1979). On generation of geomagnetic variations by acoustic oscillations during earthquakes, *Izv. AN SSSR, Fiz. Zemli* **3,** 58–68.

Berg, A.M., Tjotta, J.N. (1993). Numerical simulation of the sound pressure field from finite amplitude plane or focusing rectangular apertures, in *Proc. XII ISNA* pp. 309–14 (World Scientific, Singapore).

Brekhovskikh, L.M. (1973). *Waves in Layered Media* (Moscow, Nauka).

Catignol, D., Chapelon, J.V. (1990). High Energy Ultrasound Therapy. Part II. Shock Waves and Cavitation, in *Proc. XIII ISNA* pp. 30–35 (World Scientific, Singapore).

Christianovich, S.A. (1956). Shock wave at a large distance from the explosion point, *Prikladn. Mat. i Mekh.* **20,** 982–91.

Cole, R. (1948). *Underwater Explosions* (Princeton, New Jersey).

Fridman, V.E. (1980). A comparison of empirical and theoretical laws for explosive waves, *Izvestiya AN SSSR, Fiz. Atmosfery i Okeana* **16,** 551–9.

Fridman, V.E. (1982). Self-refraction of small amplitude shock waves, *Wave Motion* **4,** 151–61.

Gubkin, K.E. (1958). Propagation of discontinuities in sound waves, *Prikl. Mat. i Mekh.* **22,** 561–4.

Hammerton, P.W. & Crighton, D.G. (1988). Old-age behaviour of cylindrical and spherical nonlinear waves: numerical and asymptotic results, *Proc. Roy. Soc., London* **A 422,** 387–405.

Kaplan, S.A., Ostrovsky, L.A., Petrukhin, N.S. & Fridman, V.E. (1972). A calculation of self-consistent models of Sun and star chromospheres, *Astronom. Zh.* **49 (6)**, 1267–74.

Keller, J.B. (1954). Geometrical acoustics – I. The theory of weak shock waves, *J. Appl. Phys.* **25**, 938.

Kozin, D. & Saifutdinov, M.A. (1988). Generation of "fast" response of the ionosphere to the shock wave action, *Geomagnetizm i Aeronomia* **28 (4)**, 681–3.

Landau, L.D. (1945). On shock waves far from their origin, *Prikl. Mat. i Mekh.* **9**, 286–92.

Landau, L.D. & Lifshitz, E.M. (1986). *Hydrodynamics* (Moscow, Nauka).

Naugol'nykh, K.A. (1959). Propagation of spherical finite-amplitude sound impulses in a viscous thermal-conducting medium, *Akust. Zh.* **5**, 80–4.

Naugol'nykh, K.A. & Romanenko, E.V. (1959). On the dependence of the focusing system gain on the sound intensity, *Akust. Zh.* **5**, 191–5.

Naugol'nykh, K.A. (1964). On the compressional wave surrounding an expanding sphere, *JASA* **36**, 1442–4.

Naugol'nykh, K.A. (1971). Absorption of finite amplitude waves, in *High Intensity Ultrasonic Fields,* ed. L. Rozenberg (Plenum Press, New York).

Naugol'nykh, K.A. (1972). On the transition of a shock wave into an acoustic one, *Akust. Zh.* **18**, 579–83.

Naugol'nykh, K.A., Soluyan, S.I. & Khokhlov, R.V. (1963). Finite-amplitude spherical waves in a viscous thermal-conducting medium, *Akust. Zh.* **9**, 54–60.

Neighbors, T.H., Bjørnø, L. (1990). Focused finite amplitude ultrasonic pulses in liquids. Frontiers in Nonlinear Acoustics, *Proc. XII ISNA* , World Scientific, Singapore.209–14—

Ostrovsky, L.A. (1963). On the theory of waves in nonstationary compressible media, *Prikl. Mat. i Mekh.* **27**, 924–9.

Ostrovsky, L.A. (1976). Short-wave asymptotics for weak shock waves and solitons in mechanics, *Int. J. Nonlinear Mech.* **11**, 401.

Ostrovsky, L.A. & Fridman, V.E. (1985). Dissipation of intensive sound in isothermal atmosphere, *Akust. Zh.* **31**, 625–27.

Ostrovsky, L.A. & Pelinovsky, E.N. (1974). On the approximate equations for waves in a medium with small nonlinearity and dissipation, *Prikl. Mat. i Mech.* **38**, 121–4.

Ostrovsky, L.A., Pelinovsky, E.N., Fridman, V.E. (1976). Propagation of finite-amplitude acoustic waves in an inhomogeneous medium in the presence of caustics, *Akust. Zh.* **22, 6**, 914–921.

Ostrovsky, L.A. & Rubakha, N.R. (1972). Nonlinear magnetic sound in gravitational field, *Izv. Vuzov Radiofizika* **15 (9)**, 1293–9.

Ostrovsky, L.A. & Shrira, V.I. (1976). Instability and self-refraction of solitons, *Zh. ETF* **71**, 1412–20.

Ostrovsky, L.A., Pelinovsky, E.N. & Fridman, V.E. (1979). Propagation of explosive impulses in subsurface oceanic layers, *Akust. Zh.* **25 (1)**, 103–7.

Pelinovsky, E.N., Fridman, V.E. & Engelbrecht, Y.K. (1984). *Nonlinear Evolution Equations* (Valgus, Tallinn).

Petukhov, Y.V. (1983). Pressure impulse excited by a sphere expanding with constant velocity, *Akust. Zh.* **29**, 88–90.

Romanova, N.N. (1970). On vertical propagation of short acoustic waves in a real atmosphere, *Izv. AN SSSR, Fiz. Atm. i Okeana* **6 (2)**, 134–45.

Rudenko, O.V. & Soluyan, S.I. (1977). *Theoretical Foundations of Nonlinear Acoustics* (Plenum Press, New York).
Ryzhov, O.V. & Shefter, G.M. (1962). On the energy of sound waves propagating in moving media, *Prikl. Mat. i Mekh.* **26,** 854–66.
Shrira, V.T. (1980). Nonlinear refraction of solitons, *Zh. ETF* **79,** 87–98.
Taylor, G.I. (1946). The air wave surrounding an expanding sphere, *Proc. Roy. Soc., London* **A 196,** 273–92.
Whitham, G.B. (1953). The propagation of weak spherical shocks in stars, *Comm. Pure & Appl. Math* **6,** 3.
Whitham G.B. (1974). *Linear and Nonlinear Waves* (Wiley, New York).

4

Nonlinear sound beams

4.1 Paraxial beams

In this chapter we shall consider the influence of diffraction effects on finite-amplitude sound wave propagation. This generalization is adequate for the majority of problems on the emission and reception of waves by various types of antenna, wave beam propagation, plane to spherical wave transformation, wave focusing, etc.

The theoretical description of such processes runs into drastic difficulties. Even classical linear diffraction theory originally required a substantially new body of mathematics. To account for nonlinearity in such problems became an especially hot issue in view of the development of nonlinear optics. For light waves, however, because of dispersion in the medium, such a wave remains close to sinusoidal, even if the light intensity is very large, and the problem can be reduced to that of finding the space–time distribution of its two "macroscopic" parameters – amplitude and phase. In acoustics things appear still more complicated. Due to the absence of dispersion, the wave profile is constantly changing up to shock wave formation, and one needs to analyse the diffraction phenomena in such an intricately evolving field.

It was not until 1969 that problems of this type were brought under detailed consideration when a simplified equation was suggested by Zabolotskaya & Khokhlov (1969), describing the evolution of nonlinear sound beams with a narrow angular spectrum. This equation served as a basis for many further investigations. We must confess, however, that this equation still remains rather complicated and to obtain physically reliable results one has either to use computational methods or to introduce new simplifications.

In the previous chapters there was a clear-cut tendency to simplify the governing system, whenever possible, replacing it by model evolution equations of Burgers type. The effect of diffraction can be considered as a "transverse diffusion" of the field provoking energy leakage from the ray tube and can be described by an additional term in the corresponding "model" equation. The procedure is essentially the same as that described above. To explain it let us begin with a linear wave equation for, say, pressure disturbances p',

$$\frac{\partial^2 p'}{\partial t^2} - c_0^2 \nabla^2 p' = 0. \tag{1.1}$$

This equation can be written as

$$\frac{\partial^2 p'}{\partial t^2} - c_0^2 \frac{\partial^2 p'}{\partial x^2} = -c_0^2 \left(\frac{\partial^2 p'}{\partial y^2} + \frac{\partial^2 p'}{\partial z^2} \right)$$

or

$$\left(\frac{\partial}{\partial t} - c_0 \frac{\partial}{\partial x} \right) \left(\frac{\partial}{\partial t} + c_0 \frac{\partial}{\partial x} \right) p' = -c_0^2 \Delta_\perp p', \tag{1.2}$$

where

$$\Delta_\perp = \frac{\partial^2}{\partial y^2} + \frac{\partial^2}{\partial z^2}.$$

If a wave beam propagating along the x-axis is considered, then $p' = \mathcal{F}(t - x/c_0, x, y, z)$ such that the dependence on the variable $\tau = t - x/c_0$ is "fast" while that on the other variables (for τ constant) is considered to be slow. Then, as in Chapter 1,

$$\frac{\partial p'}{\partial t} + c_0 \frac{\partial p'}{\partial x} = \hat{s},$$

where $\hat{s}$ denotes the terms that are small compared with each term on the left-hand side.

Substituting this into (1.2), we obtain

$$2 \frac{\partial}{\partial t} \left(\frac{\partial p'}{\partial t} + c_0 \frac{\partial p'}{\partial x} \right) = -c_0^2 \Delta_\perp p', \tag{1.3}$$

or, using τ,

$$\frac{\partial^2 p'}{\partial \tau \partial x} = -\frac{c_0}{2} \Delta_\perp p'. \tag{1.4}$$

Equations (1.3) and (1.4) indicate that $\Delta_\perp p'$ is of the order of $\partial p'/\partial x$, i.e. the dependence of p' on x at constant τ gives second-order terms as compared with those along the transverse coordinates (if, say, $\partial/\partial x \sim \mu \ll 1$, then $\partial/\partial y, \partial/\partial z \sim \sqrt{\mu}$).

If we take a harmonic wave, setting $p' = A(x, y, z) \exp(\mathrm{i}\omega\tau)$ where

A is a slowly changing complex amplitude, then we derive from (1.4) a familiar "parabolic equation of quasi-optics":

$$\frac{\partial A}{\partial x} = -\frac{c_0}{2i\omega}\Delta_\perp A. \tag{1.5}$$

This equation was originally deduced by Leontovitch (1944), and independently by Malyuzhinets (1959).

It is worthwhile to note that there exist also more stringent and logical methods of deriving equations (1.3) to (1.5); see, for example, the book by Babich and Buldyrev (1972).

4.2 The nonlinear beam equation

Let us now consider wave beams in nonlinear viscous media. In this case a small right-hand side appears in the wave equation

$$\frac{\partial^2 p'}{\partial t^2} - c_0^2\nabla^2 p' = \hat{F},$$

where $\hat{F}$ contains nonlinear and dissipative terms (cf. Chapter 1). As $\hat{F}$ is small only derivatives with respect to τ should be retained. Thus, reducing the latter equation to one of evolution type, we should obtain for $\hat{F}$ just the same terms as those in the Burgers equation for a plane wave. Combining this equation with (1.4) yields (cf. (1.16) and (1.17) of Chapter 1)

$$\frac{\partial^2 p'}{\partial\tau\partial x} = -\frac{\partial}{\partial\tau}\left(\frac{\varepsilon}{\rho_0 c_0^3}p'\frac{\partial p'}{\partial\tau} + b\frac{\partial^2 p'}{\partial\tau^2}\right) + \frac{c_0}{2}\Delta_\perp p'. \tag{2.1}$$

Occasionally, it may be convenient to proceed to dimensionless variables:

$$u = \frac{p'}{p'_m}, \quad \theta = \omega\tau, \quad \sigma = \frac{x}{L_N} = \frac{\varepsilon\omega p'_m x}{\rho_0 c_0^3}, \quad \eta_\perp = \frac{r_\perp}{a}, \quad r_\perp^2 = y^2 + z^2, \tag{2.2}$$

where p'_m, ω are the characteristic amplitude and frequency of the perturbation, a being its typical transverse scale, and $L_N = (\varepsilon Mk)^{-1} = \rho_0 c_0^3/\varepsilon\omega p'_m$ is the distance of discontinuity formation in a plane wave with parameters p'_m and ω. Then, equation (2.1) will take the form

$$\frac{\partial}{\partial\theta}\left(\frac{\partial u}{\partial\sigma} - u\frac{\partial u}{\partial\theta} - Re^{-1}\frac{\partial^2 u}{\partial\theta^2}\right) = \frac{1}{4}N\Delta_\perp u, \tag{2.3}$$

where $N = 2/\varepsilon M(ka)^2$ is a dimensionless parameter equal to the ratio of the characteristic nonlinearity scale L_N and the diffraction scale $L_d = ka^2/2$: N is now often called the Khokhlov number. Within the brackets we have the left-hand side of the Burgers equation, which is fairly in line with our reasoning.

Equation (2.3) in the absence of losses is known as the Khokhlov–Zabolotskaya equation (see Zabolotskaya & Khokhlov, 1969). Its extension to dissipative media was performed by Kuznetsov (1970), hence (2.3) in its complete form is called the Khokhlov–Zabolotskaya–Kuznetsov (KZK) equation.

Various generalizations of equation (2.3) have been set forth. It can be formulated for a sound beam as the beam propagates within a smoothly inhomogeneous medium along separate rays (see Babich *et al.*, 1972; Pelinovsky *et al.*, 1984). This equation is offered here without deduction (for a two-dimensional case):

$$\frac{\partial}{\partial \xi}\left[\frac{\partial p'}{\partial s} + \frac{c_0\rho_0}{2}\frac{\partial}{\partial s}\left(\frac{1}{\rho_0 c_0}\right)p' + p'\frac{\partial p'}{\partial \xi} + b\frac{\partial^2 p'}{\partial \xi^2}\right]$$
$$+\frac{c}{2}\frac{\partial^2 p'}{\partial y^2} + \frac{y^2}{2c_0^2}\left(\frac{\partial^2 c_0}{\partial y^2}\right)\frac{\partial^2 p'}{\partial \xi^2} = 0 \qquad (2.4)$$

where s is now a coordinate along the ray, y is the one in the plane perpendicular to the ray, $\xi = \varphi(k, y) - t$ and φ is the eikonal.

A peculiarity of this equation lies in the presence of the last term, which contains y^2 explicitly and increases with distance from the beam axis. This means that diffraction and refraction effects are too intimately interconnected to be described by individual additive terms and are, in the long run, entangled within the last term.

In effect an equation of the type (2.3) can be derived by means of a more formalized asymptotic procedure (see Ostrovsky & Pelinovsky, 1974; Pelinovsky & Soustova, 1979). Essentially, however, it is but a record of the wave equation with small nonlinearity and dissipation in other variables; in this case one cannot even see the reduction in order of the basic equation that is attained in the Burgers equation and its "geometrical" generalizations. But the model equation (2.3) still has the advantage that it involves different effects such as nonlinearity, losses, diffraction, each one being described by a single term entering additively. Some generalizations of the equation (2.3) for solid media have also been made (Engelbrecht, 1980). The evolution equation for transient waves in solids was derived. Several types of two-dimensional evolution equations were obtained on the basis of the continuum theory of solids using the general approach of the ray method. The cases of plane and cylindrical wave beams were considered for a variety of stress–strain relationships.

The first solutions to (2.3) were obtained by its authors (Zabolotskaya

& Khokhlov, 1969), but diffraction effects were neglected. The unipolar compression pulse with an amplitude decaying abruptly towards the beam edges was shown to become divergent as it propagates, while a similar rarefaction pulse is brought into focus. It is in fair agreement with the self-refraction analysis conducted in the previous chapter, differing only in that here smooth solutions (before discontinuity formation) rather than shock waves were analysed.

When accounting for diffraction and nonlinearity effects simultaneously, one needs to employ additional approximations. Most frequently considered are the cases of small, or, on the contrary, large values of the parameter N defining the relationship between nonlinear and diffraction effects (Rudenko & Soluyan, 1975; Kunitsyn & Rudenko, 1978). In so doing, beams with an initially Gaussian amplitude distribution over radius are dealt with most often:

$$p'(x=0,r_\perp,t) = A_0 e^{-r_\perp^2/a^2}\sin\omega t. \tag{2.5}$$

If $N \gg 1$, then as a first approximation one can ignore nonlinear effects in order to bring the perturbed solution to the form of a linear diverging Gaussian beam:

$$u_1 = \frac{p'}{A_0} = \frac{\exp\left[-s^2/(1+\sigma_1^2)\right]}{1+\sigma_1^2}\sin\left(\theta - \frac{s^2\sigma_1}{1+\sigma_1^2} + \arctan\sigma_1\right), \tag{2.6}$$

where $\sigma_1 = x/L_d$, $s = r_\perp/a$.

After that, using perturbation techniques, a nonlinear correction at the second harmonic frequency is found.

Let us look for a solution of (2.3) in the form $u = u_1 + u_2$, where $u_2 \ll u_1$ and u_1 is given by (2.6). Then for u_2 we obtain

$$\begin{aligned}\frac{\partial^2 u_2}{\partial\theta\partial\sigma} - \frac{N}{4}\left(\frac{\partial^2 u_2}{\partial s^2} + \frac{1}{s}\frac{\partial u_2}{\partial s}\right) - \frac{1}{2}\frac{\partial^2 u_1^2}{\partial\theta^2}& \\ = \frac{1}{(1+\sigma_1^2)^2}\exp\left(\frac{-2s^2}{1+\sigma_1^2}\right)\cos 2\left(\theta - \frac{s^2\sigma_1}{1+\sigma_1^2} + \arctan\sigma_1\right).&\end{aligned} \tag{2.7}$$

Making use of the separation of variables technique, one can obtain the following solution of this equation:

$$\begin{aligned}u_2 = &\exp\left(\frac{-2s^2}{1+\sigma_1^2}\right)\frac{\sqrt{\ln(1+\sigma_1^2) + 4\arctan\sigma_1}}{4\sqrt{1+\sigma_1^2}} \\ &\times \cos 2\left\{\theta - \frac{s^2\sigma_1}{1+\sigma_1^2} + \frac{1}{2}\arctan\sigma_1 - \frac{1}{2}\arctan\left[\frac{2\arctan\sigma_1}{\ln(1+\sigma_1^2)}\right]\right\}.\end{aligned} \tag{2.8}$$

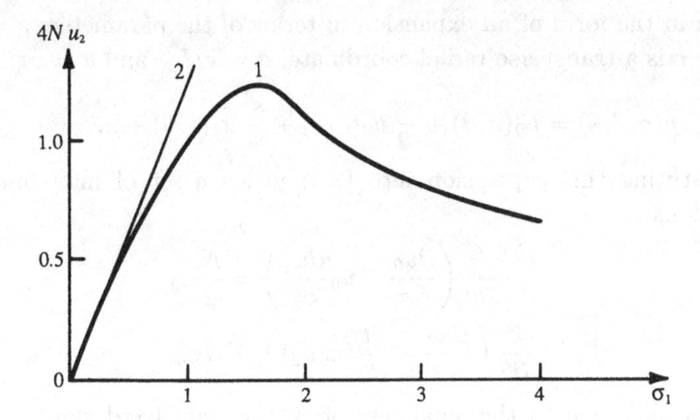

Figure 4.1. Normalized second harmonic amplitude variation; curve 1, along the beam axis; curve 2, in a plane wave, equation (2.9).

At the axis ($s = 0$) this solution has an amplitude variation

$$u_2 = \frac{1}{4N}\left[\frac{\ln^2(1+\sigma_1^2) + 4(\arctan\sigma_1)^2}{1+\sigma_1^2}\right]^{1/2}. \qquad (2.9)$$

This relationship is depicted in Figure 4.1 by curve 1. It is evident that initially the second harmonic rise is linear, as in a plane wave (curve 2), subsequently coming to a maximum and then decaying due to the divergence-related field attenuation. In this respect the diffraction effect can be compared with that of dissipation, but diffraction is pronounced at lower frequencies and does not prevent nonlinear distortions: the ratio between the second harmonic amplitude and that of the fundamental rises monotonically. The final outcome of this process will be revealed as a result of a step-by-step study to be carried out later in this chapter: discontinuities arise in a wave to be followed by the smoothing of beam directivity.

The second extreme case, when $N \ll 1$ and nonlinear effects dominate initially, seems more complicated from the viewpoint of applying the perturbation technique. When one is confined to consideration of the field shape in the vicinity of the beam axis, an additional simplification of equation (2.3) can be achieved by expanding its solutions in powers of the ratio of the current value of the radial coordinate r and the characteristic beam width a. Under this "paraxial" approximation, the combined impact of diffraction and nonlinear effects on sound beam propagation was analysed by Rudenko and Soluyan (1975). Confining ourselves to analysing the field near the beam axis we present the solution in the form of an expansion in terms of the parameters $s = r/d$, where r is a transverse radial coordinate, $\sigma = x/L_N$ and $\theta = \omega\tau$:

$$u(\sigma, \theta, s) = u_0(\sigma, \theta) + \frac{s^2}{2} u_2(\sigma, \theta) + \frac{s^4}{4} u_4(\sigma, \theta) + \dots . \tag{2.10}$$

Substituting this expression into (2.3) yields a set of interconnected equations:

$$\begin{aligned} \frac{\partial}{\partial\theta}\left(\frac{\partial u_0}{\partial\sigma} - u_0\frac{\partial u_0}{\partial\theta}\right) &= \frac{N}{2}u_2, \\ \frac{\partial}{\partial\theta}\left(\frac{\partial u_2}{\partial\sigma} - \frac{\partial}{\partial\theta}(u_0 u_2)\right) &= N u_4, \end{aligned} \tag{2.11}$$

and so on. Due to the smallness of N the right-hand sides of these equations are small, which allows us to close the set by assuming $u_n = 0$ for $n \geq 4$. The remaining set of two equations can be transformed by setting

$$u_2 = \left(\frac{2}{N}\right)\frac{\partial F}{\partial\theta}$$

and then integrating; omitting the insignificant integration constant, one obtains

$$\begin{aligned} \frac{\partial u_0}{\partial\sigma} - u_0\frac{\partial u_0}{\partial\theta} &= F, \\ \frac{\partial F}{\partial\sigma} - u_0\frac{\partial F}{\partial\theta} &= 0. \end{aligned} \tag{2.12}$$

The corresponding characteristic system has the form

$$\frac{\mathrm{d}\theta}{\mathrm{d}\sigma} = -u_0, \quad \frac{\mathrm{d}F}{\mathrm{d}\sigma} = 0, \quad \frac{\mathrm{d}u_0}{\mathrm{d}\sigma} = F. \tag{2.13}$$

Integrating gives

$$u_0 = \varphi_1(\xi) + \sigma\varphi_2(\xi), \quad F = \varphi_2(\xi), \quad \xi = \theta + \varphi_1\sigma + \frac{1}{2}\sigma^2 F,$$

which gives, taking the boundary condition (2.5) into account,

$$\begin{aligned} u &= (1 - s^2)\sin\xi + \sigma N\cos\xi, \\ \xi &= \theta + \sigma\sin\xi + \left(\frac{N}{2}\right)\sigma^2\cos\xi. \end{aligned} \tag{2.14}$$

In the domain where nonlinear effects are small ($\sigma N \ll 1$) the field at the beam axis is expressed as

$$u = \sin(\xi + \sigma N), \quad \xi = \theta + \sigma\sin\left(\xi + \frac{\sigma N}{2}\right). \tag{2.15}$$

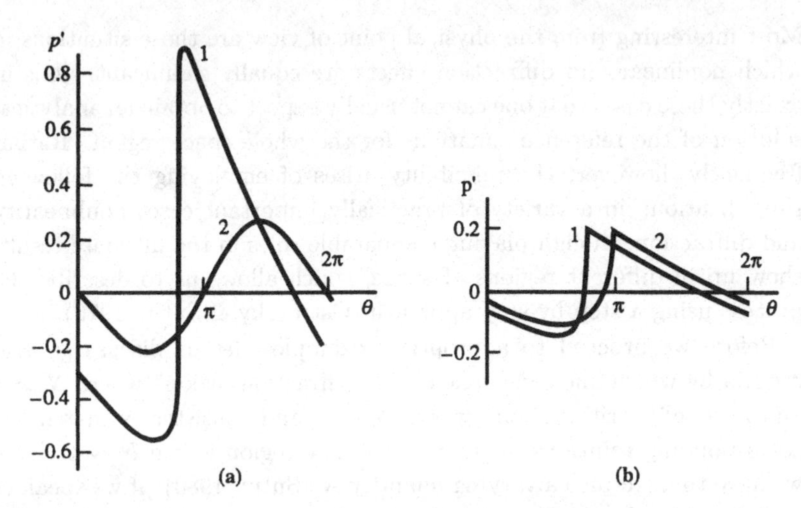

Figure 4.2. Wave profiles on the beam axis (curve 1) and at the periphery (curve 2) for significant nonlinearity ($N = \frac{1}{3}$); (a) $\sigma = 0.25$, (b) $\sigma = 1$.

From this relation it is clear that, unlike the plane wave, the profile evolution in the sound beam is asymmetric, and the displacement of profile points proceeds at different speeds in the compression and rarefaction phases. As a result of this the positive portion of the profile contracts while the negative is extended, which, in turn, leads to a higher amplitude in the positive phase than in the rarefaction phase due to the constancy of profile area (see Chapter 2).

In intermediate cases ($N \approx 1$) when the effects of nonlinear and diffraction events are comparable, numerical methods must be used to integrate equation (2.3). As an example we give the calculated result for the evolution of a Gaussian beam specified by the boundary condition (2.5) (Bakhvalov *et al.*, 1982). Figure 4.2 depicts the wave profiles for different distances from the beam axis at significant nonlinearity ($N = \frac{1}{3}$). For $\sigma = 0.25$ a discontinuity emerges only on the beam axis, whereas $\sigma = 1$ it is also observed at the periphery. The wave profile is pronouncedly asymmetric. The peculiarities of the sound beam's nonlinear evolution cited here have been observed experimentally more than once.

4.3 Step-by-step technique

Most interesting from the physical point of view are those situations in which nonlinear and diffraction effects are equally significant. It is in exactly these cases that one cannot usually expect to obtain an analytical solution of the reference equations for the whole space region. Rather frequently, however, the possibility arises of employing the following simplification: in a variety of practically important cases nonlinearity and diffraction, though playing comparable roles in the ultimate result, show up in different regions of space, which allows us to describe the process using a step-by-step approach (Ostrovsky & Sutin, 1976).

Before we proceed to appropriate examples, let us discuss general criteria by which one can single out a "diffraction region" where $N \gg 1$ and a "nonlinearity region" where $N \ll 1$ and consistently match the corresponding solutions in the intermediate region where $N \sim 1$. Here we have to introduce a varying quantity N (Sutin, 1980): if we speak of a wave beam then different characteristic parameters N are implied at different beam cross-sections. A more rigorous approach is to compare the nonlinear and diffraction terms in equation (2.3) but, in the majority of cases it suffices to consider the same parameter N as above:

$$N(x) = 2/\varepsilon M(ka)^2, \tag{3.1}$$

where for $M(x)$ and $a(x)$ characteristic beam parameters for the given cross-section are employed (x is a coordinate along the beam). From the value of $N(x)$ one can easily mark the transition from one step to another.

A more complicated problem is that of matching. When can a direct solution in the intermediate region with $N \sim 1$ be avoided? There is an evident need for the length of this region to be sufficiently small that neither nonlinear nor diffraction wave distortions have time to accumulate in it, so that a solution in this region either remains unchanged or gives way to evolution due to some third factor (e.g. geometrical divergence) that can easily be taken into account. Luckily, such cases are rather frequent.

As for the nonlinear and diffraction stages, the corresponding solutions can be achieved with comparative ease. For the former stage the nonlinear geometric acoustic (NGA) formulae inferred in the previous chapter hold true. In particular, for waves with plane, cylindrical and spherical symmetries, the solution for a simple wave can be applied in the form (Chapter 3):

$$u = F(\theta + \sigma u), \tag{3.2}$$

where $\theta = \omega\tau$, F is an arbitrary function, $u = p'r^n/p'_0R_0^n$ is the normalized pressure value and p'_0, R_0 are the initial amplitude and radius values. $n = 0, \frac{1}{2}, 1$ for plane, cylindrical, and spherical waves respectively. σ is defined by (Naugol'nykh, 1968):

$$\sigma = \varepsilon Mkr, \quad k = \omega/c_0 \tag{3.3}$$

for plane waves,

$$\sigma = 2\sigma_0 \left| \sqrt{\frac{r}{R_0}} - 1 \right|$$

for cylindrical waves, and

$$\sigma = \sigma_0 \ln r/R_0$$

for spherical waves, where

$$\sigma_0 = \varepsilon MkR_0, \quad M = p'_0/\rho_0 c_0^2.$$

Should there be discontinuities a solution to (3.2) must be supplemented by appropriate jump conditions.

At the diffraction stage the field, in the case of beams with a narrow angular spectrum, can be described using the non-stationary Kirchoff integral with a source distribution given over some beam cross-section (Morse & Feshbach, 1958),

$$p' = \frac{1}{2\pi c_0} \int_s \frac{1}{r} \frac{\partial}{\partial t} p_s \left(t - \frac{r}{c_0} \right) \mathrm{d}s, \tag{3.4}$$

where r is the distance from the observation point to a surface element $\mathrm{d}s$ in an initial beam cross-section and p_s is the acoustic pressure with respect to this element. Formula (3.4) holds true provided the surface dimensions and distance r are large compared with a wavelength.

The procedure for describing the field within the frame of a step-by-step method is now as follows: the parameter N is defined near the transmitter, and if $N \ll 1$ one can neglect diffraction at the first stage, while given $N \gg 1$ nonlinearity can be neglected. In the corresponding solution, the parameter $N(x)$ should be calculated. The region where $N \sim 1$ is considered to be the end of the first stage and the beginning of the second; as has been mentioned already, any transition region must be small compared with the nonlinear and diffraction scales in it. Several transitions from the nonlinear stage to the diffraction stage and vice versa are possible in principle.

In the ensuing sections we shall dwell upon examples of application of the step-by-step technique, and compare the approximate results obtained with experimental data and numerical results.

4.4 Divergent waves; radiation; "isotropization" of an intense acoustic beam

Let us embark on the problem of a harmonic (sinusoidal) transmitter whose parameters satisfy the relationship $N \gg 1$, i.e. nonlinearity near the transmitter can be ignored, so that well up to the far-field zone the field remains close to harmonic, and the directivity diagram formation proceeds in the standard "linear" way (Ostrovsky & Fridman, 1972; Blackstock *et al*, 1973). If, in particular, we refer to a plane transmitter with a characteristic dimension d many times exceeding the wavelength ($kd \gg 1$), then the field that it generates represents a harmonic wave beam that does not diverge until distances of the order $L_D \sim kd^2$, subsequently to form a diverging wave with a divergence angle $\alpha \sim 1/kd$. For $r \gg L_D$ (in the far-field zone) the field amplitude falls off as r^{-1} while the beam width a increases in proportion to r, so that the parameter N diminishes with distance according to

$$N = \frac{2}{\varepsilon M (ka)^2} = N_0 \frac{kd^2}{r}, \tag{4.1}$$

where N_0, d are the values of N, a at the transmitting aperture. This indicates that at the distance

$$R_s \simeq N_0 kd^2 \tag{4.2}$$

the diffraction stage is replaced by the nonlinear one. Let us suppose first that the "transition" point R_s lies within the far-field zone of a transmitter where diffraction can be disregarded, which means the wave is an inhomogeneous spherical one that is already at the linear stage. Provided the above-mentioned condition $kd \gg 1$ is fulfilled, this stage will be described by Kirchoff's formula (3.4); for instance, for a plane circular transmitter with a given displacement amplitude ξ_0 the wave field is described by the known formula (Skudrzyk, 1976)

$$p' = \frac{\mathrm{i}}{2} c\rho_0 \omega \xi_0 \frac{kd^2}{r} \left[\frac{2J_1(kd\sin\varphi)}{kd\sin\varphi} \right], \tag{4.3}$$

where φ is the angle between the radiation axis and the direction towards the observation point. Formula (4.3) holds for the wave zone. Matching with the second, nonlinear, stage must be performed within the zone, but at distances R_0 that are not too large, before the wave experiences pronounced nonlinear distortions. For $r = R_0$ an inhomogeneous spherical wave is specified:

$$p' = p'(R_0) \sin \omega t \, D(\theta, \varphi), \tag{4.4}$$

where $p'(R_0)$ is deduced from the solution to a linear problem (e.g. from (4.3)) such that D is a directivity characteristic (in the case of (4.3) it is the quantity enclosed in square brackets). To describe the further wave evolution, one can use the "spherical" solution offered in (3.2) and (3.3) with σ_0 now having an individual value for each direction: $\sigma_0 \sim M \sim D$. Thus, the wave distortion rate is different for different directions. The discontinuity formation coordinate, for example, is equal to

$$R_*(\theta, \varphi) = R_0 \exp(L_N / R_s D). \tag{4.5}$$

It is natural that R_* is minimal in the direction of a directivity diagram maximum. An inverse relationship

$$D^{-1}(\theta_*, \varphi_*) = \frac{R_0}{L_N} \ln\left(\frac{r}{R_s}\right)$$

defines the spatial angle inside which a shock wave already exists at the given distance r.

In the same way the discontinuity evolution can be tackled. Thus, the coordinate of the point at which it attains its maximum is assessed, according to (2.17) in Chapter 2, using (4.5) with an additional factor $\pi/2$ in the argument of the exponential. For large distances, when $\sigma \gg 1$, the shock amplitude is

$$p' = \frac{p'_0(R_s)}{r} \pi L_N \left(\ln \frac{r}{R_s}\right)^{-1}. \tag{4.6}$$

Here the dependence of the discontinuity amplitude on direction is lost: it is "eaten up" in the nonlinear decay. This implies that the directivity of a powerful transmitter becomes isotropic over the angular range where equation (4.6) is valid. This range obviously increases as the wave evolves. The wave profile at these angles is a sawtooth one; if viscosity is allowed for, the wave profile at still longer distances approaches sinusoidal again but isotropy is, of course, retained at this stage.

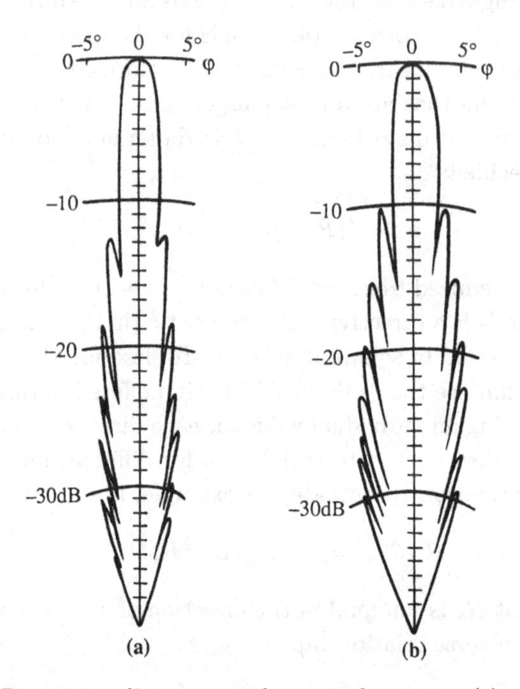

Figure 4.3. Directivity diagrams at the main frequency; (a) near the emitter ($r = 1.6$m) and (b) far from the emitter ($r = 101$m), according to the experiment by Blackstock *et al.*, (1973).

This effect was observed experimentally by Blackstock and his colleagues (Blackstock *et al.*, 1973) during experiments performed in a lake using a powerful hydroacoustic transmitter with diameter 7.6 cm and operating frequency 450 kHz. Angular properties of the field were measured at the fundamental frequency and its harmonics (it is useful to recall that at the sawtooth stage the relation between the amplitude harmonics is kept fixed). Figure 4.3 depicts the angular dependence of the field at the basic frequency at distances of 1.64 m (this coincides approximately with the beginning of the wave zone) and 101m. Broadening of the directivity diagram with distance can be clearly observed.

Later on a series of works appeared concerned with the study of intense ultrasound transmitter properties (see, for example, Aanonsen *et al.*, 1984; Hennion, 1978). Theory and experiments (Hamilton *et al.*, 1985) showed that a fine multilobe field structure appears with a different amount of sidelobes for different harmonic numbers.

Diffraction of a nonlinearly distorted wave

In what follows it seems quite natural to consider the opposite cases, when the nonlinear stage is followed by one of diffraction. To begin with let us analyse the problem of radiation by means of plane wave diffraction at an opening in a plane screen (the simplest version of the "aperture" antenna) (Ostrovsky & Sutin, 1976). Let a plane harmonic transmitter with diameter $d \gg \sqrt{\lambda \ell}$ be situated at a distance ℓ from the screen; then a wave incident on the opening, of diameter $a < d$, remains plane but before approaching the opening may, in general, become nonlinearly distorted. Therefore, for the transmitter we have $N \ll 1$ whereas, as we shall suppose, $N \gg 1$ for the opening. If $\ell \ll L_N$ then the situation previously considered arises: a harmonic beam coming out of the opening initially suffers linear diffraction, eventually to become essentially nonlinear in the far-field zone. If, however, $\ell \geq L_N$, then within the region between the transmitter and the screen the nonlinear stage is set up and the wave is distorted, while the opening brings about diffraction of the already distorted wave of type (3.2) which may be described by Kirchoff's integral (3.4).

Initially we suppose that the wave at the opening has no discontinuities. The far-field zone for the whole wave is formed at a distance $L_D \sim ka^2$. For $r > L_D$ the field has an appreciably different character far from the axis as compared with that in the near-axis cylinder (i.e. for angles $\theta < a/2r$ where a is the aperture radius). At a distance $r > L_D$ one can neglect the variation of retarded time in the Kirchoff integral for points lying inside this cylinder, and then the solution acquires the form

$$p' = \frac{a^2}{2c_0 r}\frac{\partial}{\partial t}p'_s\left(t - \frac{r}{c_0}\right), \tag{4.7}$$

where r is the distance from the screen plane and p'_s is a simple wave (3.2). Therefore, near the axis a wave with a steep front transforms into a sequence of unipolar "peaks" (see Figure 4.4 (a)). The peak value of the field equals $p_1/(1-\sigma)$, where $p_1 = p'_0 ka^2/2r$ is the pressure amplitude at the axis that would be realized for a diffracted harmonic wave with amplitude p'_0. The intensity of such a wave, averaged over a period, is given by

$$I = \frac{2}{\sigma^2}\left(\frac{1}{\sqrt{1-\sigma^2}} - 1\right) I_1, \tag{4.8}$$

where $I_1 = p_1^2/2\rho_0 c_0$ is the intensity corresponding to a harmonic wave.

Therefore, both the peak pressure and the field intensity increase with

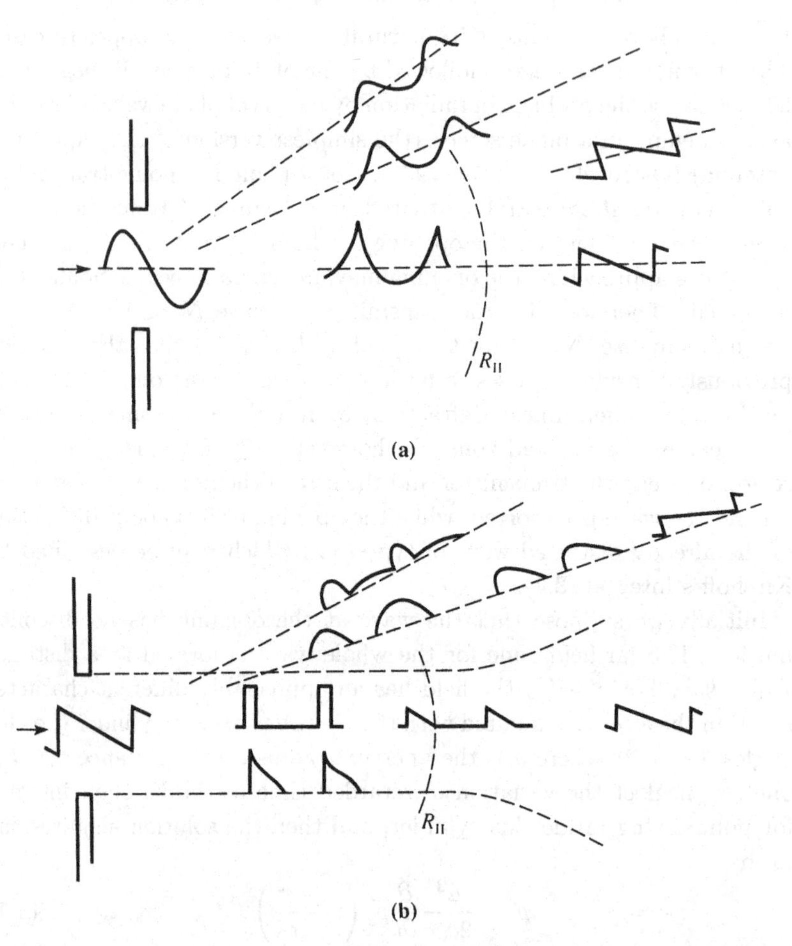

Figure 4.4. Schematic pictures of waveforms after diffraction for a simple wave (a) and for a sawtooth wave (b).

profile steepening in the diffracted wave being the result of more effective energy concentration and weak divergence of higher harmonics arising as the nonlinear wave evolves. And, naturally, the asymmetry between the positive and the negative half periods mentioned above is expressed here in full.

Another picture can be observed outside the "projector cylinder" ($\theta \gg a/2r$). Here the Kirchoff integral can be made one-dimensional by dividing the radiating aperture into "strips" with width $\mathrm{d}y$ normal to the xz-plane passing through a normal to the screen and the observation point; the variations of retarded time between various points of the strip can be disregarded. In this case, as can easily be seen, a steep front gives rise to a pulse whose width increases with the angle θ, so that at angles exceeding $1/ak$ the pulses from various periods start to overlap.

Diffraction of a sawtooth wave

If a sawtooth wave ($\sigma > 1$) is observed at the opening, the solution is simplified (Ostrovsky & Sutin, 1976). The Kirchoff integral can be written as

$$p' = \sum_{n=-\infty}^{\infty} \frac{p_{0s}}{\pi c} \int_s \frac{\delta(t - nT)}{r} ds - \frac{p_{0s} sk}{2\pi^2 r}, \tag{4.9}$$

where δ is a Dirac delta function, p_{0s} is the initial shock magnitude, $s = \pi a^2$, T is the wave period and n is an integer (the "period number"); the last term reflects the transformed continuous parts of the wave, whereas the first refers to the discontinuous fronts (see also Kharkevich, 1950). It can be noted that (4.9) implies a radiated pulse area that is independent of direction: this area equals $sp'_s/\pi rc$. For points outside the "projector cylinder" we derive pulses of the form

$$p'_n = \frac{p_{0s}}{\pi r \theta} F\left(\frac{c(t - nT) - r}{\theta}\right), \tag{4.10}$$

where $F(z)$ is a local opening width ("strip length") in the xz-plane. The duration of these pulses is equal to $a_z\theta/c$ where a_z is the maximum value of F.

Thus, here the temporal pulse shape repeats the dependence on z of the radiating surface width in the direction normal to z. If, in particular, the radiating structure has a symmetry axis, then the pulse shape in a plane passing through this axis coincides with the shape of the contour itself. For example, a circular opening produces semicircular pulses while an aperture of "dumb-bell" type yields a two-hump pulse in a plane oriented along the "dumb-bell" axis.

For large angles, as was already mentioned, an overlapping of several pulses occurs. As a result of this the wave profile is smoothed and the angular dependence of the field reveals a "lobed" structure, though the amplitude vanishes nowhere (except on the planes parallel to the sides of a rectangular aperture).

Within the projector zone cylinder the field consists of pulses with discontinuous fronts, amplitude $2p'_s$ and duration τ lying within the interval $a^2/8cr < \tau < a^2/2cr$, i.e. the pulse energy decreases with distance due to pulse shortening rather than amplitude decay.

In the wave zone a nonlinear stage again sets in, for which one can make use of the NGA formulae. Within the angle range $\theta < \lambda/a$ where separate pulses are observed (note that at larger angles, the manifestation of nonlinearity in a real environment does not seem at all feasible, because of viscous action) each pulse will be transformed into a triangular one. For large distances the parameters of such pulses are defined only by the initial pulse area, which, as was already mentioned, does not depend on direction. Therefore, field "isotropization" sets in at this stage. As time evolves, the pulses expand and overtake one another to result in the wave's transformation into a sawtooth one. The general structure of the radiated field described here is depicted in Figure 4.4 (b).

We have already emphasized that the matching conditions between the linear and nonlinear stages should be dealt with most attentively. Let us now assess the variation of the characteristic parameter N with distance. Inserting the peak amplitude value in the projector cylinder into the expression for N yields

$$N = (1 - \sigma^2)N_0ka^2/2r, \tag{4.11}$$

which indicates that the diffraction stage boundary is defined by the coordinate

$$R_s \simeq \frac{ka^2}{2}(1 - \sigma^2)N_0. \tag{4.12}$$

This expression contains a factor $(1 - \sigma^2)$ the meaning of which is quite plain: the new nonlinear stage sets in earlier because of wave front contraction at the first stage (before the screen). As we supposed that nonlinearity begins to show up only in the wave zone of the opening, i.e. $R_s > L_D \approx ka^2/2$, the applicability condition for our approximation will appear as

$$N_0(1 - \sigma^2)/2 \gg 1.$$

Under this condition a step-by-step procedure gives a correct result everywhere.

In this case, however, a difficulty arises in considering a sawtooth discontinuous wave when $\sigma \geq 1$. For the region outside the projector cylinder, one can use a radiated pulse length (θa) instead of $2\pi/k$ in the expression for N, while using the following quantity for M:

$$M = p_s d / 4\pi r \theta \rho c^2.$$

As a diffraction length (for angles $\theta > a/2r$) we shall simply use the aperture size a, and then $N \approx r\theta^2/\varepsilon a M_0$; should the condition $\theta^2 \gg \varepsilon M_0$ be satisfied then N is large even for $r \approx a$, and nonlinear effects will only show up in the wave zone, as was actually expected.

Things go less well for a projector cylinder where for discontinuities the wave zone does not exist at all. Because of this one cannot, strictly speaking, pass again to a nonlinear stage and, therefore, one has to resort to additional assumptions.

One possible method is to account for the shock front self-refraction considered in the previous chapter. Initially a discontinuity decays more rapidly at the projector cylinder edge, where the pulses have sharper peaks compared with those at the centre, resulting in faster travel of the middle compared with the edges and causing the plane front to change to a spherically diverging one. Diverging pulses are quickly transformed to triangular ones and the field becomes more isotropic; later on the wave becomes sawtooth.

Now let us briefly discuss a free sound beam, as produced by a plane harmonic transmitter. If here $N_0 \gg 1$ then we embark on a problem dealt with at the beginning of this section: the linear wave behaviour in the wave zone is replaced by nonlinear behaviour and the isotropization effect is observed. If, however, $N_0 \ll 1$ then even the first stage appears to be nonlinear, and the sawtooth wave is formed in the near-field zone. Eventually the wave enters the diffraction stage when $x > L_D$.

These situations, however, are not typical: the state-of-the-art nonlinear transmitters usually have N only slightly below unity. In this case one cannot strictly separate the nonlinear and diffraction stages. Nevertheless, even here a step-by-step approach makes it possible to obtain correct qualitative and partially quantitative results (Sutin, 1980) that are in fair agreement with experimental results such as those offered by (Browning & Mellen, 1968); comparison with their data is presented in Figure 4.5. Other experiments concerning this problem were performed by Theobald *et al.* (1976), and later by Lucas *et al.* (1983), Aanonsen *et al.* (1984b) and Andreev *et al.* (1985).

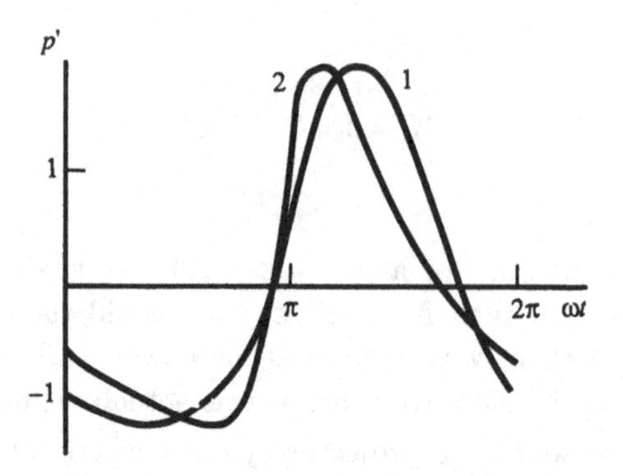

Figure 4.5. Theoretical (curve 1) and experimental (curve 2) waveforms in the wave zone of a radiator. The frequency is 150 kHz and the pressure amplitude is 0.2 MPa.

4.5 Foci and caustics

There are interesting problems related to the field behaviour in regions that are singular in the geometrical acoustics approximation; these are specified as foci and caustics. In such regions the field is subject to the joint action of nonlinearity and diffraction. Here a step-by-step approach may also be useful. Actually, in spite of the field amplitude growth in these regions they may turn out to be linear regions because of the more rapid growth of diffraction effects so that nonlinear distortions fail to accumulate within such small regions (Ostrovsky & Sutin, 1975).

Focusing

Let a radiating surface (concentrator) create a harmonic convergent beam with convergence angle α, focal length F and aperture $d = F\alpha$ (Figure 4.6 (a)); as was the case previously, a small-angle approximation is considered to be valid, at least for $\alpha < 30°$. If the initial wave amplitude p_a is large enough and α is not too small, so that the initial Khokhlov number $N_0 = L_N/L_D$ is small, then everywhere up to the immediate vicinity of the focus the geometrical acoustics solution for convergent waves holds true:

$$p' = \left(\frac{p_a F}{r}\right) \sin\left(\omega\tau_F + q\sigma_0 \ln\left(\frac{F}{r}\right)\right), \tag{5.1}$$

where $\sigma_0 = \varepsilon k M F$, $M = p_a/\rho_0 c_0^2$, $q = p' r / p_a F$, $\tau_F = \tau - F/c_0$.

The radius R_s at which the nonlinearity and diffraction scales become comparable may be found from the condition $N(R_s) = 1$. It can easily

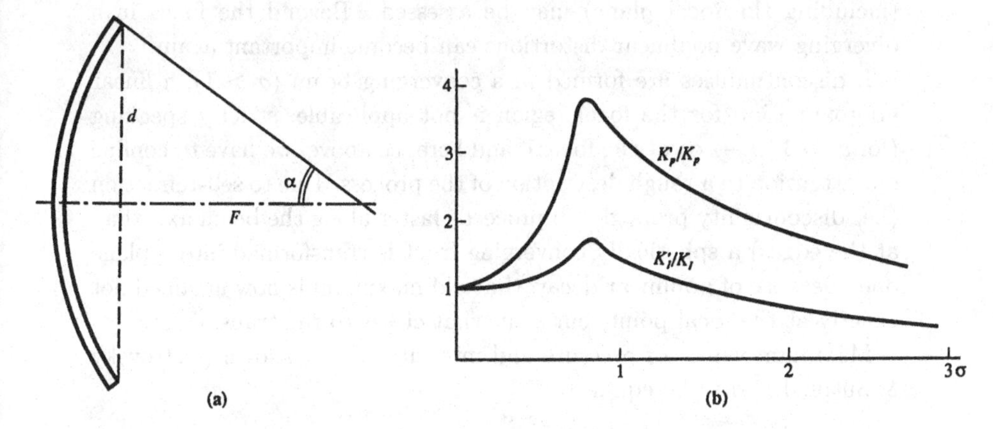

Figure 4.6. (a) Wave focusing: d is the focusing system aperture size, and F is the focal distance. (b) Dependence of the focusing system amplification coefficients for sound amplitude and intensity on the reduced distance.

be seen that $N = N_0 F/r$ and hence $R_s \simeq N_0 F$. For $r < R_s$ we shall describe the wave propagation through the solution of a linear diffraction problem for which, as before, the Kirchoff integral may be employed. This yields the same formulae as those obtained for a plane beam while the field at infinity is now replaced by that in a focus; such a "conversion" is well known in diffraction theory. Namely, at a focal point ($r = 0$) the field pressure appears as (cf. e.g. (4.7))

$$p'_F = \frac{R_s \alpha^2}{2} \frac{\partial p'}{\partial t}\bigg|_{t-R_s/c_0}. \tag{5.2}$$

Thus the peak value of p'_F is

$$p_m = \frac{kF\alpha^2 p_a}{2[1-\sigma(R_s)]}. \tag{5.3}$$

One can introduce the gain of the focusing system as the ratio of the pressure peaks K_P or of the mean intensities K_I at the focus $r = 0$ and at the transducer $r = R_0$. For the linear case, obviously $K_P = \sqrt{K_I} = \alpha^2 kF/2$. As σ increases to unity the gain increases due to the stronger concentration of high-frequency harmonics that arises because of nonlinearity (here $\sigma = \sigma_0 \ln F/r$). The average intensity I_m is given in

(4.8) where now I_1 is an intensity value at the focus with no allowance for nonlinearity.

As in the previous section the field distribution in a beam cross-section (including the focal plane) may be assessed. Beyond the focus in a diverging wave nonlinear distortions can become important again.

If discontinuities are formed in a converging beam ($\sigma > 1$), a linear approximation for the focal region is not applicable, strictly speaking (for $\sigma \to 1$, $p' \to \infty$ at the focus), and here, as above, we have to confine our attention to a rough description of the process. Due to self-refraction (i.e. discontinuity propagation proceeds faster along the beam axis than at the edges) a spherically converging front is transformed into a plane one. Because of nonlinear decay, the field maximum is now attained not exactly at the focal point, but somewhat closer to the transmitter.

Maximum values of pressure and intensity can be shown (Ostrovsky & Sutin, 1975) to be equal to

$$p_m \simeq \frac{2K_{p\ell}p_a}{\sigma_0 \ln \eta}, \quad I_m \simeq \frac{3}{\pi}\xi K_{pl}^2 I_0, \tag{5.4}$$

where $K_{p\ell} = k\alpha^2 F/2$ is the amplification factor of a linear harmonic wave; the parameter η is defined by the equation $\sigma_0 \eta \ln \eta = 2\pi/\xi$ where $\xi = 1$ for $\sigma_0 \leq \pi/2$ and $\xi = pr/p_0 F$ for $\sigma_0 > \pi/2$. When σ_0 is large enough, then for a sawtooth wave $\xi \simeq \sigma_0^{-1}$ and the quantity η (along with p_m and I_m) ceases to depend on the initial amplitude p_a, i.e. everything proceeds as in a spherically symmetric converging wave; this case was analysed in the studies by Naugol'nykh & Romanenko (1959), and Naugol'nykh (1968). It is clear that within this limit the amplification factor of the system decreases due to nonlinear damping; thus, the relations

$$K_p = K_{p\ell}\left(1 + \sigma_0 \ln K_{p\ell}\right)^{-1}, \quad K_I = \frac{2}{3}\left(\frac{p_m}{p_a}\right)^2, \tag{5.5}$$

depict the dependence of the concentrator amplification factor, normalized to the linear system amplification factor, on σ. The normalized dependence of the amplification factor K_p and K_I on σ is presented in Figure 4.6*b*.

As one should expect, there exists within the range $\sigma_s \approx 1$ a maximum of the field amplification associated with an increased higher harmonic energy concentration at the focus.

Numerical calculations referred to in the book by Bakhvalov *et al.* (1982) confirm, for the most part, the results of the approximate theory given above. Nonlinear effects in focused beams were discussed also in Tjotta and Tjotta (1984) and Du and Breazeale (1987). The possibility of additional field amplification due to the joint action of nonlinearity and diffraction may become significant for real acoustic concentrators, as has been demonstrated experimentally more than once (Borisov & Gyunkina, 1973; Smith *et al.*, 1969; Zaretsky, 1968).

Caustics

Another important example involving the use of a step-by-step procedure is given in a study of nonlinear wave refraction with the formation of caustics (Ostrovsky *et al.*, 1976). The geometrical acoustics aspect of this problem was tackled in the preceding chapter. A near-acoustic domain can be examined (at least locally) in a "plane" formulation: plane wave oblique propagation in an inhomogeneous medium is described within the framework of NGA (we shall resort to a linear-ray approximation), while in the vicinity of a caustic we use a linear description. This, certainly, has a qualitative basis. In fact, when "approaching" a caustic, in a domain close to it one can neglect the variation of the medium parameters ρ and c in a nonlinearly distorted wave, taking into account only the rapid change of the ray tube cross-sectional area $Q(\ell)$, as at the caustic we have $Q = 0$ and nearby $Q \sim |\ell_C - \ell|^{1/2}$ where ℓ is the ray coordinate and ℓ_C corresponds to the caustic point.

The reduced coordinate z entering the geometrical acoustics formulae in Chapter 3 therefore varies as $|\ell_C - \ell|^{3/4}$, remaining finite up to the caustic. Thus, even within the framework of geometrical acoustics an unbounded increase of the field amplitude in the near-caustic region is not accompanied by an unbounded waveform distortion due to nonlinearity: the reduced distance z remains finite. At the same time, distortions due to diffraction near a caustic are rather pronounced, which justifies the application of a step-by-step procedure: the relation between an incident wave and that reflected from a caustic will be given by a linear approximation.

It is well known that in a linear medium a caustic is equivalent to a phase converter shifting the phase of each of the spectral components by $-\pi/2$. For a non-monochromatic signal this obviously results in strong distortions of its shape (Brekhovskikh, 1973). Such a phase-shifting conversion can be formulated in integral form, as a Hilbert transform:

$$p'_t(t, z_C) = \frac{\omega}{2\pi} \int_{-\pi/\omega}^{\pi/\omega} p'_i(t', z_C) \cot \frac{\omega(t - t')}{2} \, dt' \qquad (5.6)$$

where z_C is the reduced coordinate for the caustic, p_i' is the pressure of a wave reaching a caustic, p_t' is the pressure of the transmitted wave and ω is the basic frequency of the wave (supposed periodic). Strictly speaking, (5.6) should contain the current coordinate of a typical point lying near a caustic, but within the framework of the geometrical acoustics application, due to finiteness of the reduced path this coordinate can be replaced with z_C and the "geometrical" value of p_i' at this point can be employed. The problem is that p_i' is an implicit function (a simple wave). However, passing over to the variable $\xi = \omega t + z_C p_i'/p_{i0} L_N$ (p_{i0} is the incident wave amplitude and L_N is the shock formation distance corresponding to it) on which p_i' depends, one can write (5.6) as an explicit integral (Ostrovsky *et al.*, 1976),

$$p_t'(t, z_C) = \frac{p_{i0}}{4\pi} \left[\int_0^{\pi - \xi_C} \psi(\xi', \xi)\, \mathrm{d}\xi' + \int_{\pi + \xi_C}^{2\pi} \psi(\xi', \xi)\, \mathrm{d}\xi' \right], \qquad (5.7)$$

where

$$\psi(\xi, \xi') = \sin \xi' \left(1 + \frac{z_C}{L_N} \cos \xi' \right) \cot \frac{1}{2} \left[\xi - \xi' + \frac{z_C}{L_N} (\sin \xi - \sin \xi') \right],$$

and $\xi_C = 0$ for $z_C \leq L_N$ and $\xi_C = p_i(z_C) z_C / p_{i0} L_N$ for $z_C > L_N$ (the last factor in φ is the coefficient of transformation from $\mathrm{d}t'$ to $\mathrm{d}\xi'$ as was the case in Chapter 1).

Integral (5.4) can easily be calculated using a computer. As one would expect, for weak nonlinearity ($z_C \ll L_N$) the incident and reflected waveforms are close to sinusoidal, while at dramatic distortions a negative peak occurs for an incident wave with sharp front. For a discontinuous wave the peak value is formally infinite (a logarithmic divergence is observed), and for a sawtooth wave ($z_C \gg L_N$) with amplitude p_s, the following analytical expression for p_t results:

$$p_t'(t, z_C) = p_s(z_C) \left(\ln 2 + \frac{1}{2\pi} \ln \left| \sin \frac{\omega t}{2} \right| \right). \qquad (5.8)$$

(Note that this peak diminishes very rapidly in the course of further propagation.)

As regards the field evolution after the caustic, it is again described by geometrical acoustics formulae, whereas the initial waveform is now deduced from the linear solutions (5.7) or (5.8). In the initially continuous wave discontinuities may arise again (in the negative phases).

The applicability conditions for the step-by-step procedure are essentially the same as those discussed in the previous sections.

References

Aanonsen, S.I., Barkve, T., Tjotta, J.N. & Tjotta, S. (1984*a*). Distortion and harmonic generation in a narrow field of a finite amplitude sound beam, *JASA* **75,** 749–68.

Aanonsen, S.I., Hamilton, M.F., Tjotta, J.N. & Tjotta, S. (1984*b*). Nonlinear effects in sound beams, in *Proc. X ISNA,* ed. Kobe (A. Nakamura, Osaka Univ.).

Andreev, V.G., Karabutov, A.A. & Rudenko, O.V. (1985). Experimental investigation of nonlinear sound beam propagation in free space, *Akust. Zh.* **31,** 423–8.

Babich, V.M. & Buldyrev, V.S. (1972). *Asymptotic Methods in Problems of Short-wave Diffraction* (Nauka, Moscow).

Bakhvalov N.S., Zhileikin, Y.M. & Zabolotskaya, E.A. (1982). *Nonlinear Theory of Sound Beams* (Nauka, Moscow).

Blackstock, D.T., Lockwood, J.C. & Muir, T.G. (1973). Directive harmonic generation in the radiation field of circular piston, *JASA* **53,** 1148–53.

Borisov, Y.Y. & Gynkina, L.M. (1973). Investigation of elliptic sound concentrators, *Akust. Zh.* **19,** 616–8.

Brekhovskikh, L.M. (1973). *Waves in Layered Media* (Nauka, Moscow).

Browning, D.G. & Mellen, R.H. (1968). Finite-amplitude distortion of 150 kHz acoustic waves in water, *JASA* **44,** 644–6.

Du, G. & Breazeale, M.A. (1987). Theoretical description of a focused Gaussian ultrasonic beam in a nonlinear medium, *JASA* **81,** 51–77.

Engelbrecht, J.K. (1981). Two-dimensional nonlinear evolution equations. The derivation and transient wave solutions, *Int. J. Nonl. Mech.* **16(2),** 199–212.

Hamilton, M.F., Tjotta, J.N. & Tjotta, S. (1985). Nonlinear effects in the farfield of a directive sound source, *JASA* **78,** 202–16.

Hennion, P.Y. (1979). Etude de l'émission parametrique en champ proche par analyse de Fourier: Application à un emetteur focalisant, *Proc. VIII ISNA,* Paris, *J. de Phys.* **41 Suppl. 11,** 126–36.

Kharkevich, A.A. (1950). *Transient Wave Effects* (Gostekhizdat, Moscow-Leningrad).

Kunitsin, V.E. & Rudenko, O.V. (1978). Generation of the second harmonic in a field of piston transmitter, *Akust. Zh.* **24,** 549–55.

Kuznetsov, V.P. (1970). Equations of nonlinear acoustics, *Akust. Zh.* **16,** 548–53.

Leontovitch, M.A. (1944). On a method of problem solution concerning the electromagnetic wave propagation, *Izv. An. SSSR Ser. Fiz.* **8,** 8–24.

Lucas, B.G., Tjotta, J.N.& Muir, T.G. (1983). Field of parametric focusing source, *JASA* **74,** 1966–71.

Malyuzhinets, G.D. (1959). Developing concepts on diffraction phenomena, *UFN.* **69,** 321–34.

Morse, P.M. & Feshbach, H. (1953). *Methods of Theoretical Physics,* Vol. I (McGraw-Hill, New York, London, Toronto).

Naugol'nykh, K.A. (1968). Absorption of finite amplitude sound waves, in *Powerful Ultrasonic Fields,* ed. L. Rozenberg (Nauka, Moscow).

Naugol'nykh, K.A. & Romanenko, E.V. (1959). On the focusing system amplification factor dependence on the sound intensity, *Akust.Zh.* **3,** 294–6.

Ostrovsky, L.A. (1976). Short-wave asymptotic for weak shock waves and solitons in mechanics, *Int. J. Nonl. Mech.* **11,** 401–16.
Ostrovsky, L.A. & Fridman, V.E. (1972). On the powerful acoustic radiation directivity, *Akust. Zh.* **18,** 584–9.
Ostrovsky, L.A. & Pelinovsky, E.N. (1974). On the approximate equations for waves in media with small nonlinearity and dissipation, *PMM.* **38,** 121–4.
Ostrovsky, L.A. & Sutin, A.M. (1975). Focusing of finite amplitude acoustic waves, *Akust. Zh.* **22,** 1300–3.
Ostrovsky, L.A. & Sutin, A.M. (1976). Diffraction and radiation of sawtooth acoustic waves, *Akust. Zh.* **22,** 93–100.
Ostrovsky, L.A., Pelinovsky, E.N. & Fridman, V.E. (1976). Propagation of finite amplitude acoustic waves in inhomogeneous medium with caustics, *Akust. Zh.* **22,** 914–21.
Pelinovsky, E.N. & Soustova, I.A. (1979). Structure of nonlinear sound beam in inhomogeneous medium, *Akust. Zh.* **25,** 631–3.
Pelinovsky, E.N., Fridman, V.E. & Engelbrecht, J.K. (1984). *Nonlinear Evolution Equations* (Valgus, Tallinn).
Rozenberg, L.D. (1949). *Sound Focusing Systems* (Izd. AN SSSR, Moscow-Leningrad).
Rudenko, O.V. & Soluyan, S.I. (1977). *Theoretical Foundations of Nonlinear Acoustics* (Plenum Press, New York).
Skudrzyk, E. (1976). *Fundamentals of Acoustics,* Vol. 2 (Mir, Moscow).
Smith, C.W. & Beyer, R.T. (1969). Ultrasonic radiation field of a focusing source at finite amplitudes, *JASA* **46,** 806–13.
Sutin, A.M. (1980). Diffraction effects in intense sound beams, in *Nonlinear Acoustics, Theory and Experiment,* ed. V. Zverev & L. Ostrovsky (IPF AN SSSR, Gorky).
Theobald, M.A., Webster, D.A. & Blackstock, D.T. (1976). The importance of finite-amplitude distortion in outdoor propagation experiments, in *Abstr. Pap.* 7th *ISNA Blackburg, USA,* ed. A. Nayfeh & J. Kaiser (?, ?).
Tjotta, J.N. & Tjotta, S. (1984). Sound field of parametric focusing source, *JASA* **75,** 1392–4.
Zabolotskaya, E.A. & Khokhlov, R.V. (1969). Quasiplane waves in nonlinear acoustics of bounded beams, *Akust. Zh.* **15,** 40–7.
Zaretsky, A.A. (1968). Measurement of the sound pressure distribution in radiator focus using location technique, *Akust. Zh.* **14,** 471–2.

5
Sound–sound interaction (nondispersive medium)

5.1 One-dimensional interactions

In a nonlinear medium the waves interact with each other, giving rise to harmonics and waves at combination frequencies. In the case of weak nonlinearity the most efficient energy exchange between various spectral components of the field is observed when the synchronism conditions are satisfied, i.e. the medium response at a combination frequency propagates with the velocity of a free wave in the system at this frequency. In other words, a space–time resonance should take place.

When describing such processes in acoustics certain difficulties arise that are associated with lack of dispersion. In this case one can seldom deal with two- or three-wave interaction, as the synchronism conditions are simultaneously satisfied at many frequencies. As was mentioned in the first chapter, the process of nonlinear distortion of the profile on an initially harmonic wave can be treated as the interaction of a large number of synchronously propagating harmonics; the Bessel–Fubini series and its extension to the discontinuous stage are adequate to describe this interaction pattern.

In more general cases, provided that quadratic nonlinearity is the most significant, a wave triplet may be considered as "basic" for interactions, i.e. a triad of waves with frequencies $\omega_{1,2,3}$ and wave vectors $\mathbf{k}_{1,2,3}$ for which the following resonance conditions are satisfied:

$$\omega_1 + \omega_2 = \omega_3, \quad \mathbf{k}_1 + \mathbf{k}_2 = \mathbf{k}_3 \tag{1.1}$$

(it is known that these conditions, in the quantum description, express the laws of conservation of energy ($\hbar\omega$) and of momentum ($\hbar\mathbf{k}$) in the disintegration and merging of quanta, which are field quasi-particles; here $\hbar$ is the Planck constant).

In the absence of dispersion, when $k_i = \omega_i/c$, the resonance relations

(1.1) will be fulfilled only for collinearly propagating waves, when all $\mathbf{k}_i$ are parallel. Under this restriction, however, these equalities hold true for triads of any harmonics with frequencies $n\omega$ provided n is the same for all waves. Furthermore, cascade processes can easily arise in which, for example, a wave with frequency ω_3 gives birth to a new one with frequency $\omega_4 = \omega_3 + \omega_1$, and so on. All this not only renders the problem more complicated but, which is more important, changes the physical results: energy transfer toward the higher wave numbers, into the small-scale part of the spectrum, leads to a pronounced nonlinear damping; this is exactly the case of discontinuity formation. Here the variety of processes that are observed goes far beyond the framework of a classical triplet of waves with periodic energy exchange.

Along with this, the losses due to higher harmonics may frequently be avoided by introducing phase velocity dispersion, or selective losses. Finally, many applications (indeed, those most highly developed nowadays) are connected with interactions within a bounded space region and are, as a rule, of low energy efficiency, but allow one to create a field of a given frequency outside this region ("parametric radiation") or to produce a signal associated with the influence of external radiation on the field within the region ("parametric reception"). In this case one can frequently (albeit not always) confine one's attention to the interaction of a small number of specified waves, since the fields at combination frequencies and harmonics remain small.

In this chapter we shall discuss some of the various types of wave interaction in the "classical" acoustics of nondispersive media.

As was noted above, in such a medium resonant relations can be fulfilled only in the case of collinear propagation of all interacting waves in the same direction, but then they hold true for all the wave harmonics. Therefore, a space–time description based on the Burgers equation ((1.29), Chapter 1) often turns out to be more efficient here:

$$\frac{\partial v}{\partial x} - \alpha v \frac{\partial v}{\partial y} = \delta \frac{\partial^2 v}{\partial y^2}, \tag{1.2}$$

where $\alpha = \varepsilon / c_0^2$. As a boundary condition we impose a pair of sinusoids with different frequencies ω_1 and ω_2; in the course of propagation higher harmonics of these frequencies and other combination frequencies arise. As always, our concern will be mostly with the case of large acoustic Reynolds numbers (though we shall complete this section with a brief discussion of the opposite limit).

At large Re the total field is close to the simple wave

$$v = F(\xi), \quad \xi = y + \alpha v x. \tag{1.3}$$

Then the problem is reduced to the following one: specifying a field on the boundary, one needs to define the required field component from (1.3), for example combination spectral components at sum or difference frequencies. Let us discuss the most typical cases.

Weak signal interaction with a simple wave

The evolution of a weak (linear) disturbance in the presence of a simple wave of arbitrary amplitude can be generally solved by linearization of the one-dimensional gas dynamics equations (Ostrovsky, 1963). In the case of unidirectional propagation considered here it suffices to linearize the solution (1.3). Let us present v in two ways:

$$v = F_0(\xi) + F'(\xi) = F_0(\xi_0) + v',$$

where $\xi_0 = y + \alpha x F_0(\xi_0)$, $F_0(\xi_0)$ being the unperturbed solution in the form of a simple wave and F' a small perturbation specified at $x = 0$. It is clear that

$$F_0(\xi) = F_0(\xi_0 + \alpha x v') \simeq F_0(\xi_0) + \alpha x v' \frac{\mathrm{d}F_0}{\mathrm{d}\xi_0}.$$

As a result we have

$$v' = F'(\xi_0) + \alpha x v' \frac{\mathrm{d}F_0}{\mathrm{d}\xi_0},$$

or

$$v' = F'(\xi_0) \Big/ \left(1 - \alpha x \frac{\mathrm{d}F_0}{\mathrm{d}\xi_0}\right). \tag{1.4}$$

(The argument ξ in F' can be replaced by ξ_0 due to the smallness of v'.)

Hence, it can immediately be seen that at the point $x = x_*$ where a discontinuity emerges in the basic wave, the quantity v' diverges. This is perfectly clear: perturbation results in a small displacement of the wave profile in space, which implies substantial changes of the field at a given point if the front is steep.

If, for $x = 0$, two harmonic waves are specified, i.e. $F_0 = A_1 \sin \omega_1 y$, $F' = A_2 \sin \omega_2 y$ ($A_2 \ll A_1$), then

$$v' = \frac{A_2 \sin \omega_2 \xi_0}{1 - \alpha x A_1 \omega_1 \cos \omega_1 \xi_0}. \tag{1.5}$$

As was mentioned above, one frequently needs to define a field arising at the sum or difference frequencies $\Omega_{\pm} = \omega_1 \pm \omega_2$. Amplitudes of these components are given by the expression†

$$A_{\pm} = \frac{\Omega_{\pm} A_2}{\pi} \int_{y}^{y+2\pi/\Omega_{\pm}} \frac{\sin \omega_2 \xi_0 \, \sin \Omega_{\pm} y}{1 - \alpha x \omega_1 A_1 \cos \omega_1 \xi_0} \, \mathrm{d}y.$$

Here again, as when deducing the Bessel–Fubini series, we make use of the "implicit argument technique": using the expression

$$\mathrm{d}\xi_0 = \mathrm{d}y(1 - \alpha x \omega_1 A_1 \cos \omega_1 \xi_0)^{-1}$$

we obtain

$$A_{\pm} = \frac{\Omega_{\pm} A_2}{\pi} \int_{-\pi/\Omega_{\pm}}^{\pi/\Omega_{\pm}} \sin \omega_2 \xi_0 \, \sin \Omega_{\pm} (\xi_0 - \alpha x A_1 \sin \omega_1 \xi_0) \, \mathrm{d}\xi_0,$$

or, after elementary transformations

$$A_{\pm} = \pm \frac{\Omega_{\pm} A_2}{2\pi \omega_1} \int_{-\pi\omega_1/\Omega_{\pm}}^{\pi\omega_1/\Omega_{\pm}} \cos(\gamma - \sigma \sin \gamma) \, \mathrm{d}\gamma, \tag{1.6}$$

where $\sigma = \alpha \Omega_{\pm} A_1 x$.

Rather typical here are the two cases presented below.

The first is concerned with the weak low-frequency field interaction with a strong high-frequency wave ($\omega_2 \ll \omega_1$), which brings about side-band combination tones with frequencies $\Omega_{\pm}$ close to ω_1. In this case the integration limits in (1.6) are near to $\pm\pi$, and, as is well known

$$\int_{-\pi}^{\pi} \cos(\gamma - \sigma \sin \gamma) \, \mathrm{d}\gamma = 2\pi J_1(\sigma) \tag{1.6a}$$

where J_1 is the Bessel function of the first order.

The second case deals with the interaction of waves of nearly equal frequencies, which produces a field at the difference frequency $\Omega_- \ll \omega_2$. In this case the limits of the integral in (1.6) can be written as $n\pi + s$, where $n \gg 1$ and $|s| < \pi$. Then, similar to (1.6*a*), this integral approximately equals $2\pi n J_1, (\sigma)$, the contribution due to s being negligible.

Accordingly, in both cases (1.6) yields the same expressions

$$A_{\pm} = \pm A_2 J_1(\alpha \Omega_{\pm} A_1 x) \tag{1.7}$$

valid up to the point $x_* = (\alpha \omega_1 A_1)^{-1}$ where discontinuity formation would be observed in the basic wave. It is noteworthy that in the case

† It should be noted that the function (1.5) is not necessarily periodic in ξ_0, although it does possess a discrete spectrum.

of low Ω_- the Bessel function argument is small everywhere, and the low-frequency field amplitude has the form

$$A_- = -\frac{\alpha}{2}\Omega_- A_1 A_2 x. \tag{1.8}$$

Interaction of low-frequency and high-frequency waves

The case of $\omega_2 \ll \omega_1$ makes possible a simpler consideration, valid at finite amplitudes of the low-frequency signal as well (Gurbatov & Malakhov, 1977). This is based on the same general expression for a simple-wave field,

$$v = F_h(\xi) + F_\ell(\xi), \quad \xi = y + \alpha x(v_h + v_\ell), \tag{1.9}$$

where h and ℓ stand for the waves of high and low frequency, respectively. Furthermore, F_ℓ is considered to be (locally) a constant and, therefore, $\alpha x v_\ell$ simply constitutes a phase shift for the high-frequency wave,

$$v_h = F_h(\xi_0 + \alpha x v_\ell), \tag{1.10}$$

where $\xi_0 = y + \alpha x v_\ell$ has the same meaning as above. Under a harmonic boundary condition, the Bessel–Fubini expansion remains valid for this function, so that

$$v_h = \sum_{n=1}^{\infty} \frac{2J_n(n\sigma)}{n\sigma} \sin \omega_1 n(y_1 + \alpha x v_\ell) \tag{1.11}$$

where, as above, $\sigma = \alpha\omega_1 A_1 x$. If nonlinear distortions of the low-frequency field are disregarded, then by setting $v_\ell = A_2 \sin \omega_2 y$ it is not difficult to perform the expansion of (1.11) into a spectrum in the combination frequencies $\Omega_\pm^{(n)} = n\omega_1 \pm \omega_2$. Given a weak low-frequency signal ($v_\ell \to 0$) the expansion takes the form

$$A_\pm^{(n)} = \pm A_2 J_n(n\alpha\Omega_\pm^{(n)} A_1 x), \tag{1.12}$$

which for the lowest combination frequencies $\Omega_\pm = \omega_1 \pm \omega_2$ coincides completely with (1.7). However, it is necessary to note that here also the solution (1.12) is not an exact one, even for a weak signal, since even at a low-frequency ω_2 the function $v_\ell(\xi)$ contains a high-frequency component at the points where σ is close to unity. Actually, the argument ξ depends on the high-frequency field v_h and is subject to abrupt changes at these points, bringing about, therefore, abrupt changes in v_ℓ as well (this is quite obvious in the expression (1.5)). At low ω_2, however, these abrupt changes of v_ℓ are small (they appear as "steps" whose value would not exceed the variation of the function $v_\ell(0, y)$ over a half period at the high frequency ω_1); the error here is of the same order as that which arose when the integration limits in (1.6) were replaced with $\pm\pi$.

The same approach is adequate for the discontinuous stage as well if, instead of (1.11), one employs the appropriate spectral expansion. For the sawtooth wave stage, when the periodic "pumping" harmonic amplitudes are given as $A_s^{(n)} = 2A_1/n\sigma = 2/\alpha n\omega_1 x$, we obtain

$$A_\pm^{(n)} = \pm A_2, \tag{1.13}$$

that is, all combination components have the same amplitudes. In fact this is not valid for all n, being restricted to $n < Re$, where $Re = \alpha A_{1s}/\delta\omega_0$ is the Reynolds number, that is to frequencies that are low compared with the limiting frequency of the wave spectrum determined by the shock front thickness (cf. Chapter 2). On the other hand, to make linearization with respect to the signal possible, the amplitude of the n-th pumping harmonic should exceed A_2. This leads to the condition $n < A_s/A_2 \approx (\alpha\omega_1 A_2 x)^{-1}$. The dominance of one or other of these restrictions depends on the Reynolds number for the emerging satellite frequency, $Re_\Omega \simeq \alpha A_2/\delta\omega_1$: provided Re_Ω is low, the restriction is due to the first condition while in the opposite case it is due to the second.

Formula (1.12) implies that any combination tone amplitude is always below that of the low-frequency signal A_2. It can be shown that in the absence of dispersion, even for comparable frequencies ω_1 and ω_2, a significant amplification at a selected frequency is impossible (see also Gol'dberg, 1972). The physical reason for this is that in a nondispersive medium the energy is spent predominantly on generating higher spectral components of a signal. Notice that, if all satellites – and, according to (1.13) many of them attain the same amplitude – can be summed up then the total energy efficiency of such a transformation will rise considerably (Gurbatov, 1980).

The interaction of strong low-frequency and weak high-frequency waves, with generation of the sum and difference frequencies, was studied experimentally by Burov *et al.* (1978). An experiment was conducted in water; the low-frequency wave had frequency $f_2 = 1.35\,\text{MHz}$ and pressure amplitude $p_2 = 10^5\,\text{Pa}$, while the high-frequency wave had $f_1 = 11.5\,\text{MHz}$ and $p_1 = 1.2 \times 10^3\,\text{Pa}$. Figure 5.1 depicts the high-frequency signal amplitude dependences on distance due to interaction with a low-frequency wave (curves 1,2), and without the low-frequency field (curve 3). It can be seen that in the first case additional signal decay is observed (above that associated with linear dissipation), and

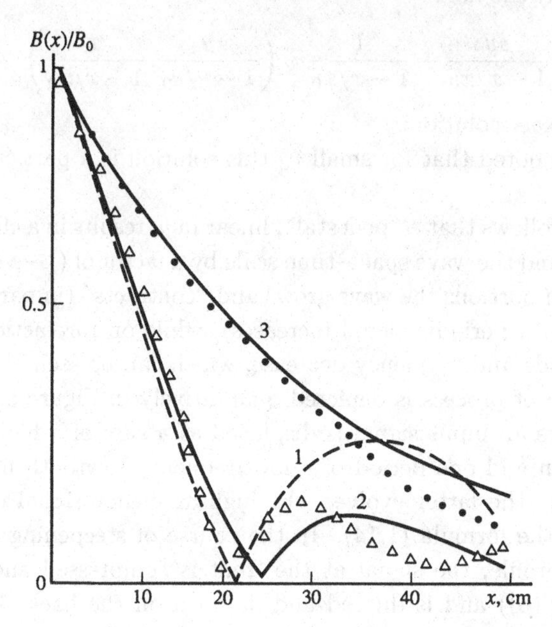

Figure 5.1. The spatial dependence of the amplitude of the high-frequency wave when interacting with a low-frequency wave (curves 1, 2) and in the absence of a low-frequency wave (curve 3). Curve 1 gives an approximate solution and curve 2 gives a numerical result. Triangles and circles represent the experimental data (Tagunov, 1981).

seems to be connected with energy transfer to the side-band components with frequencies $f_{\pm} = f_1 \pm f_2$; also observed are oscillations due to interaction of all these components. Theoretical curves were obtained as a result of analyzing interactions of small perturbations with a low-frequency field, taking damping into account (Tagunov, 1981). Symbols represent experimental results.

These results refer to the case when the nonlinear distortions of the low-frequency field can be neglected. There exists, however, a case when an exact solution can be achieved to the Burgers equation which describes an arbitrary nonlinear wave interaction with a compression or rarefaction wave possessing a linear profile in y (Moffett, 1978; Gurbatov *et al.*, 1983). It can easily be seen that the solution $v_1 = sy/(1 - x/x_0)$ (where s is constant, $x_0 = (\alpha s)^{-1}$) satisfies the Burgers equation. Let any other solution $v_2(y, x)$ be known; then (as can be proved by direct substitution) the field

$$v = \frac{sy}{1 - x/x_0} + \frac{1}{1 - x/x_0} v_2 \left(\frac{y}{1 - y/y_0}, \frac{x}{1 - x/x_0} \right) \tag{1.14}$$

is also an exact solution.

It can be noted that for small v_2 this solution is a particular case of (1.4).

Hence it follows that a "pedestal", linear in y, results in a change in the amplitude and the wave space–time scale by a factor of $(1-x/x_0)^{-1}$. For compression portions the wave grows and "contracts" (in particular, the frequency of a periodic signal increases), while on rarefaction portions the amplitude and frequency decrease, which can be seen also in (1.4).

This kind of process is depicted qualitatively in Figure 5.2 (Moffett, 1978), where an input signal is displayed as a sum of a high-frequency perturbation and one period of a low-frequency sawtooth field (Figure 5.2 (a)). As the latter evolves, the high-frequency signal changes according to the formula (1.14). In the course of steepening of the low-frequency profile, the signal at the front is compressed and amplified (Figure 5.2 (b)) and is dilated and damped on the back slope, which leads to modulation of the high-frequency signal; the evolution is shown in Figure 5.2c. This example illustrates the behavior of the wave, described by the general equation (1.4).

Detection of modulated waves

Now let us address the problem of "detection", i.e. generating a low-frequency field as a product of the interaction of neighbouring-frequency waves or, in general, of the propagation of a modulated high-frequency wave for which the boundary condition is

$$v(y, 0) = A(y) \sin \omega y \tag{1.15}$$

with a slowly varying amplitude $A(y)$. Here it seems convenient, as previously, to formulate a solution as $v = v_h + v_\ell$ where v_h is the oscillating signal and v_ℓ is a small low-frequency component, to be found for a boundary condition $v_\ell(y, 0) = 0$. The high-frequency component has the form of a simple wave,

$$v_h = A(\xi) \sin \omega \xi, \quad \xi = y + \alpha x v_h, \tag{1.16}$$

and further, neglecting distortion of the envelope A, we set $A = A(y)$ (in this case the error is the same as that discussed previously with respect to v_ℓ). The low-frequency portion can be defined by an averaging

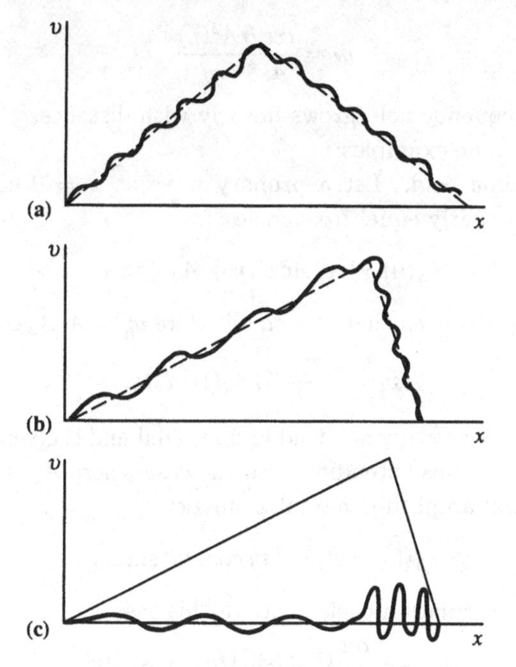

Figure 5.2. Interaction of a high-frequency signal with a sawtooth low-frequency wave: (a) initial profile of wave; (b) distorted profile; (c) low-frequency and high-frequency signals having separated after interaction.

technique: substituting $v = v_h + v_\ell$ into the reference equation (1.1) (for $\delta = 0$) and averaging over a high-frequency period we easily obtain

$$\frac{\partial v_\ell}{\partial x} = \frac{\alpha}{2} \frac{\partial}{\partial y} \left(\overline{v_h^2} \right), \tag{1.17}$$

where a bar designates averaging over y, so that slowly varying quantities are differentiated.

The averaging of v_h^2 is reduced to calculation of the integral

$$\begin{aligned} \overline{v_h^2} &= \frac{A^2}{2\pi} \int_{-\pi}^{\pi} \sin^2 \omega\xi \, \mathrm{d}y \\ &= \frac{A^2}{2\pi} \int_{-\pi}^{\pi} \sin^2 y'(1 + \alpha\omega A x \cos \omega y') \, \mathrm{d}y' = \frac{A^2}{4\pi}. \end{aligned} \tag{1.18}$$

It should be noted that averages over ξ and y for any periodic simple wave are equal to each other (see the last section in Chapter 2). Therefore, to the given approximation,

$$v_\ell = \frac{\alpha x}{4}\frac{\partial A^2(y)}{\partial y}, \tag{1.19}$$

i.e. the low-frequency field grows linearly with distance.

Let us cite some examples:

(i) Biharmonic field. Let a primary wave at $x = 0$ consist of two sinusoids with nearly equal frequencies:

$$v_h(0) = A_1 \sin\omega_1 y + A_2 \sin\omega_2 y,$$

where $\omega_1 = \omega_2 + \Omega$ such that Ω is small. Here $\overline{v_h^2} = A_1A_2\cos\Omega y$, so that

$$v_\ell = -\frac{\alpha x}{2}A_1A_2\Omega\sin\Omega y. \tag{1.20}$$

In this case the low-frequency field is sinusoidal and is given by the same formula as (1.8), thus extending it to the case where A_2 is not small.

(ii) Sinusoidal amplitude modulation. Let

$$v_h(0) = A(1 + m\cos\Omega y)\sin\omega y,$$

where m is a modulation coefficient. In this case

$$v_\ell = -\frac{\alpha x}{2}\Omega A^2(\sin\Omega y + \cos 2\Omega y), \tag{1.21}$$

i.e. a two-frequency field is excited.

(iii) Pulse modulation. If the primary field appears as a localized high-frequency pulse, then according to (1.19) the low-frequency field is an alternating one such that

$$\int_{-\infty}^{\infty} v_\ell \,\mathrm{d}y = 0.$$

Such a process has been named "self-demodulation", though this term can be applied to essentially all processes considered in this section.

All the above refers to the region $x < L_N$, L_N being the discontinuity formation length, before discontinuities emerge in a wave. For $x > L_N$ equation (1.17) still holds good, while the averages need to be calculated accounting for the energy decay at discontinuities, which, moreover, arise at different distances for various portions of a modulated wave. As, locally (in the vicinity of a given point of the envelope profile $A(y)$), everything develops as in a periodic wave, one can make use of the appropriate formulae of Chapter 2. A discontinuity appears at the point $L_N = (\alpha\omega x A(y))^{-1}$ while the average amplitude changes for $x > L_N$ according to the approximate formula (Sutin, 1978)

$$\frac{\overline{v_h^2}}{\overline{v_0^2}} = \left(1 - \frac{\sigma v_s}{\pi}\right) \pm v_s\sqrt{1 - v_s^2} + \frac{2\sigma}{3\pi}v_s^2. \tag{1.22}$$

Here $\overline{v_0^2} = A^2/2$, $\sigma = x/L_N$ and v_s is a normalized discontinuity amplitude defined by the relation $v_s = \sin\sigma v_s$. In this case the positive sign is employed for $\sigma < \pi/2$, the negative for $\sigma > \pi/2$ (when the discontinuity lies at the front and the back of the reference wave, respectively). Here the quantities $\overline{v_0^2}$, σ, v_s are all slowly varying functions of y.

Integrating (1.17) we assume, to simplify calculations, that the wave is a sawtooth one over the whole discontinuous stage ($x > L_N$). Then, if we divide the integral

$$\int_0^x \overline{v_h^2}\,\mathrm{d}x$$

arising on averaging (1.17) into two portions: from 0 to L_N where $\overline{v_h^2}$ is constant, and from L_N to x where the wave has a sawtooth wave profile, we obtain for $x > L_N$

$$v_\ell \simeq \frac{1}{\omega}\frac{\partial A}{\partial y}\left\{1 + \frac{\pi^2}{\sigma}\left[1 - \frac{2}{(1 + 2\alpha\omega x A)^2}\right]\right\}. \tag{1.23}$$

This equation is valid when $\alpha\omega x A \geq 1$. Therefore $|v_\ell|$ increases monotonically with x, and as $x \to \infty$, $|v_\ell|$ tends to an asymptotic value (Naugol'nykh *et al.*, 1963)

$$v_{\ell\infty} \simeq \frac{\pi}{2}\frac{\Omega}{\omega} \tag{1.24}$$

(the latter expression corresponds to 100 per cent harmonic modulation of the carrier wave of frequency Ω). It has to be emphasized that at this stage the field is linearly rather than quadratically related to A. Another important conclusion is as follows: the ratio v_ℓ/A is of the order of Ω/ω, where Ω is the modulation frequency. Therefore, the efficiency of the low-frequency wave generation is here of the order of $(\Omega/\omega)^2$ (Sutin, 1978). We note that in dispersive media it may attain the order of Ω/ω (see the ensuing chapter).

It seems worthwhile to underline that, as was shown in Chapter 2, even in an unmodulated nonlinear wave strictly periodic in y, the average value of v (as of other physical quantities) is, rigorously speaking, different from zero: this average is, however, of the order of v^2/c_0^2 and does not increase with distance; formula (1.19) does not take this into account.

Low Reynolds numbers

Let us consider a simple but sometimes practically important case, that of a viscous medium when the Reynolds number is small at high frequency. Then the high-frequency field is described by the linearized Burgers equation, whereas the low-frequency one is described by the same equation, but with a right-hand side, as follows:

$$\frac{\partial v_\ell}{\partial z} - \delta\frac{\partial^2 v_\ell}{\partial y^2} = \overline{\alpha v_h \frac{\partial v_h}{\partial y}}. \tag{1.25}$$

For example, let a pumping at $x = 0$ have the form of the sum of two waves with close frequencies. Then

$$v_h = A_1 \exp(-\delta\omega_1^2 x)\sin\omega_1 y + A_2\exp(-\delta\omega_2^2 x)\sin\omega_2 y. \tag{1.26}$$

Substituting this into (1.25) one can determine the field at the difference frequency (Naugol'nykh *et al.*, 1963) as

$$A_\Omega = \{\exp[-\delta(\omega_1^2+\omega_2^2)x] - \exp(-\delta\Omega^2 x)\}\,\frac{\alpha\Omega A_1 A_2}{2\delta(\omega_1^2+\omega_2^2-\Omega^2)}. \tag{1.27}$$

This implies that A_Ω grows initially, to attain its maximum at the point $x_m = \delta(\omega_1^2+\omega_2^2-\Omega^2)\ln(\omega_1^2+\omega_2^2)/\Omega^2$ and decrease thereafter (at large x it is proportional to $\exp(-\delta\Omega^2 x)$). The relation of the maximum amplitude A_Ω to the characteristic amplitude of the high-frequency pumping at $x = 0$ is equal to

$$\frac{A_\Omega}{\sqrt{A_1 A_2}} = \frac{\Omega}{\omega}Re = \frac{\alpha\sqrt{A_1 A_2}}{4\omega\delta} \ll 1. \tag{1.28}$$

Thus, the maximum amplitude transformation factor is far less here than the value of Ω/ω reached (according to (1.24)) at large Reynolds numbers.

Here comparatively much attention has been devoted to one-dimensional problems, for they form a basis for the different problems arising for parametric transmitters and receivers considered in what follows. Thus, nonlinear "detection" is at the root of parametric transmitter operation, and the interaction of a weak low-frequency signal with a high-frequency "pumping" (with side-band component generation) is central to parametric receivers.

5.2 Parametric acoustic transmitters

A region of sound wave interaction serves as a source of waves at combination frequencies and, primarily, at the sum and difference frequencies.

If this region is bounded in space, then a "secondary" field of combination frequencies can be radiated from it. This can be clearly observed even when solving the one-dimensional problem considered above: at large distances, where primary waves decay, the low- (difference-) frequency field amplitude tends to a nonzero constant value (if viscosity is taken into account it of course decays, but at a much slower rate than the primary waves).

A more complicated matter is that of the interaction of sound beams having finite cross-sections. In this case the interaction area is localized in space and "sound scattering by sound" is observed, accompanied by radiation at combination frequencies.

The problem of the possibility and efficiency of such scattering provoked animated discussions back in the 1960s. Nowadays it does not seem helpful to reproduce the details of those discussions, for the main ideas are rather simple and the theoretical aspects of the problem have been cleared up to a great extent.

On noncollinear interaction, when the resonance conditions are not satisfied, the fields from different portions of the interaction domain interfere in such a manner that, at points outside this domain, the field is typically small even if the domain size is many times greater than the primary beam wavelength.† Because of this, in the case of bounded beams as well, the most interesting situation turns out to be the collinear interaction when primary and secondary waves are in synchronism. Then the interaction domain acts as if it were a virtual, "bodiless" travelling wave antenna. Of special concern here is the case of close frequency waves, or that of a modulated wave, bringing about a low-frequency field. This is used in the parametric sound array (PA) proposed in the 1960s by Westervelt (1963) and Zverev and Kalachev (1970).‡

Allowing for the fact that parametric arrays are now widely known (see, e.g., the book by Novikov *et al.*, 1980) we shall confine ourselves here to discussing merely the most illuminating physical models.

The theory of PA is usually based upon an approximation to a given pumping field according to which one disregards the influence of the

† Note that nowadays this problem is being considered in connection with the "nonscattering of sound by sound" (Westervelt, 1987).

‡ introduced by P. Westervelt's mechanism does not exactly correspond to the notion of parametric resonance (see, however, the next chapter) adopted in radio-engineering and the theory of oscillations. It would have been more precise to speak of a nonlinear antenna. But for about three decades this term has penetrated so deeply into the scientific soil that it seems senseless to try to change it now.

low-frequency field on the primary beams which may be described in a linear approximation. Let us proceed from the nonlinear acoustics equations introduced in the first chapter and take the pressure p' as the unknown variable in what follows (as is the accepted procedure for PA in the literature). Let a high-frequency "pumping" be prescribed in the form of a wave beam with time-modulated pressure amplitude,

$$p'_h = \frac{1}{2} A(\mathbf{r}, t) \exp[i(\omega t - kx)] + c, \tag{2.1}$$

where A is a slowly-varying pressure amplitude here, and x is a coordinate along the beam axis. A solution is sought in the form $p' = p'_h + p'_\ell$ where p'_ℓ is a weak low-frequency component. As for the nonlinear term, the pumping field ($p' \simeq p'_h$) alone will be substituted; moreover, inasmuch as the wave is close to a plane wave, one can use the relations $p' = \rho_0 c_0 v = c_0^2 \rho'$ in these terms. The dissipation is taken into account only in consideration of the pump wave. Finally, considering p'_ℓ to be a slowly-varying function, let us average the Westervelt equation (1.15) of Chapter 1 over a time interval of the order of several pumping periods, as was done above. Then for the slowly varying component a linear inhomogeneous wave equation results,

$$\nabla^2 p'_\ell - \frac{1}{c_0^2} \frac{\partial^2 p'_\ell}{\partial t^2} = - \frac{\alpha}{\rho_0 c_0^2} \frac{\partial^2}{\partial t^2} \langle p_h'^2 \rangle = - \frac{\alpha}{2 \rho_0 c_0^2} \frac{\partial^2 A^2}{\partial t^2}, \tag{2.2}$$

where the brackets again designate averaging over "fast" time. The right-hand side of equation (2.2) comprises a distribution of "virtual" sources for the low-frequency field, associated with the detection of high-frequency oscillations due to the medium nonlinearity.

The general solution of the wave equation (2.2) is known to have the form

$$p'_\ell = - \frac{\alpha}{8 \pi \rho_0 c_0^2} \int \frac{1}{r} \frac{\partial^2 A^2 (t - r/c_0)}{\partial t^2} \, \mathrm{d}V, \tag{2.3}$$

where the integral is taken over the whole volume occupied by sources, and r is a distance from the element $\mathrm{d}V$ to the observation point.

In the general case the quantity A^2 is specified by the solution of a nonlinear diffraction problem for the pumping wave (in the spirit of the problems treated in Chapter 4). Both finding A^2 and calculating the integral in (2.3) can then turn out to be rather complicated matters, which promotes the adoption of additional simplifying assumptions based on a qualitative analysis of the primary beam behaviour, depending on the

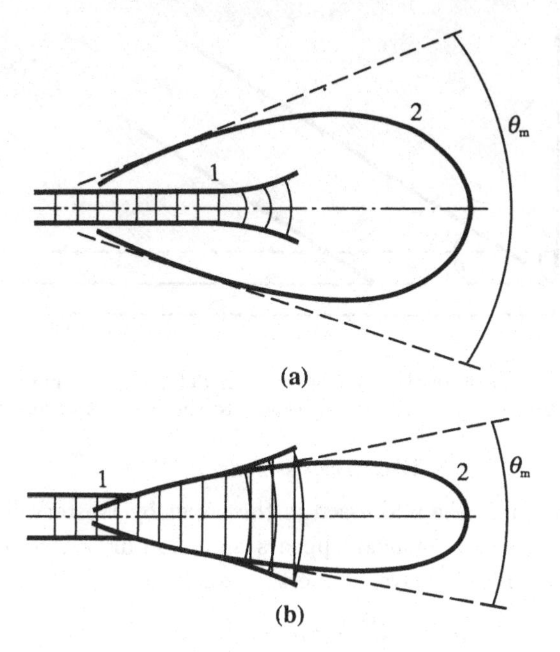

Figure 5.3. Two regimes of parametric array performance: interaction in (a) the near-field zone and (b) the far-field zone.

relation between the characteristic scales L_N for nonlinearity, L_D for diffraction, L_d for dissipation.

Let us begin with a relatively simple but rather realistic instance of "linear" pumping, when $L_N \gg L_D$, i.e. nonlinear distortions are small in the pumping field so that it can be taken from the linear theory of acoustic beams with damping. The formation of a low-frequency field proceeds differently, depending on the ratio L_d/L_D determining the real pumping field geometry. From this viewpoint one can single out two limiting cases: (i) $L_d \ll L_D$, when the primary wave field decays significantly over the near-field "projecting" zone (collimated beam) (Figure 5.3(a)); (ii) $L_d \gg L_D$, when the principal contribution to the low-frequency wave generation is that of the far-field zone, where the primary waves diverge (Figure 5.3(b)). Curve 1 on these schematic figures delineates the pumping field, whereas curve 2 represents the directivity diagram of the low-frequency radiation.

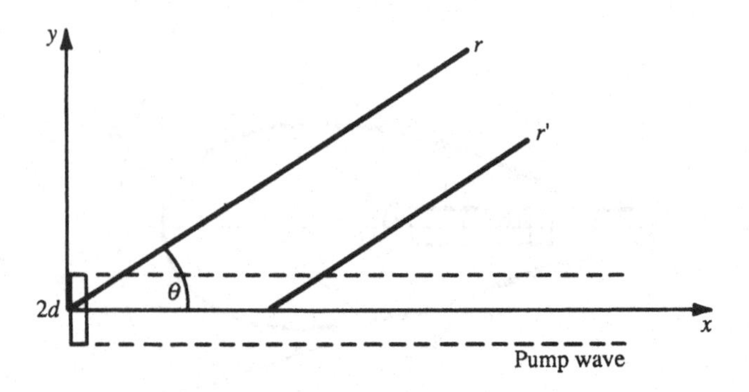

Figure 5.4. Schematic of parametric array model. $2d$ is the pump wave transducer apperture, and r and r' are distances to the point of observation.

Westervelt's model of PA

Consider the first case addressed in the paper by Westervelt (1963). In this case, the pumping field appears as a non-diverging but damped beam, with a finite transverse aperture, such that

$$A = p_h(r_\perp)E(y)\exp(-ax), \quad y = t - x/c_0, \tag{2.4}$$

where a is a damping factor at the carrier (pump) frequency (so that $L_d = a^{-1}$) and $E(y)$ is a modulation function. $p_h(r_\perp)$ is a function describing the pump wave amplitude distribution over the cross-section of the beam, and the function $p_h(r_\perp)$ is required to be integrable in the whole y, z- plane.

Let us substitute (2.4) into (2.3) and calculate the field in the far-field zone of a beam (at distances $r > L_d$) where, as usual, $1/r$ can be moved outside the integral sign:

$$p'_\ell = \frac{\alpha}{4\pi c_0^2 r \rho_0} \int p_h^2(r_\perp) \frac{\partial^2}{\partial t^2} E^2(y - r'/c_0)\exp(-2ax)\,\mathrm{d}V. \tag{2.5}$$

The integration is supposed to be made over all the region of interaction V and the variables r', r are the distances from the source element and the coordinate system origin to the observation point (Fig. 5.4). In the case of an axially symmetric beam with a homogeneous amplitude distribution over its cross-section, $\int p_h^2 dydz = \pi p_h^2 d^2$, where d is the beam radius.

For the case of a biharmonic pump wave of equal amplitudes p_h at each of the components at frequencies $\omega \pm \Omega/2$, where ω is the carrier frequency and Ω is the difference frequency, $E = 2\sin(\frac{\Omega y}{2})$. Therefore one gets

$$\frac{\partial^2}{\partial t^2} E^2 \left(y - \frac{r'}{c_0} \right) = \Omega^2 \exp[i\Omega \left(y - \frac{r'}{c_0} \right)],$$

First we shall consider the beam to be narrow compared to the wavelength at the difference frequency. Then in the far zone $r' \simeq r - x\cos\theta$ (Fig. 5.4) and $y - \frac{r'}{c_0} = t - \frac{r}{c_0} - \frac{x}{c_0}(1 - \cos\theta)$ so that from (2.5) one obtains

$$\begin{aligned} p'_\ell &= \frac{\alpha p_h^2 k_\Omega^2 d^2}{4r\rho_0 c_0^2} \int \exp[i(\Omega t - k_\Omega r) - ik_\Omega x(1 - \cos\theta) - 2ax] \mathrm{d}x \\ &= \frac{\alpha p_h^2 k_\Omega^2 d^2}{8r\rho_0} \frac{\exp[i\Omega(t - \frac{r}{c_o})]}{ik_\Omega \sin^2\frac{\theta}{2} - a}, \end{aligned} \tag{2.6}$$

where $k_\Omega = \frac{\Omega}{c_0}$. Separating the real part of the amplitude we get the Westervelt formula

$$P_\ell = \frac{\alpha p_h^2 k_\Omega^2 d^2}{8\rho_0 a r} D_1(\theta), \tag{2.7}$$

where

$$D_1(\theta) = \left[1 + \left(\frac{k_\Omega}{a} \right)^2 \sin^4 \frac{\theta}{2} \right]^{-1/2}. \tag{2.8}$$

In the case of a broader pump wave beam the integral over transverse coordinates entering (2.5) yields an additional aperture factor $D_t(\theta)$. For the case of homogeneous distribution of p_h over the beam cross-section

$$D_t(\theta) = \frac{2J_1(kd\cos\theta)}{kd\cos\theta}$$

where J_1 is the Bessel function and d the beam radius.

This implies that the radiation maximum is directed along the primary beam axis and the characteristic angular width of the directivity pattern is equal to

$$\theta_m^w \simeq \sqrt{\frac{a}{k_\Omega}} \sim (k_\Omega L_d)^{-1/2}, \tag{2.9}$$

where $L_d = 1/a$ is the characteristic dissipation length. As in any travelling-wave antenna, θ_m^w turns out to be proportional to $1/\sqrt{L_d}$ rather than to $1/L_d$.

In a similar way, solutions for other modulation laws can be derived. In particular, the general expression for the field on the PA axis was derived by Berktay (1965). Since the travel speed of the source $E^2(\tau)$ coincides with that of sound, the source moves along the axis in synchronism with the wave, so that at a given point the contributions from all profile points are summed in phase, although account is taken of decay through the factor $\exp(-ax)$. Therefore

$$\int_0^\infty E^2(\tau)\exp(-ax)\,\mathrm{d}x = E^2(\tau)\int_0^\infty \exp(-ax)\,\mathrm{d}x = \frac{1}{a}E^2(\tau), \quad (2.10)$$

and eventually (2.5) yields

$$p'_\ell = \frac{\alpha p_h^2 d^2}{16a\rho_0 c_0^2 r}\frac{\partial^2 E^2}{\partial t^2}. \quad (2.11)$$

For a biharmonic wave this certainly gives formulae (2.6) and (2.7) with $\theta = 0$. If, however, $E = 1 + m\cos\Omega\tau$ (sinusoidal amplitude modulation), then the field on the axis will prove biharmonic (cf. (1.19)).

Frequently employed also is a pump in the form of a finite-length pulse (wave packet), which makes the low-frequency radiation take the form of a "video" pulse of complex shape. As was mentioned before, in such instances one can speak of a "self-demodulation" regime. Note that at non-zero θ the pulse shape may differ from that on the beam axis (similar behaviour of a diffracted wave was discussed in the previous chapter).

Let us consider now the problem of PA efficiency, i.e. the relation

$$K = \frac{\sqcap_\ell}{\sqcap_h} \approx \frac{\overline{p_\ell'^2}\theta_m^2 r^2}{\overline{p_h^2}d^2}, \quad (2.12)$$

where d is the beam diameter, $\sqcap_h$ is the primary beam power while $\sqcap_\ell$ is that of the secondary beam (integration is carried out over the entire solid angle Σ where the radiation field is present). The formulae derived above readily yield here

$$K = \left(\frac{p_h}{\rho_0 c^2}\right)^2 (k_\Omega d)^2 (k_\Omega/64\pi a). \quad (2.13)$$

Hence it is clearly seen that to this approximation the transmitter efficiency is proportional to p_h^2, i.e. the pump intensity. In this sense a powerful PA proves to be most efficient, but only up to certain limits defined by nonlinear dissipation of the primary beam.

Berktay model of PA

Consider now the case of weak damping of the pump wave, when $L_D \gg L_d$. Then without significant dissipation a wave will transform into a diverging beam, i.e. an inhomogeneous spherical wave with a directivity angle θ_ω of the order of $(kd)^{-1}$. Then the contribution to the low-frequency radiation is both that of a collimation zone and that of the far-field zone. If then the condition $\theta_\omega < \theta_m$ is satisfied, that is

$$\left(\frac{k}{k_\Omega}\right)^{1/2} < (kd)^{-1},$$

i.e. radiation from the beam side surface estimated above (see formula (2.9)) does not escape beyond the angular limits of the primary beam, then the low-frequency radiation "falls within" the pump directivity diagram. In achieving this, the secondary sources form something like a horn antenna (Berktay regime). The low-frequency field here can be derived just as was done previously in Section 1 for the plane waves, taking into account only the primary field variation due to ray tube divergence. Then the field amplitude on the axis at large distances is obviously proportional to

$$\int_{r_0}^{\infty} \exp(-a(r-r_0))\,\frac{\mathrm{d}r}{r},$$

where r_0 is the initial radius (it can be taken as of the order of L_D). If, as was supposed, $ar_0 \ll 1$, then this integral is approximately equal to $\ln(ar_0)^{-1}$, and the entire field in the wave zone has the form (Berktay & Leahy, 1974)

$$p'_\ell \simeq \frac{\varepsilon p_h^2 k_\Omega r_0^2}{4\rho_0 c_0^2 r}\left(\frac{\partial E^2}{\partial \tau}\right)\ln(ar_0)^{-1} D_h^2(\theta), \tag{2.14}$$

where $D_h(\theta)$ is the directivity function of the primary source, $r_0 \simeq L_D \simeq kd^2/2$, and p_h is the pump wave amplitude. It seems of interest that here the directivity function for low frequency is equal to D_h^2, i.e. it proves even narrower than that of the primary beam. Another salient feature of (2.14) is that the field on the beam axis is proportional to the first derivative of the modulation function squared rather than to the second, so that the radiated pulse shape, for example, will be different from that in the Westervelt regime. In this case the transmitter efficiency (for a biharmonic pump) is equal to

$$K = \left[\frac{k_\Omega L_d p_h}{4\rho_0 c_0^2}\ln(2dL_d\Omega/\omega)\right]^2. \tag{2.15}$$

This quantity exceeds (2.13), since the beam radius enters it only under the logarithmic sign.

Of course, cases of practical concern are not always described by these approximate formulae. For an intermediate case, when $L_d \sim L_D$, contributions of both near- and far-field zones of the primary beam are comparable and as a rule some additional approximation should be made. Summing up the fields of the near- and far-field zones of the primary beams, one obtains an interpolation formula for the radiator axis pressure, as follows (Dunina *et al*, 1976):

$$p'_\ell = \frac{\alpha p_h^2 (kd)^2 (k_\Omega d)^2}{16 \rho_0 k_\Omega r} \int_0^L \frac{\exp(-ar)\,\mathrm{d}r}{(k^2 d^2/2k_\Omega) + r}, \tag{2.16}$$

uniting relations (2.7) (for $\theta = 0$) and (2.14). Numerous calculations have also been carried out within the framework of the Khokhlov–Zabolotskaya–Kuznetsov equation. This problem also has been given a detailed computer-aided investigation. All those interested are referred to the aforementioned book by Novikov *et al.*

Pump saturation regime

As was shown above, PA efficiency rises with primary beam intensity. However, in a strong primary wave, nonlinear distortions may show up, and one can expect that following discontinuity formation nonlinear losses will restrict the further growth of efficiency.

A solution of this problem may, as previously, be based on the "given pump field" approximation, although when considering this field one has to allow, along with diffraction, for nonlinearity according to the suggestions set forth in Chapter 4. Let us again consider the extreme situations treated above. Assume initially that $L_N < L_d < L_D$. Then the wave turns out to be a sawtooth one, even at the near-field (projector) stage, after which it decays rapidly (so that the inequality $L_d < L_D$ has to be noted, taking losses at discontinuities into account). Here formula (1.22) from Section 1 is valid, and will be employed later. In order to consider the field on the axis we need, in analogy with (2.5), to calculate an integral, setting p_h proportional to $E(\tau)$, and take the integral over the beam portion before the discontinuity formation point. As a result we obtain (for a circular transmitter)

$$p'_\ell = \frac{p_h d^2}{2\omega c_0 r} \frac{\partial^2 E}{\partial \tau^2}. \tag{2.17}$$

It should be noted that the contribution of the discontinuous stage is also essential so the equation (2.17) gives only an estimate for the p'_ℓ.

This formula differs from (2.11) first in that it does not involve the nonlinearity parameter and second in that p'_ℓ is linearly related to p_h and to E. In this case a biharmonic pump gives rise to a non-sinusoidal field: differentiating the modulation function $|\sin \Omega\tau/2|$ twice, we shall observe sharp peaks in the vicinity of the envelope zeros (although near such zeros (2.14) becomes inapplicable, because here pumping cannot develop to the sawtooth stage). Such peaks have been observed experimentally (Willette *et al.*, 1976).

On the other hand, a sinusoidally modulated pump produces a harmonic secondary field on the axis.

As for the angular field dependence in the saturation regime, the calculations are more complex, and usually require the use of a computer. However, estimation of the angular width of the directivity diagram can again be accomplished (for a narrow antenna) using (2.9), where for the antenna length L a nonlinear pump decay length L_N can be substituted. Then

$$\theta \sim \frac{1}{\sqrt{k_\Omega L_N}} = \frac{1}{c_0}\left(\frac{\varepsilon \omega p_h}{k_\Omega \rho_o c_0}\right)^{1/2}, \tag{2.18}$$

Here again an alternative regime is possible, when discontinuity formation and the following rapid decay of the pumping field take place in the far-field zone and something like a horn radiator is formed at low frequency. To describe the primary field, here one can also use formula (1.22) provided σ is understood as a reduced distance for a spherical wave (see Chapter 3) and the diffraction length $L_D \approx d^2/\lambda$ is used for r_0. Then integrating over r_0 yields the following formula for the field on the beam axis (Gurbatov *et al.*, 1979):

$$p'_\ell = \frac{p_h d^2}{2c_0 r}\frac{\partial E}{\partial t}. \tag{2.19}$$

Thus, p'_ℓ is again linearly dependent on E, but this time it is proportional to the first derivative of E, so that the field shape in the case of non-harmonic $E(\tau)$ will be different from that given by (2.17).

An important problem is that of the transmitter efficiency. If, using (2.12), we recover the efficiency from (2.17) and (2.19), then it can easily be seen that in the first case $K = K_1 \approx (\Omega/\omega)^4$, while in the second $K = K_2 \approx (\Omega/\omega)^2$, i.e. $K_2/K_1 \approx (\omega/\Omega)^2$. Strictly speaking, a value of the order of K_2 can be obtained not only by elongating the interaction region ("virtual antenna") but by shortening it as well: if the intensity is increased until the length $L \sim z_*$ becomes less than the low-frequency wavelength then the width of the directivity diagram will be determined simply by the aperture, $\theta_m \sim K_\Omega d$, whereas the efficiency becomes of the order of $(p'_\ell \theta_m / p_h \theta_h)^2 \approx (\Omega/\omega)^2$ (θ_h being the angular width of the primary radiation). In this case, however, all the advantages of PA, stemming from its narrow and "lobeless" diagram, fade away. Therefore, the "horn" regime should be considered as optimal in this sense, with the maximum efficiency being of the order (Sutin, 1978) of

$$K_{\max} \approx (\Omega/\omega)^2. \tag{2.20}$$

Accordingly, only a minor portion of the pumping wave energy is transferred into low-frequency radiation (causes for this have been discussed above).

A more elaborate investigation (Naugol'nykh & Ostrovsky, 1982) demonstrates that the maximum pressure value on the axis, for a narrow radiation diagram (width of the order of $\theta_h \sim 1/kd$) and efficiency conforming to (2.20), is realized for the regime transient to the "horn" regime, when discontinuities appear at distances of the order of L_D.

Applications of parametric transmitters

As was said above, a parametric transmitter is known to have some unique merits: it has a high level of directivity and practical absence of the diagram sidelobes for small dimensions of the primary wave transmitter, as well as wide bandwidth; moreover, to substantially change the radiation frequency a rather small pumping frequency variation suffices in practice. The major disadvantage of a PA is its low energy efficiency. However, the efficiency value, no matter how important it may be, is not always a decisive factor; on some occasions it may be compensated by a field gain along the radiator axis due to high directivity as compared with the isotropic low-frequency transducer.

Nowadays PAs are still not very widely used, but they are usually employed when one needs to avoid reverberation interference of various kinds. In ocean acoustic they are utilized to perform investigations in measuring acoustics basins, in shallow seas, in fishery, in near-to-bottom area sounding, in study of sound scattering layers, etc.

The results of elaborate analysis of PA radiation behaviour in water are in fair agreement with the the theory. Some comparative data regarding the PA directivity diagram width are shown in Figure 5.5 (Hobaek & Vestrheim, 1976). The figure resents the dependence of the actual directivity diagram width (normalized to that of the Westerwelt model, $\bar{\theta} = \theta/\theta^w$, θ^w being given by equation 2.8), on the parameter $N_a \approx \sqrt{\Omega L_d / \omega L_D}$ characterizing the ratio of the primary wave diagram width to the parametric radiation diagram width θ^w. Line 1 corresponds

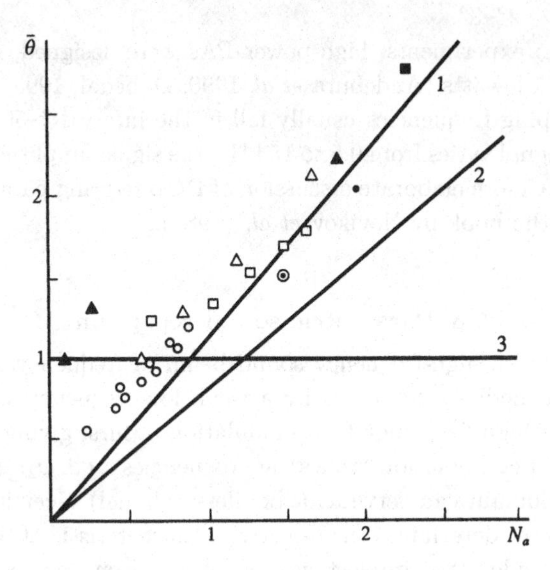

Figure 5.5. The theoretical and experimental dependences of the directivity pattern width of a parametric array. $N_a = \sqrt{\Omega L_d/\omega L_D}$ is the ratio of the primary and the secondary wave directivity pattern widths; $\overline{\theta}$ is the low frequency radiation directivity pattern half-width normalized to its value calculated from Westervelt formula (2.9), $\overline{\theta} = \theta_m/\theta_m^w$. Line 1 – Berktay's model, parametric generation in a far field, equation (2.14); line 2 – Westerwelt's model, parametric generation in a near field, equation (2.8).

to Berktay's model of PA; according to equation (2.14), the directivity pattern of low-frequency radiation in this case can be presented as $\mathrm{D}(\theta) = \mathrm{D}_h^2(\theta)$, where $\mathrm{D}_h(\theta)$ is the directivity characteristic of the high-frequency (pump) wave. The horizontal line 2 obviously corresponds to the Westerwelt model (2.9). The symbols represent various experimental data.

Apart from marine investigations, PAs are discussed for use in atmospheric soundings and in air turbulence level inspection in the vicinity of airports, as well as in medical acoustics.

In laboratory installations and in electroacoustic transducer calibration systems (the advantages of PAs here are associated with their wide bandwidth and high directivity) small pumping transducers are typically used (a few centimetres in diameter) working at high pumping frequencies (≈ 1 MHz). In this case the pump power is of the order of 10 W, the radiation frequency amounts to 1–100 kHz, the signal amplitude (normalized to 1 m distance) is about $1\,\mathrm{Pa\,m^{-1}}$, and the directivity pattern width is $\theta_m \sim 2°$–$3°$.

For marine experiments, high-power PAs were designed, ranging up to dozens of kilowatts (Andebura *et al*, 1990; Dybedal, 1993; Muir *et al*, 1980). Pumping frequencies usually fall in the interval 5–50 kHz while that of the signal varies from 0.5 to 15 kHz, the signal amplitude reaching $10\,\mathrm{Pa\,m^{-1}}$. A more elaborate discussion of PA operating parameters can be found in the book by Novikov *et al.* (1980).

5.3 Parametric sound reception

When an intense high-frequency sound beam of frequency ω_1 is propagating in a medium disturbed by a weak low-frequency wave of frequency ω_2, a high-frequency field modulation occurs, giving rise to the appearance of combination "tones" at frequencies $\omega_1 \pm \omega_2$. Thereby an interaction domain can serve as a bodiless (virtual) receiving antenna whose geometry determines the directivity characteristic of the receiver.

Let us consider the simplest version of a parametric receiver (Figure 5.6), consisting of a high-frequency field (pump) transmitter and a receiving array selecting combination frequencies excited when a low-frequency wave is incident on the interaction domain (Zverev & Kalachev, 1970). Assume first that the distance (path length) L between these transducers is less than that of the pumping transmitter projector zone, so that a high-frequency wave (in the absence of a signal) can be treated as plane,

$$p_1' = p_1 \exp[-i\omega_1(t - x/c_0)]. \tag{3.1}$$

Let a plane low-frequency wave propagate at an angle θ with respect to the x-axis,

$$p_s' = p_2 \exp(-i\omega_2 t + ik_2 x\cos\theta - ik_2 y\sin\theta), \tag{3.2}$$

the amplitude p_2 being small compared with that of the high-frequency pumping wave.

The problem of the combination frequency wave generation can in principle be solved by substituting the sum of the fields (3.1) and (3.2) into the nonlinear wave equation. There is, however, a still more elementary approach which yields the same result. The medium perturbation due to the low-frequency wave leads to a variation in the pump wave propagation velocity of $\Delta c = \Delta c_1 + \Delta c_2$, where $\Delta c_1 = v_s\cos\theta$ is the sound velocity variation due to the medium motion, and

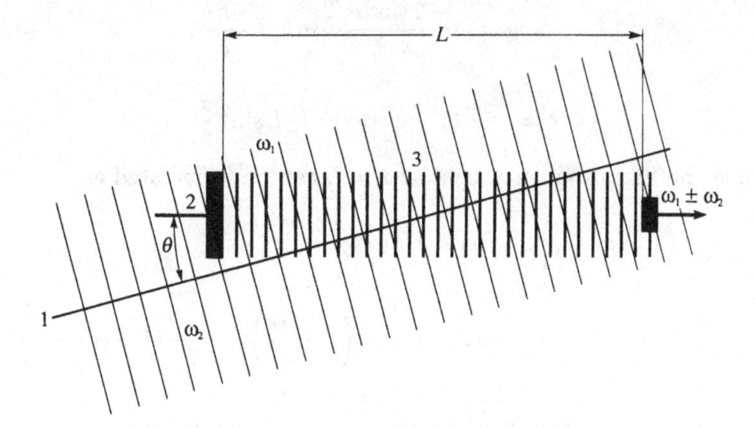

Figure 5.6. The parametric receiver of sound. 1 – signal wave; 2 – pump wave transducer; 3 – region of interaction.

$\Delta c_2 = [(\gamma - 1)/2\rho_0]c_0\rho'_s$ is that due to the medium density change, $v_s = p_2/\rho_0 c_0, \rho'_s = p_2/c_0^2$. Therefore

$$\Delta c = \frac{\gamma - 1 + 2\cos\theta}{2\rho_0 c_0} p_2 \cos(\omega_2 t - k_2 x \cos\theta). \tag{3.3}$$

The high-frequency beam diameter d is assumed here to be small compared with the low-frequency wavelength. Because of this (3.3) is formulated for the pumping beam axis ($y = 0$).

The sound velocity modulation leads to phase modulation of the high-frequency wave at the reception point, which can be assessed as follows. The wave front leaving point $x = 0$ at time t will arrive at point $x = L$ at time

$$t + \int_0^L \frac{\mathrm{d}x}{c_0 + \Delta c} \simeq t + \frac{L}{c_0} - \int_0^L \frac{\Delta c}{c_0^2}\,\mathrm{d}x,$$

and the phase shift due to the sound velocity variation is equal to

$$\Delta\varphi(t) \simeq -\int_0^L \frac{\omega_1}{c_0^2}\Delta c\left(t + \frac{x}{c_0}, x\right)\,\mathrm{d}x \tag{3.4}$$

(to this approximation the time delay can be considered as equal to x/c_0 for all fronts).

Substituting here (3.3), we obtain after elementary transformations,

$$\Delta\varphi = \Delta\varphi_0 \cos\left[\omega_2 t + \frac{k_2 L}{2}(1 - \cos\theta)\right], \tag{3.5a}$$

where

$$\Delta\varphi_0 = \frac{\varepsilon - 2\sin^2\theta/2}{\rho_0 c_0^2} p_1 k_1 L \frac{\sin\delta}{\delta}, \tag{3.5b}$$

$$\delta = \frac{k_2 L}{2}(1-\cos\theta) = k_2 L \sin^2\frac{\theta}{2}.$$

Hence the high-frequency field at $x = L$ can be formulated as

$$\begin{aligned} p_1(t, L) =& p_1 \exp\Bigg\{ -\mathrm{i}\omega_1(t - L/c_0) \\ & - \mathrm{i}\Delta\varphi_0 \cos\left[\omega_2\left(t - \frac{L}{c_0}\right) + \frac{k_2 L}{2}(1-\cos\theta)\right]\Bigg\} \\ =& p_1 \exp\left[-\mathrm{i}\omega_1\left(t - \frac{L}{c_0}\right)\right]\sum_{-\infty}^{\infty}(-\mathrm{i})^n J_n(\Delta\varphi_0) \\ & \times \exp\left\{-\mathrm{i}\left[\omega_2\left(t - \frac{x}{c_0}\right) + \frac{k_2 L}{2}(1-\cos\theta)\right]\right\}. \end{aligned} \tag{3.6}$$

Thus the effect of the low-frequency wave is to give rise to combination frequencies $\omega_1 \pm n\omega_2$ in the pump wave spectrum.

For $\Delta\varphi_0 \ll \pi$, using an approximate expression for J_1, one can find the amplitudes of the combination signals $\omega_1 \pm \omega_2$ traditionally employed at reception as

$$p_+ \simeq p_- = p_c = \frac{\varepsilon - 2\sin^2\theta}{\rho_0 c_0^2} p_1 p_2 k_1 L \frac{\sin\delta}{\delta}. \tag{3.7}$$

The directivity diagram of the parametric receiver signals is given by the factor $\sin\delta/\delta$. The diagram half-width (at the level $1/\sqrt{2}$ with respect to the maximal amplitude) is determined, for $K_2 L \gg 1$, by the quantity

$$\theta_m \simeq 4.7/\sqrt{k_2 L} \tag{3.8}$$

(cf. (2.9) for a transmitter).

The basic disadvantage of a parametric receiver is the smallness of the combination component amplitude, which contains the factor $p_1/\rho_0 c_0^2 = M$, the low-frequency wave Mach number. As in the case of a parametric transmitter, the transformation factor grows with the pumping amplitude p_1. At large p_1, however, the pumping wave is again transformed into a sawtooth one as it propagates from the transmitter to the receiver. Then each harmonic of the sawtooth pumping wave undergoes a phase modulation under the effect of the low-frequency wave. Taking formula

(1.13) into consideration for the amplitude of the satellite components having frequencies $n\omega_1 \pm \omega_2$, we obtain

$$p_c^{(n)} = p_1^{(n)} \Delta\varphi_0 = \left(\varepsilon - 2\sin^2\frac{\theta}{2}\right) p_2 \frac{\sin\delta}{\delta}. \tag{3.9}$$

Therefore, the satellite amplitudes of the sawtooth pumping wave harmonics generated by low-frequency modulation are equal (Gurbatov, 1980) until they become sensitive to viscosity. Such a circumstance implies, in principle, the possibility of increasing the parametric receiver effectiveness in the nonlinear regime by extracting the satellites of various harmonics and then summing their intensities.

An important property of any receiver is the signal-to-noise ratio. A parametric receiver "collects", in principle, noises at both low and high frequencies. It can be shown that, allowing for directivity effects, the signal-to-noise ratio for it is equal (McDonough, 1975) to

$$\frac{S}{N} = \frac{G^2 W_S}{N_p D_p + G^2 N_S D_S}. \tag{3.10}$$

Here $G = p_c/p_1$ is the combination tone to signal amplitude ratio; N_p and N_S are the powers of the isotropic noise in the reception band for pumping and signal frequencies; D_p, D_S are factors defining the directivity of the system at pump and signal frequencies; W_S is the signal power in the reception frequency band. For large coefficients G of parametric transformation the ratio S/N would approach a value corresponding to a linear antenna with the same directivity; however, in real cases $G \ll 1$ and a parametric receiver turns out to have considerably less satisfactory signal-to-noise characteristics.

Reception over a diverging and curved pumping beam

Let us consider now a more intricate case, in which for the pumping beam $L_d > L_D$, so that along with the "projector" zone the wave zone of a beam also "works". We shall proceed from the wave equation for a wave of the sum frequency,

$$\nabla^2 p_k + k_+^2 p_k = \frac{\varepsilon\omega_+^2}{\rho_0 c_0^4} p_1 p_2, \tag{3.11}$$

where p_1, p_2 are complex amplitudes of the pumping and signal waves and $k_+ = \omega_+/c_0$. This equation holds within a "quasi-one-dimensional" approximation when the angle θ between the directions of the pump and signal wave propagation is small enough. Note that for the difference frequency, p_2 in (3.11) should be replaced with p_2^*.

The solution to (3.11) has the form of a scattering integral,

$$p_k(\mathbf{r}) = -\frac{\varepsilon\omega_+^2}{4\pi\rho_0 c_0^4}\int \frac{p_1 p_2 \exp(ik_+|\mathbf{r}-\mathbf{r}'|)}{|\mathbf{r}-\mathbf{r}'|}\,\mathrm{d}\mathbf{r}'. \tag{3.12}$$

Calculation of (3.12) includes integration over a beam cross-section s, to be followed by that over the x-coordinate along the beam axis:

$$p_k = \frac{\mathrm{i}\varepsilon\omega_+}{2\rho_0 c_0^3}\int_0^L u(x)\,\mathrm{d}x, \tag{3.13a}$$

where L is the interaction domain length and

$$u(x) = \frac{\mathrm{i}k_+}{2\pi}\int \frac{p_1 p_2 \exp[\mathrm{i}(k_1+k_2)|\mathbf{r}-\mathbf{r}'|]}{|\mathbf{r}-\mathbf{r}'|}\,\mathrm{d}s, \tag{3.13b}$$

($k_1 = \omega_1/c_0$, $k_2 = \omega_2/c_0$). When the amplitude and phase of the low-frequency signal do not change significantly over the cross-section s, the factor $p_2\exp(\mathrm{i}k_2|\mathbf{r}-\mathbf{r}'|)$ can be considered as depending only on the x-coordinate. In addition, taking into account that $k_+ \simeq k_1$ and that the reception point is located at $x = L$, we obtain

$$u(x) = p_2(x)\exp[\mathrm{i}k_2(L-x)]\cdot\left(\frac{\mathrm{i}k_+}{2\pi}\int \frac{p_1\exp(ik_1|\mathbf{r}-\mathbf{r}'|)}{|\mathbf{r}-\mathbf{r}'|}\,\mathrm{d}s\right).$$

A further simplification of (3.13) can be performed (Donskoy & Sutin, 1981). According to the Huygens–Fresnel principle, the integral (the Huygens–Rayleigh integral) in brackets, for any cross-section of a wave beam, is equal to the value $p_1(x = L)$ at the observation point, i.e.

$$u(x) = p_1(L)p_2(x)\exp[\mathrm{i}k_2(L-x)]. \tag{3.14}$$

Substituting (3.14) into (3.13a) yields an expression for the combination frequency wave amplitude,

$$p_k = \frac{i\varepsilon\omega_+ p_1(L)}{2\rho_0 c_0^2}\int_0^L p_2(x)\exp[\mathrm{i}k_2(L-x)]\,\mathrm{d}x. \tag{3.15}$$

Typically, a signal wave can be considered as plane so that its amplitude is constant within the interaction domain. As has been mentioned, the signal phase variation over a characteristic beam width d can be neglected, i.e. the condition $k_2\theta d \ll \pi$ is supposed to be fulfilled. Note that instead of d one can take the width of the first Fresnel zone for a pump beam, since the sources lying outside this zone are compensated by one another. Substituting $d_F = \sqrt{L\lambda}$ yields the condition

$$k_2\sqrt{2\pi L/k_1}\theta \ll \pi. \tag{3.16}$$

Then the signal wave amplitude can be displayed as $p_2 \exp(\mathrm{i}k_2 x \cos\theta)$ and, according to (3.15),

$$p_k = \frac{\varepsilon\omega_+ L p_1(L) p_2}{2\rho_0 c_0^3} D(\theta) \exp[\mathrm{i}(k_2 L - \delta)], \qquad (3.17)$$

where

$$D(\theta) = \frac{\sin\delta}{\delta}, \qquad \delta = k_2 L \sin^2\theta/2.$$

Formula (3.17) extends relation (3.7) for a narrow beam (as compared with the signal wavelength) to the case of an arbitrary width beam. The directivity angle coincides with that defined previously in (3.8). Therefore, for a parametric receiver as distinct from a transmitter, no significant difference is observed between regimes in which interaction occurs in the near- or far-field zones of the pumping transmitter.

Formula (3.17) can be extended to the case of an inhomogeneous medium, when reception is performed on a curved pumping beam and even on several beams (as in a layered waveguide, pumping can reach a reception point following different paths). In the latter case, the amplitude of the combination-frequency wave resulting from an interaction of a signal with a pumping wave is given by the relation

$$p_k = \sum_s \frac{\mathrm{i}\varepsilon\omega_+ p_1(L_s)}{2\rho_0 c_0^3} \int_0^{L_s} p_2(\ell_s) \exp\left[\mathrm{i}k_2(\psi(L_s) - \psi(\ell_s))\right] \mathrm{d}\ell_s. \quad (3.18)$$

Here $\psi(\ell_s)$ is an eikonal along the s-th ray, L_s being the length of the interaction path along the s-th ray (beam) and ℓ_s a coordinate along the ray.

Experimental investigations of parametric receivers have been carried out more than once, although not as extensively as for transmitters. We could refer to, as an example, reception directivity diagram measurements (Truchard, 1975). The distance between the transmitter and the receiver is 15 m, the pump frequency is 90 kHz, and the signal frequency varies within the range 3–6 kHz. Since, according to (3.7), the sensitivity of such a receiver grows with its base length, possibilities have been considered for increasing the former. A parametric receiver with base length equal to 340m was created (Goldsberg *et al.*, 1978). The signal transmitter operated at frequencies in the interval 40–80 Hz, with a pump frequency of 65 kHz and pump amplitude up to about 105 Pa m. Experiments were performed at a depth of 23 m. Possibilities of parametric reception on a superlong base up to 26 km were also considered (McDonough, 1975). Certainly, the relatively low efficiency of such systems makes one think of them not as rivals with linear antennae in their typical application field, but rather as a convenient means to analyse the spatial structure of an acoustic field. In an ocean waveguide, for instance, parametric reception is feasible on a base curved along a ray, which allows one to accomplish a rather sensitive selection of sound beams in an inhomogeneous ocean (Donskoy *et al.*, 1983).

Nonlinear acoustic tomography

The methods of nonlinear acoustic tomography can also be referred to, in the general sense, as "parametric reception". The term "tomography" (a well-known one in modern medical X-ray diagnostics, ocean acoustics, etc.) designates image recovery, or, in general, a recovery of the spatial distribution of any parameter, using the output data obtained from the intersections of the domains under investigation by a probing wave field (e.g. rays) at various levels and/or directions. In the case considered here, we speak of the recovery of the space distribution of the nonlinearity parameter ε (or of $B/A = 2(\varepsilon - 1)$; see Chapter 1) which, generally speaking, is associated with the temperature or other important properties of a medium. Such methods are of special interest for cases where there are no significant variations of the linear acoustic impedance – which renders standard methods of ultrasonic echo-sounding (tomographic versions included) inapplicable.

A method advocated by Japanese scientists (see Ichida *et al.*, 1983; Sato *et al.*, 1985; and others) is based on the interaction of a weak probing high-frequency signal with a relatively strong low-frequency pump (continuous or, more frequently, of a "video-pulse" type) which modulates the signal phase such that, as above, the phase shift $\Delta\varphi$ is proportional to the nonlinearity parameter. Different schemes for the realization of such a method were considered where the pumping and the signal propagate collinearly, or perpendicularly, or oppose each other. Let us briefly discuss the case where a continuous probing signal of frequency ω and a pumping pulse propagate in opposing directions (along the x-axis). The phase shift of the signal is described by an expression similar to (3.4). Let the pumping pulse enter the system at point $x = 0$ and propagate in the positive direction. This changes the sound speed to $\Delta c = \Delta c(t - x/c_0)$. The signal portion which at $t = 0$ was situated at the point $x = x_0$ will, at time t, reach the point $x = x_0 - c_0 t$, interacting, therefore, with that portion of the pump pulse where $\Delta c = \Delta c[(x_0 - 2x)/c_0]$. Consequently, the total phase shift of this portion of the signal, which has run in the negative direction to the point $x = 0$ (the system end) from the point $x = L$ (the system origin) amounts to (cf. (3.4))

$$\Delta\varphi(x_0) = -\frac{\omega}{c_0^2}\int_0^L \Delta c\left(\frac{x_0 - 2x}{c_0}\right)\,dx. \tag{3.19}$$

But the quantity Δc can be readily related to the pump pulse pressure via

$$\Delta c = \frac{B/A}{2\rho_0 c_0} p_p', \tag{3.20}$$

and if $p_p' = p_p f(t - x/c_0)$, where p_p is the amplitude and f is a known dimensionless function defining the pulse shape, then (3.19) will furnish

$$\Delta\varphi(x_0) = -p_p\omega\int_0^L N(x) f\left(\frac{x_0 - 2x}{c_0}\right)\,dx, \tag{3.21}$$

where

$$N(x) = \frac{B/A}{2\rho_0 c_0^3} \tag{3.22}$$

is the phase shift parameter, determined by the distribution of B/A along the x-axis which we seek to define.

Introducing a new function g, instead of f, according to the equation $f(z) = g(c_0 z/2)$, puts (3.21) into the form

$$\Delta\varphi(x_0) = -p_p\omega\int_0^L N(x) g\left(\frac{x_0}{2} - x\right)\,dx. \tag{3.23}$$

Note that the phase shift observed at the system output at time $t = x_0/c_0$ corresponds to the value of N at point $x = x_0/2$.

Since the function g is known beforehand and $\Delta\varphi$ is measured, function $N(x)$ can be determined by a deconvolution operation, namely

$$N(x) = -\frac{1}{p_p\omega}F^{-1}\left[\frac{F(\Delta\varphi(x))}{F(g(x))}\right], \tag{3.24}$$

where F and F^{-1} stand for the direct and inverse Fourier transforms, respectively.

Thus the problem has been solved of the recovery of the nonlinear parameter distribution along the probing signal propagation direction. Scanning the probing ray and the pump mechanically or electronically, in a transverse direction y one can recover a two-dimensional distribution $N(x, y)$.

Such a system, together with the others, was realized by Sato *et al.* (1985). The probing signal frequency was 5 MHz, whereas the pump appeared as a pulse in the form of one period of a sinusoid with frequency 500 kHz. The signal and pump powers were equal to 0.01 mW cm^{-2} and 10 mW cm^{-2}, respectively. The imaged surface size was 20×20 cm^2 with a resolution of 2 mm. Apart from specially fabricated models, an image of the cross-section of a rodent *in vivo* was obtained.

An attractive aspect of such methods, in our opinion, is the diversity of their realizations which offers, it seems, some new possibilities for nonlinear acoustical diagnostics. At the same time it should be admitted that no real practical application of nonlinear tomographic methods has been reported as yet.

References

Andebura, V.A., Donskoy, D.M., Naugol'nykh, K.A., Stepanov, Y.S., & Sutin, A.M. (1990). Sound field of a powerful transmitter in a parametric regime, *Akust. Zh.* **36,** 548–50.

Berktay, H.O. (1965). Possible exploitation of nonlinear acoustics in underwater transmitting applications, *J. Sound Vib.* **2,** 435–61.

Berktay, H.O. & Leahy, D.J. (1974). Farfield performance of parametric transmitters, *JASA* **55,** 539–46.

Burov, A.K., Krasil'nikov, V.A. & Tagunov, E.Y. (1978). Experimental investigation of collinear interaction of weak ultrasonic signal with intense low-frequency perturbations in water, *Vestn. MGU Ser. 3 Fiz. Astr.* **19,** 53–8.

Donskoy, D.M. & Sutin, A.M. (1981). Parametric reception of acoustic signals in nonhomogeneous media, *Akust. Zh.* **27,** 876–80.

Donskoy, D.M., Zverev, V.A. & Kalachev, A.I. (1983). Parametric selection of sound beams in inhomogeneous media, *Akust. Zh.* **29,** 181–6.

Dunina, T.A., Esipov, I.B. & Kozyaev, E.F. (1976). On the theory of parametric radiation, *Proc. VI ISNA, Moscow MGU* **1,** 296–307.

Dybedal, J. (1993). Topas: Parametric end-fire array used in offshore applications, *Proc. XIII ISNA,* ed. H. Hobaek, Bergen, 264–69.

Gol'dberg, Z.A. (1972). Efficiency estimation for parametric amplification of acoustic waves in liquid, *Physics, Proc. Kazakh. Univ., Alma-Ata* **3,** 38–43.

Goldsberg, T.G., Reeves, C.R. & Rohde, D.F. (1978). Experimental measurements with a large aperture PARRAY, *JASA Suppl.* **64,** 125.

Gurbatov, S.N. (1980). On interaction of waves in media with high-frequency decay, *Akust. Zh.* **26,** 467–9.

Gurbatov, S.N., Malakhov, A.N. (1977). On statistical characteristics of random quasimonochromatic waves in nonlinear media, *Akust. Zh.* **29,** 569–75.

Gurbatov, S.N., Demin, I.Yu. & Sutin, A.M. (1979). Interaction of nonlinearly bounded beams in parametric transmitters, *Akust. Zh.* **25,** 515–20.

Gurbatov, S.N., Saichev, A.I. & Yakushkin, I.G. (1983). Nonlinear waves and one-dimensional turbulence in non-dispersive media, *UFN.* **141 (2),** 221–25.

Hobaek, H. & Vestrheim, M. (1976). Properties of the parametric acoustic array in different parametric regimes, *Proc. VI ISNA, Moscow, MGU* **1,** 308–19.
Ichida, N., Sato, T. & Linzer, M. (1983). Imaging of the nonlinear ultrasonic parametric of a medium, *Ultrasonic Imaging* **5,** 295–9.
McDonough, R.N. (1975). Long-aperture parametric receiving arrays, *JASA* **57,** 1150–5.
Moffett, M.B., Konrad, W.L. & Carlton, L.F. (1978). Experimental demonstration of the absorption of sound by sound in water, *JASA* **63,** 1048–51.
Muir, T.G., Thomson, L.A., Cox, L. & Frey, H.G. (1980). A low frequency parametric research tool for ocean acoustics, in *Bottom-interacting ocean acoustics,* ed. W.A. Kuperman & F.B. Jensen (Plenum Press, New York).
Naugol'nykh, K.A. & Ostrovsky, L.A. (1982). On nonlinear effects in ocean acoustics, in *Akust. Okeana. Sov. Sostoyanie.,* ed. L. Brekhovskih & I. Andreeva (Nauka, Moscow).
Naugol'nykh, K.A., Soluyan, S.I. & Khokhlov, R.V. (1963). On nonlinear interaction of sound waves in an absorbing medium, *Akust. Zh.* **9,** 192–7.
Novikov, B.K., & Timoshenko, V.I. (1980). *Parametric arrays in hydrolocation* (L. Sudostroenie, Leningrad).
Ostrovsky, L.A. (1963). To the theory of waves in non-stationary compressible media, *Prikladnaya mathematka i mechanika* **27,** 924–9.
Sato, T., Fukusima, A., Ichida, N., Ishikava, H., Miwa, H., Igarashi, Y., Shimura, T. & Muzakami, K. (1985). Nonlinear parameter tomography system using counterpropagating probe and pump waves, *Ultrasonic Imaging* **7,** 49–59.
Sutin, A.M. (1978). On a limiting regime of parametric array action, *Akust. Zh.* **24 (1),** 104–7.
Tagunov, E.Y. (1981). Investigation of nonlinear interactions of weak ultrasonic signals with powerful low-frequency acoustic perturbations, *Moscow MGU.,Thesis* **14,** .
Truchard, J.J. (1975). Parametric acoustic receiving array, *Experiment. JASA* **58,** 1146–50.
Westervelt, P.J. (1963). Parametric acoustic array, *JASA* **35,** 535–7.
Willette, J.G., Moffett, M.B. & Konrad, W.L. (1976). Difference-frequency harmonics from saturation-limited parametric acoustic sources, *Proc. VI ISNA, Moscow, MGU* **1,** 272–9.
Zverev, V.A. & Kalachev, A.I. (1970). Modulation of sound by sound under acoustic wave intersection, *Akust. Zh.* **16,** 245–52.

6

Nonlinear acoustic waves in dispersive media

6.1 Media with selective absorption

Sound speed dispersion may prevent the cumulative energy transfer to higher harmonics or other frequency components that arises in the case of intense wave propagation unless the special resonance (synchronism) conditions relating the wave frequencies and the wave numbers are satisfied. Thus, the velocity of perturbation at the sum or the difference frequency emerging due to interaction of two waves should coincide with the velocity of one of the system's free waves; then the latter is excited in a resonant way. In other words, for three resonantly interacting waves propagating collinearly the conditions (1.1) of Chapter 5 are to be satisfied:

$$\omega_1 \pm \omega_2 = \omega_3, \quad k_1 \pm k_2 = k_3. \tag{1.1}$$

For a nondispersive medium the first of these relations automatically implies the second one. Because of this, the initial signal energy spreads over a wide frequency band.

In dispersive media the resonance conditions can be satisfied only for selected triads, but the efficiency of interaction between them will be higher. Such effects have long been employed, for example in radiophysics and optics.

The sound dispersion mechanisms are quite diverse. First there is relaxation which is, however, directly associated with losses (see Chapter 1). Second, as was also shown in Chapter 1, media with internal temporal or spatial scales are typically characterized by a pronounced dispersion. Finally, the same effects are caused by "geometrical" dispersion typical, in particular, of bounded systems such as waveguides and resonators. A selective suppression of individual components in the wave spectrum is also possible.

In this and in the next chapter we shall consider various nonlinear acoustic effects arising in dispersive media; along with the models considered in Chapter 1 we shall introduce some new models of dispersive media in acoustics.

Let us begin by considering a rather abstract situation in which one allows the interaction of only a finite number of spectral components and, on the other hand, in which at certain frequencies an increased selective absorption is observed. Let, for example, a periodic wave be described by the nonlinear integro-differential equation (Rudenko, 1983)

$$\frac{\partial v}{\partial x} - \alpha v \frac{\partial v}{\partial y} = -\sum_{n=1}^{\infty} D_n A_n(x) \sin n\omega y. \qquad (1.2)$$

Here

$$A_n = \frac{2}{\pi} \int_0^{\pi} v(x, y') \sin n\omega y' \mathrm{d}(\omega y')$$

are the amplitudes of the appropriate harmonics; D_n are their decay coefficients. The values of D_n depend on the dissipation mechanism. Then, it is obvious that for the Burgers equation when the right-hand side of (1.2) is equal to $b\frac{\partial^2 v}{\partial y^2}$

$$D_n \sim n^2.$$

If the decay can be neglected (all $D_n \to 0$) then the solution of (1.2) will evidently appear as a Riemann simple wave. Of interest here are the cases when $D_n \to \infty$ for a certain $n = k$; then the corresponding harmonics in the wave spectrum are suppressed. Indeed, let us represent the field as a Fourier series with equal phases of the harmonics:

$$v = \sum_{n=1}^{\infty} A_n(x) \sin n\omega y. \qquad (1.3)$$

Then (1.2) can be written as a set of equations in ordinary derivatives,

$$\frac{\mathrm{d}A_n}{\mathrm{d}x} + D_n A_n = S_n, \qquad (1.4)$$

where

$$S_n = \frac{2}{\pi} \int_0^{\pi} \alpha v \frac{\partial v}{\partial y'} \sin n\omega y' \, \mathrm{d}(\omega y')$$

is the Fourier component of a nonlinear term containing, generally speaking, pairwise products of all A_n.

Therefore it follows, in particular, that provided some D_n are large enough, then within the relaxation period, of order of D_n^{-1}, the corresponding value of A_n will amount to

$$A_n = S_n/D_n \to 0,$$

showing, as one could anticipate, that rapidly decaying harmonics "drop out" of the interaction. Consider now several important particular instances.

Resonant frequency doubling and parametric amplification

First we discuss the interaction of two synchronously propagating harmonics at frequencies ω and 2ω. Substituting the sum (1.3) into (1.2), disregarding the losses at these frequencies and assuming that $D_n \to \infty$ for $n \geq 3$, we obtain for A_1 and A_2 the equations

$$\begin{aligned} \frac{\mathrm{d}A_1}{\mathrm{d}x} &= \frac{\alpha\omega}{2} A_1 A_2, \\ \frac{\mathrm{d}A_2}{\mathrm{d}x} &= -\frac{\alpha\omega}{2} A_1^2. \end{aligned} \tag{1.5}$$

These yield in particular the energy integral

$$A_1^2 + A_2^2 = A_0^2 = \text{constant}, \tag{1.6}$$

and after that the solution of (1.5) will take the form

$$\frac{A_2}{A_0} = \frac{A_{20}\cosh\beta x - A_0\sinh\beta x}{A_0\cosh\beta x - A_{20}\sinh\beta x}, \tag{1.7}$$

where $A_{20} = A_2(0)$, $\beta = \alpha\omega A_{20}/2$.

It is obvious that as $x \to \infty$ we always have $A_1 \to 0$, $A_2 \to -A_0$, i.e. for the properly phased components the energy is completely transferred into the second harmonic (which, of course, would be impossible in a medium without dispersion and losses).

Of particular interest is the case when the initial dominance is that of the second harmonic, i.e. $A_1(0) \ll A_2(0)$. Then in the initial stage the quantity $A_2 = A_{20}$ can be assumed constant, and (1.5) or (1.7) yields

$$A_1 = A_{10}\exp(\hat{\beta}x), \tag{1.8}$$

where $\hat{\beta} = \alpha\omega A_{20}/2$. This is the standard process of parametric amplification of the subharmonic field (at frequency ω) by a "pump" wave (with frequency 2ω). If $A_{20} > 0$ then A_1 will grow exponentially (while for $A_{20} < 0$ it will decay, i.e. the process depends on the "pumping" phases). At later stages of the process A_2 decreases in absolute value, changes sign and approaches $-A_0$ while A_1 attains its maximum value and then eventually vanishes.

These results, first obtained by Khokhlov (1961), played an important role in the development of nonlinear optics. Khokhlov (1961) and Bloembergen (1965) also approached the more complicated problem of two- and three-wave interaction, under deviations from exact synchronism and with arbitrary phase detuning at $x = 0$ between the waves of different frequencies.

In this case full energy transfer into the second harmonic is impossible, and partial periodic energy exchange between the harmonics always occurs (see also Sections 2 and 3 of this chapter).

Media with finite dissipation

Perfect blocking of all unwanted harmonics is usually a difficult matter, although the introduction of an increased selective absorption even at one or two frequencies will distinctly improve the energy exchange conditions as between the lower spectral components (Woodsum, 1981; Moffett & Mellen, 1981).

In order to analyse the possibility of controlling the energy transfer between the harmonics by means of removing selected spectral components from the wave field, a phenomenological integrodifferential equation of type (1.2) was investigated (Rudenko, 1983) retaining also the viscous "Burgers" term:

$$\frac{\partial u}{\partial \sigma} - u\frac{\partial u}{\partial \theta} - Re^{-1}\frac{\partial^2 u}{\partial \theta^2} = D_3 A_3(\sigma)\sin 3\theta - D_4 A_4(\sigma)\sin 4\theta. \qquad (1.9)$$

Here $u = v/v_0$, $\theta = \omega y$, $\sigma = x/L_N = \varepsilon\omega v_0 x/c_0^2$, $Re = 2\varepsilon v_0 \rho_0 c_0/\omega b$, v_0 is the amplitude of the wave velocity, ω its frequency, while the $A_n(\sigma)$ are the amplitudes of these harmonics derived from (1.2) for $n = 3, 4$.

Consider a process of second harmonic generation under the propagation of an initially harmonic wave of frequency ω, so that $u(0,\theta) = \sin\theta$. In Figure 6.1 the range dependences of the first harmonic amplitude (broken line) and the second (solid lines), for $Re^{-1} = 0$ and for different values of D_1, deduced by numerical integration of equation (1.9), are depicted. Here curve 1 corresponds to a weak selective absorption ($D_3 = D_4 = 2$); curve 2 embodies a weak absorption of the third harmonic ($D_3 = 2$) and a strong one of the fourth harmonic ($D_4 = 20$) while curve 3 corresponds to the reverse case ($D_3 = 20$, $D_4 = 2$); and, finally, curve 4 demonstrates the case of strong absorption of both the third and fourth harmonics ($D_3 = D_4 = 20$).

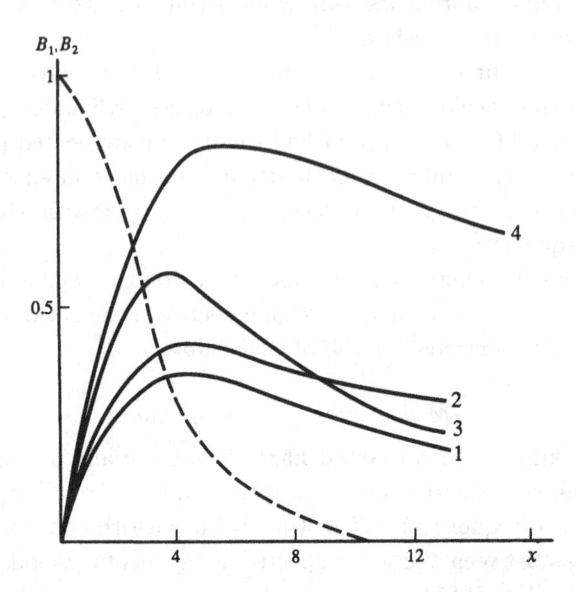

Figure 6.1. The range dependence of the first harmonic amplitude (dashed line) and of the second (solid curves) for different values of the selective absorption coefficient D_n. Curve 1, weak selective absorption ($D_3 = D_4 = 2$); curve 2, weak absorption of the third harmonic ($D_3 = 2$) and strong absorption of the fourth harmonic ($D_4 = 20$); curve 3, strong absorption at the third harmonic ($D_3 = 20$) and weak absorption at the fourth ($D_4 = 2$); curve 4, strong absorption at the third and fourth harmonics ($D_3 = D_4 = 20$). (Rudenko, 1963.)

As is seen from Figure 6.1, the second harmonic variation turns out to be different for the various selective absorption regimes. When absorption of the third harmonic dominates (curve 3) a more rapid growth of A_2 is observed in which A_2 attains large values as compared with those in the case when absorption is significant at the fourth harmonic (curve 2). On the other hand, after that A_2 decreases with distance more slowly in the case of selective absorption of the fourth harmonic. Generally speaking, as one might expect, the influence of the third harmonic on the process proves more effective than that of the fourth.

As Re^{-1} increases, which corresponds to a decrease in pumping amplitude or dissipation growth, the second harmonic generation intensity will certainly diminish.

Andreev *et al.* (1985) performed the numerical integration of equation (1.9) subject to another boundary condition,

$$u(0, \theta) = m \sin \theta - \sin 2\theta,$$

where $m \ll 1$ is the amplitude of the weak signal amplified parametrically by the pump at frequency 2ω. The result indicates that, in this case also, the introduction of selective absorption at the third and fourth harmonics intensifies the energy exchange between the low-frequency spectrum components, and increases the parametric amplification coefficient up to values ranging from 10^2 to 10^3.

Thus, selective absorption allows one, in principle, to improve the efficiency of parametric sound amplification. It should be noted that in a nondispersive (and thus nonselective) medium the parametric amplification coefficient cannot rise far above unity, even if there were perfect synchronism (Gol'dberg, 1972; Rudenko & Soluyan, 1975). Such a selection can be accomplished, for example, in a plane resonator (acoustic interferometer) that has a wall in the form of a slab of finite thickness L whose acoustic impedance differs from that of the environment. At its own resonance frequencies the plate would not reflect the waves. It is this fact that can be used in order to avoid energy transfer into the unwanted harmonics (Zarembo & Serdobol'skaya, 1974). Exciting the resonator at frequency $\omega_m = \pi c_0/mL$, we observe that at the m-th harmonic and at the higher ones of frequency ω_n, the resonator quality Q is low (it is "open"), whereas at the basic frequency and its harmonics with numbers less than m the value of Q remains high because the slab is equivalent to a rigid wall and the frequency spectrum of such a resonator remains equidistant.

To conclude this section, we mention the problem of mode interaction in resonators for the case when the number of modes may exceed three. Such a problem has been considered by Gol'dberg (1983) for a plane resonator with rigid walls excited by harmonic oscillations of one of the walls at frequency ω. The result depends strictly on the number of interacting modes. When only three modes at frequencies ω, $\omega_1 < \omega$, and $\omega_2 = \omega - \omega_1$ are present, the parametric generation of oscillations corresponding to ω_1 and ω_2 occurs after the ω-mode (pump) amplitude exceeds a threshold value A_1. It may be shown that A_1 is a minimum when ω_1 coincides with the lowest eigenfrequency of the resonator (i.e. ω_1 and ω_2 are maximally separated); see also the next section. For the four-mode interaction, however, when the mode at frequency $\omega_4 = \omega + \omega_1$ is also present from the very beginning the result is different: the lowest threshold will be achieved for the subharmonic value $\omega_1 = \omega_2 = \omega/2$.

The three-mode case will be considered below in more detail in applications to waveguide resonators.

6.2 Nonlinear waves in a waveguide

In addition to the models of dispersive media considered in the first chapter, another quite simple and general class of systems is widely known, namely that of waveguides, in which dispersion is caused by the presence of boundaries (or, in the more general case, by smooth inhomogeneities) in a nondispersive medium. Then excitation of various waveguide modes allows us to perform a resonant energy transfer among waves of different frequencies.

Before writing the basic equations of "waveguide nonlinear acoustics" let us clarify, using elementary examples, the nature of the mode synchronism. Consider a re-reflection of a plane wave between rigid boundaries (the Brillouin concept of waveguide modes). On reflection from such a boundary the wave phase does not change, and propagation along a broken line between the boundaries takes place in a way very similar to that along an appropriate straight path, the only difference being that the effective velocity of the field energy propagation along the waveguide axis x (the group velocity) turns out to be smaller than the sound velocity c_0. In the case of a finite amplitude wave the harmonic generation, discontinuity formation, etc., are essentially the same as in a plane wave (the interaction of noncollinearly propagating waves, i.e. the incident and reflected waves, proves insignificant within the "quadratic" approximation). In this sense a waveguide with solid walls has no meaningful distinction from free space.

Now let one of the walls (say, $z = 0$) be free, i.e. the condition $p'(z = 0) = 0$ holds. Then upon reflection the pressure phase is changed by π. Evidently, in this case rarefaction portions in a wave are replaced by those of compression, and vice versa. This testifies to the fact that nonlinear distortions do not accumulate when propagating along a broken line, i.e. over a cycle the reflected wave profile (ideally) re-acquires its initial shape (if no discontinuity was formed in it), or, in spectral terms, the second harmonic phase with respect to the first changes its sign on reflection. For such a process, wave energy transfer to higher harmonics cannot be cumulative.

Nevertheless, a synchronous interaction of two waves can be realized again provided that they propagate at different angles with respect to the waveguide axis such that the difference in the path length is compensated for by a different number of reflections from the walls, each reflection producing a phase shift by π.

Consider now the details of the combination-component generation process for sound propagation in a homogeneous waveguide, proceeding from the nonlinear wave equation in the form ((1.13), Chapter 1)

$$\nabla^2\varphi - \frac{1}{c^2}\varphi_{tt} = \frac{1}{c^2}\frac{\partial}{\partial t}\left[(\nabla\varphi)^2 + \frac{\gamma-1}{2c^2}(\varphi_t)^2\right]. \tag{2.1}$$

Here c is assumed constant.

In order to demonstrate the role of dispersion, let us confine our attention to second harmonic generation alone (Keller & Miltman, 1971; Hamilton & Ten Cate, 1987). To the first approximation, let us formulate the field as a single mode:

$$\varphi_0 = Af_m(z)\exp[\mathrm{i}(\omega t - kx)] + c.c., \tag{2.2}$$

where x and z are longitudinal and transverse coordinates, respectively (for the present the waveguide is assumed two-dimensional; the case of a three-dimensional waveguide, e.g. a tube or a channel, is essentially similar and will be discussed in what follows). The functions f_m, where m is the mode number, are orthogonal eigenfunctions of the equation

$$f_{mzz} + \mathcal{X}_m^2 f_m = 0 \tag{2.3a}$$

with the appropriate boundary conditions imposed on the waveguide walls. The value of $\mathcal{X}_m$, the transverse wave number, is related to the longitudinal wave number through the equation

$$\mathcal{X}_m^2 + k_m^2 = \omega^2/c^2 = k_0^2. \tag{2.3b}$$

Let us write the solution in the form $\varphi = \varphi_0 + \varphi'$, making the approximation $\varphi = \varphi_0$ in the nonlinear terms; thus

$$\nabla^2\varphi' - \frac{1}{c^2}\varphi'_{tt} = \frac{1}{c^2}\frac{\partial}{\partial t}\left[(\nabla\varphi_0)^2 + \frac{\gamma-1}{2c^2}\varphi_{0t}^2\right]. \tag{2.4}$$

Substituting the first-approximation expression (2.2) into (2.4) yields

$$\nabla^2\varphi' - \frac{1}{c^2}\varphi'_{tt} = \frac{\mathrm{i}\omega}{c^2}A^2\left[f_{mz}^2 - f_m^2\left(k_m^2 + \frac{\gamma-1}{2}\frac{\omega^2}{c^2}\right)\right]\exp\left[2\mathrm{i}(\omega t - k_m x)\right]. \tag{2.5}$$

Expanding φ' in the waveguide natural modes we have

$$\varphi' = \sum_{n=1}^{\infty} f_n(z)\exp(2\mathrm{i}\omega t)[a_n\exp(-2\mathrm{i}k_m x) + b_n\exp(-\mathrm{i}k_n x)]. \tag{2.6}$$

The first term in square brackets corresponds to a forced solution, the second to a free solution in which k_2 is related to 2ω via the dispersion equation (2.3b).

Denoting the right-hand side of (2.5) as $\mathrm{i}S(z)\exp[2\mathrm{i}(\omega t - k_m x)]$, multiplying by f_m, integrating over the waveguide height $(0, H)$ and taking the orthogonality condition

$$\frac{1}{H}\int_0^H f_m f_n \,\mathrm{d}z = \delta_{mn}$$

into account, we obtain for the forced part the solution

$$a_n = \frac{\mathrm{i}H^{-1}}{4\mathcal{X}_m^2 - \mathcal{X}_n^2}\int_0^H s(z) f_n(z)\,\mathrm{d}z. \tag{2.7}$$

If a boundary problem is being solved in which a harmonic field is specified at the waveguide input: $\varphi(x=0) = \varphi_0$, $\varphi'(x=0) = 0$, then $b_n = -a_n$, and the perturbation will appear as

$$\varphi' = \exp(2\mathrm{i}\omega t)\sum_{n=1}^{\infty} a_n f_n(z)\left(\exp(-2\mathrm{i}k_m x) - \exp(-\mathrm{i}k_n x)\right), \tag{2.8}$$

where a_n is determined from (2.7). Hence, it follows that the second-harmonic perturbation is a multi-mode one and the amplitude variation of each mode (the m-th one included) along the x-axis has the form of beats with period $\Lambda_{mn} = 2\pi/|2k_m - k_n|$. The entire pattern of the spatial field distribution at frequency 2ω can of course be complicated.

Special attention should be devoted to the case of synchronism, when $k_n(2\omega) = 2k_m(\omega)$ and thus $\mathcal{X}_n^2 = 4\mathcal{X}_m^2$. In this situation the forced solution (2.7) diverges, whereas the full solution (2.8) remains finite and grows in space. Indeed, let us set $k_n = 2k_m + \Delta$ where $\Delta \to 0$. Then $4\mathcal{X}_m^2 - \mathcal{X}_n^2 = 4k_m$, and expanding the exponent differences in (2.8) we get

$$\varphi_n' = -\frac{f_n(z)x}{4k_m}\tilde{a}_m, \quad \tilde{a}_m = \int_0^H s(z) f_m(z)\,\mathrm{d}z. \tag{2.9}$$

Such resonance cases are of special interest. Whether they arise or not depends on the wave mode dispersion properties. Let us restrict ourselves here to two examples (Ostrovsky & Papilova, 1973).

(a) A waveguide with rigid walls. In this case $\partial\varphi/\partial z(0, H) = 0$ and, obviously, $f_n = \sqrt{2/H}\cos(n\pi z/H)$. Therefore, $\mathcal{X}_n = n\pi/H$ and, as can readily be observed, synchronism (i.e. satisfaction of the condition $k_n(2\omega) = 2k_m(\omega)$) is realized for $n = 2m$ where m is an arbitrary integer. This is quite natural: as was shown earlier in this section, on re-reflection from a rigid wall nonlinear distortions are accumulated like those in a homogeneous medium, and according to the Brillouin concept an m-th mode at frequency ω and an n-th one at frequency 2ω propagate at the same angle with respect to the x-axis.

(b) A waveguide with free walls. Here $\varphi(0, H) = 0$ and $f_n = \sqrt{2/H}\sin(n\pi z/H)$. In this instance the condition $n = 2m$ provides synchronism again. However, an elementary calculation of the integral in (2.9) indicates that here $\tilde{a}_n = 0$: although a "nonlinear" force in (2.5) is in resonance with the perturbation its total effect (over the vertical plane) is zero because of its spatial orthogonality to the field of an m-th mode. This is also clear from what was said above: owing to the phase shift of π on reflection, there is no real synchronism of the basic frequency wave and its second harmonic along the x-axis. This prevents the upward transfer of energy (and, hence, discontinuity formation).

A similar discussion can be given for three-dimensional waveguides (pipes). For a waveguide with a rectangular cross-section relations analogous to (2.2) and (2.3) hold true as well, though this time in (2.3*b*) we have

$$\mathcal{X}^2 = \left(\frac{m\pi}{a}\right)^2 + \left(\frac{n\pi}{b}\right)^2,$$

where a and b are the waveguide's side dimensions and m, n are integers. Thus for the second harmonic (and subharmonic) generation the resonance conditions are

$$\frac{m_2^2}{a^2} + \frac{n_2^2}{b^2} = 4\left(\frac{m_1^2}{a^2} + \frac{n_1^2}{b^2}\right), \tag{2.10}$$

where m_1, n_1 are the mode numbers at the fundamental frequency and m_2 and n_2 are those at the second harmonic.

These requirements are met, for example, for any mode pair with $m_2 = 2m_1$, $n_2 = 2n_1$. As above, in a waveguide with rigid walls the resonance conditions are satisfied for all higher harmonics of a signal, so that eventually a discontinuity emerges, whereas in a waveguide with free boundaries no interaction occurs for such mode pairs. However (as distinct from the two-dimensional case), other possibilities now exist for a free-wall waveguide as well. Thus, condition (2.10) is satisfied provided that $m_1 = n_1 = 1$, $m_2 = 1$, $n_2 = 3$ and the waveguide dimensions are such that $b^2/a^2 = 5/3$. In these circumstances the phase change on reflection from the boundaries is compensated for by a difference in the number of reflections for the two modes. Therefore, a resonant energy transfer upwards into selected frequencies is possible without discontinuity formation (the third harmonic is no longer in synchronism with the first one). It should be noted that here the possibility still exists for some higher mode generation at the second harmonic frequency (in the case considered it may be, say, a mode with $m_2 = 2$, $n_2 = 6$).

One more version can be considered – a waveguide in the form of an open rectangular channel with rigid walls and bottom, containing water. This gives instead of (2.10) the formula

$$\frac{m_2^2}{a^2} + \frac{(2n_2+1)^2}{4b^2} = 4\left[\frac{m_1^2}{a^2} + \frac{(2n_1+1)^2}{4b^2}\right]$$

(here n and m are the "vertical" and "horizontal" mode numbers, respectively). This condition is satisfied, for example, if $b^2/a^2 = 3/4$ and $m_1 = 2$, $n_1 = 0$, $m_2 = 1$, $n_2 = 0$.

In analogy with this, one can analyse more intricate situations when the interaction of two waves of different frequencies gives rise to the appearance of signals at sum and difference frequencies.

Back in the 1970s and 1980s a series of experiments was performed on mode interaction in waveguides. Let us quote the result of an investigation (Hamilton & Ten Cate, 1987) of the nonresonant interaction of acoustic waves in an air-filled waveguide having cross-section $(a \times b)$ $= 7 \times 3.8\,\mathrm{cm}^2$. Two primary modes were excited in it – one being a purely longitudinal mode ($m_1 = n_1 = 0$) at frequency $f_1 = 165\,\mathrm{Hz}$, the other corresponding to a mode with $m_2 = 1$, $n_2 = 0$ at frequency $f_2 = 3200\,\mathrm{Hz}$. The pressure amplitude levels in these modes amounted to $p_1 = 129.8\,\mathrm{dB}$, $p_2 = 120.7\,\mathrm{dB}$, respectively (i.e. $p_1 = 60\,\mathrm{Pa}$, $p_2 = 20\,\mathrm{Pa}$). Figure 6.2 depicts the wave amplitude variations along the waveguides at the sum ($f = 3365$ Hz) and difference ($f = 3035$ Hz) frequencies. The solid lines represent the results obtained using formulae similar to (2.8) but allowing for a slight attenuation of the waves. Fair agreement between the theory and experiment is observed. Note that the spatial oscillation period at the difference frequency is less than that at the sum frequency (due to a stronger dispersion influence at lower frequencies which are closer to a critical frequency of the mode $\omega_{cr} = c\mathcal{X}$).

6.3 Parametric amplification and generation of sound

The possibility of "sound generation by sound", i.e. direct transformation of the acoustic oscillation frequency without electromagnetic field participation, is of considerable interest. In many aspects, as was stated previously, such a process for selected frequencies cannot be effective in the absence of dispersion. That is why, while in microwave radio-engineering and optics, parametric amplifiers and generators have been

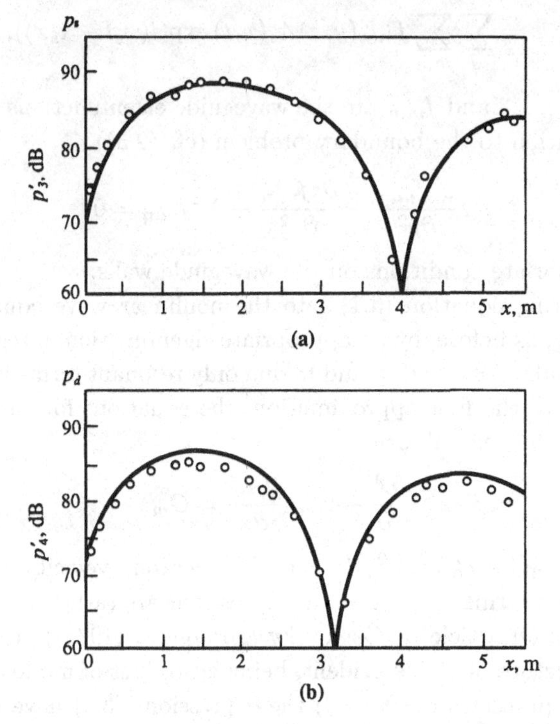

Figure 6.2. The variation in wave amplitude along the waveguide for (a) the sum frequency wave and (b) the difference frequency wave. Solid curve, theory; open circles, experiment (Hamilton & Ten Cate, 1987).

constructed and used for a long time, frequency transformation in acoustics has been performed, as a rule, electroacoustically. Of course, we have just considered parametric antennae in Chapter 5 (as we have mentioned already, the term "parametric" for them does not strictly correspond to its traditional meaning), but they have an extremely low transformation efficiency. As will be seen below, substantial parametric amplification and effective parametric sound generation *are* possible in dispersive systems.

Travelling waves in waveguides

Let us first consider resonant parametric interactions in a waveguide of rectangular cross-section. The x-axis is directed along that of the waveguide, the y- and z-axes in the cross-sectional plane, with dimensions $a \times b$. A solution in the form of three interacting modes at frequencies

$\omega_{1,2,3}$ will be sought:

$$\varphi = \sum_s \sum_{m,n} f_{m,n}(y,z) A_s(x,t) \exp(i(\omega_s t - k_s x)), \tag{3.1}$$

where $s = 1, 2, 3$ and $f_{m,n}$ are the waveguide eigenfunctions determined by the solution to the boundary problem (cf. (2.3))

$$\frac{\partial^2 f_{m,n}}{\partial y^2} + \frac{\partial^2 f_{m,n}}{\partial z^2} + \mathcal{X}_s^2 f_{m,n} = 0 \tag{3.2}$$

with appropriate conditions on the waveguide walls.

Substituting equation (3.1) into the nonlinear wave equation (2.1), multiplying, as before, by an appropriate eigenfunction, integrating over the waveguide cross-section and taking only resonant terms into account we obtain, to the first approximation, the equations for the mode amplitudes in the general form

$$\frac{\partial A_m}{\partial t} + v_{\text{gr}} \frac{\partial A_m}{\partial x} = G_m. \tag{3.3}$$

Here $v_{\text{gr}} = c_0(1 - c_0^2 \mathcal{X}^2/\omega^2)^{1/2}$ is the mode group velocity, and the G_m are nonlinear terms whose general expressions are rather unwieldy (they are given in an article by Ostrovsky & Papilova (1973)). However, the overall structure of G_m is evident, being a product of mode amplitudes. For steady processes ($\partial/\partial t = 0$) the expressions (3.3) have the form of the triplet equations

$$\begin{aligned} \mathrm{i}\frac{\partial A_1}{\partial x} &= -B_1 A_2^* A_3, \\ \mathrm{i}\frac{\partial A_2}{\partial x} &= -B_2 A_1^* A_3, \\ \mathrm{i}\frac{\partial A_3}{\partial x} &= B_3 A_1 A_2, \end{aligned} \tag{3.4}$$

where B_1, B_2, B_3 can be considered as real.

As is well known, this system has exact solutions expressed via the Jacobi elliptic functions (Bloembergen, 1965) and satisfying the following simple equalities (a particular case of the familiar Manley–Rowe relations):

$$\frac{|A_1|^2 - A_{10}^2}{B_1} = \frac{|A_2|^2 - A_{20}^2}{B_2} = -\frac{|A_3|^2 - A_{30}^2}{B_3}, \tag{3.5}$$

where the subscript 0 stands for initial values of the appropriate quantities. Characteristic solutions of (3.4) conform to a periodic variation of the amplitudes of all waves, such that the first and the second waves change in phase, while the third one, the "pump", does so in antiphase, so as to conserve the total energy flux.

Let us analyse in more detail the case of a "degenerate" interaction of two modes at frequencies ω and 2ω. In extending the formulae of Section 1 we allow for a slight detuning between the waves, i.e. we set $\omega_2 = 2\omega_1 + \Delta\omega$, $k_2 = 2k_1 + \Delta k$, where $\Delta\omega \ll \omega_1$ and $\Delta k \simeq \Delta\omega \mathrm{d}k_2/\mathrm{d}\omega$. Then instead of (3.4) we obtain

$$\mathrm{i}\frac{\partial A_1}{\partial x} = -A_1^* A_2 B_1 \exp\mathrm{i}(\Delta\omega t - \Delta k\, x)$$
$$\mathrm{i}\frac{\partial A_2}{\partial x} = A_1^2 B_2 \exp\mathrm{i}(-\Delta\omega t + \Delta k\, x). \tag{3.6}$$

In terms of real amplitude and phase such that $A_{1,2} = a_{1,2}\exp(i\theta_{1,2})$, this system acquires a form generalizing (1.5),

$$\frac{\partial a_1}{\partial x} = -a_1 a_2 \mathrm{Im}\left[B_1 \exp(\mathrm{i}\theta)\right],$$
$$\frac{\partial a_2}{\partial x} = a_1^2 \mathrm{Im}\left[B_2 \exp(\mathrm{i}\theta)\right],$$
$$\frac{\partial \theta}{\partial x} = \Delta k + \mathrm{Re}\left[B_1 \exp(\mathrm{i}\theta)\left(a_2 - \frac{a_1^2}{a_2}\right)\right], \tag{3.7}$$

where $\theta = \theta_2 - 2\theta_1 - \Delta k\, x$.

We begin with a study of the linear stage of parametric amplification, when the "pump" amplitude a_2 can be considered as given, and discuss the behaviour of "signal" a_1 having a given initial amplitude $a_{10} \ll a_2$.

Provided $\Delta = 0$ (exact synchronism) the amplitude a_1 grows exponentially in full compliance with formula (1.7):

$$a_1 = a_{10}\exp(q), \quad q = a_2 \,\mathrm{Im}[B_1 \exp(\mathrm{i}\theta) x]. \tag{3.8}$$

In particular, with real $B_1 > 0$ a maximum increment is attained at $\theta = \pi/2$, equal to $B_1 a_2$.

To assess the frequency band $2\Delta k_0$ of amplification one needs to solve the system (3.7) with $\Delta k \neq 0$. The complete expression for $a_1(x)$ in this case is rather awkward. Therefore, we shall consider the case of comparatively large detuning, when $\Delta k \gg B_1 a_2$ and in the third equation of (3.7) one can set $\theta = \Delta k\, x + \theta_0$. In so doing a_1 then varies according to the law (Akhmanov & Khokhlov, 1964)

$$a_1 = a_{10}\exp\left(B_1 a_2 \frac{\sin \Delta k\, x/2}{\Delta k/2}\right), \tag{3.9}$$

i.e. a_1 is changing along the x-axis periodically, with period $\Lambda = 2\pi/\Delta k$.

If one requires that amplification at the boundaries of this band achieves its maximum over a system length L_0, then $\Delta k = \Delta k_0 = \pi/L_0$.

It can be noted that when the frequency ω_2 approaches a critical mode frequency, where $k_1 \to 0$, the amplification increment rises, whereas its bandwidth narrows.

Waveguide systems possess a salient feature as regards pulse propagation. In the general case, when the phase velocities v_{ph} are equal, the group velocities v_{gr} differ and the pulses at pump and signal frequencies will be separated, and as a result amplification will only be observed over a limited range. In our case, however, as equation (2.3) implies, the relation $v_{ph}v_{gr} = c_0^2$ holds true for each mode, so that the phase velocity equality means that of the group velocity also, and until the wave packet becomes spread out due to dispersion its amplification will proceed in the same way as that of a continuous signal.

Let us make some estimates concerning the cases addressed in the previous section (Ostrovsky & Papilova, 1973).

(a) For a waveguide with rigid walls, when the interaction is that of modes with indices $0, 1$ (at frequency ω) and $0, 2$ (at frequency 2ω) we can estimate the maximum increment q in accordance with (3.8). Let $d_1 = 0.7\,\text{cm}$, $d_2 = 2.8\,\text{cm}$; the critical frequency for the $0, 1$ mode is equal to 268 kHz. We take $f_1 = 300$ kHz, $f_2 = 600$ kHz ($f = \omega/2\pi$). If the pump intensity $I_2 = 5\,\text{W}\,\text{cm}^{-2}$ (maximum Mach number $M = 3\times10^{-4}$) then $q = 0.84\,\text{m}^{-1}$ and the amplification factor $K = \text{e}^q$ amounts to 2.3. The amplification band is about 0.6 per cent of the carrier frequency. Note that for a homogeneous plane wave (zero mode) one would have $K = 1.3$. The difference can be attributed to the fact that for a non-zero mode in a waveguide the effective wave path is elongated (v_{gr} is decreased) as compared with free space (zero mode) which corresponds, at equal pumping intensities ($I_2 = v_{gr}a_2^2$) to a higher velocity amplitude (which more than compensates for the weakening of the interaction due to the different transverse structures of the modes engaged).

(b) Consider a waveguide with free boundaries in the form of a rectangular channel with thin walls, filled with water and placed in air. As was specified earlier, the synchronism conditions are satisfied for the $1, 3$ (pump) and $1, 1$ (signal) modes provided $b^2/a^2 = 5/3$.

For $d_1 = 0.7\,\text{cm}$, $d_2 = 0.87\,\text{cm}$ and $I_2 = 6\,\text{W}\,\text{cm}^{-2}$ (i.e. $M = 3\times10^{-4}$) we obtain the amplification factor and $q = 1.74\,\text{m}^{-1}$. The amplification band is of the order of 2 per cent.

(c) For a waveguide in the form of an open channel filled with water the following estimates are available. Let $d_1 = 0.7\,\text{cm}$, $d_2 = 0.35\,\text{mm}$,

$f_1 = 200\,\text{kHz}$; then at $I_2 = 5\,\text{W}\,\text{cm}^{-2}$ ($M = 3 \times 10^{-4}$) one obtains $q = 0.28\,\text{m}^{-1}$, i.e. $K = 1.4$. Strictly speaking one has to account for a possible capillary wave excitation on the surface. At high frequencies, however, this effect seems to be unimportant.

Parametric sound generation in resonators

The estimates given above imply that in the sonic and ultrasonic frequency ranges high amplification for a travelling wave is difficult to obtain, for the nonlinearity is relatively small. The amplification factor is nevertheless usually far in excess of the decay decrement due to losses. This allows the use of multiple wave interactions arising in bounded systems of resonator type with reflecting ends, such that, as was mentioned in Chapter 2, both wave reflections should proceed in phase. In such systems not only is amplification observed, but also, thanks to a feedback mechanism, an instability, i.e. parametric sound generation.

Let us discuss the interaction of three modes in a resonator formed by a waveguide of length L with reflecting ends (Ostrovsky *et al.*, 1978). Note that this consideration involves the plane parallel resonator (acoustic interferometer) case as well, provided that the interaction is that of zero (longitudinal) modes. We write the solution as a superposition of three single modes of standing waves:

$$\varphi = \sum_{s=1}^{3} A_s(t) f_s(y, z) \cos(k_s x + \psi_s) + c, \tag{3.10}$$

where the ψ_s are phases defined by the boundary conditions at $x = 0, L$; k_s are the wave numbers of the longitudinal modes, so that at resonance we have $k_s L + \psi_s = n_s \pi$ (n_s being integers). The frequencies of the interacting waves are again related through the condition $\omega_1 + \omega_2 = \omega_3$.

Substituting (3.10) into (2.1) and proceeding in the same way as for the travelling waves in a waveguide, we obtain for the complex amplitudes the equations

$$\begin{aligned}
\frac{\mathrm{d}A_3}{\mathrm{d}t} &= -\mathrm{i}A_3\Delta\omega_3 - \frac{B_3}{\omega_3} A_1 A_3 - \delta_3 A_3 - A_0\omega_3, \\
\frac{\mathrm{d}A_1}{\mathrm{d}t} &= -\mathrm{i}A_1\Delta\omega_1 + \frac{B_1}{\omega_1} A_2^* A_3 - \delta_1 A_1, \\
\frac{\mathrm{d}A_2}{\mathrm{d}t} &= -\mathrm{i}A_2\Delta\omega_2 - \frac{B_2}{\omega_2} A_1^* A_3 - \delta_2 A_2,
\end{aligned} \tag{3.11}$$

where $\Delta\omega_s$ is a detuning between ω_s and the nearest natural frequency ω_{s0} of the resonator, the δ_s are decrements due to losses in the resonator, the B_s are the nonlinearity coefficients in proportion to those in (3.4), and A_0 is a prescribed amplitude of an external exciting source related, for example, to the oscillations of a terminal reflecting wall at frequency ω_3. Owing to the presence of this source the equation for A_3 turns out to be inhomogeneous.†

Passing to real variables ($A_s = a_s \exp(i\theta_s)$), we get four equations

$$\begin{aligned} \omega_{s0}\dot{a}_s &= -\omega_{s0}\delta_s a_s - B_s a_k a_\ell \sin\alpha - A_0\omega_3^2 \sin\varphi_3 \delta(s-3), \\ \dot{\alpha} &= \Delta\omega + \left(\frac{B_1 a_1 a_3}{\omega_{10} a_1} + \frac{B_2 a_1 a_3}{\omega_{20} a_2} - \frac{B_3 a_1 a_2}{\omega_{30} a_3}\right)\cos\alpha \\ &\quad + \frac{A_0 \omega_3^2 \cos\theta_3}{a_3 \omega_0}, \\ \alpha &= \theta_2 + \theta_3 - \theta_1. \end{aligned} \tag{3.12}$$

Here s, k, $\ell = 1, 2, 3$ such that $k, \ell \neq s$ and $\delta(s-3)$ is a delta function.

Equations of this kind are exhaustively studied in connection with the theory of parametric generators in radio-electronics (Kaplan *et al.*, 1966). They always have an obvious equilibrium state when $a_1 = a_2 = 0$ and

$$a_3 = A_0 \frac{\omega_3}{\delta_3}\left(1 + \frac{\Delta\omega_3^2}{\delta_3^2}\right)^{-1/2}. \tag{3.13}$$

At a rather strong excitation, however, when $A_0 > A_{\text{th}}$, and where the threshold value A_{th} is equal to

$$A_{\text{th}} = \left[(\omega_{10}\omega_{20}\delta_1\delta_2)\left(1 + \frac{\Delta\omega_1^2}{\delta_1^2}\right)\left(1 + \frac{\Delta\omega_2^3}{\delta_2^2}\right) B_1^{-1} B_2^{-1} \delta_3 \omega_{30}^{-1}\right]^{1/2}, \tag{3.14}$$

this state becomes unstable. Simultaneously, at values of

$$A_0 > \bar{A} = A_{\text{th}}\left(\frac{\Delta\omega_1}{\delta_1} + \frac{\Delta\omega_2}{\delta_2}\right),$$

there exists another equilibrium state where all the a_s are different from zero:

$$a_3^2 = \omega_{10}\omega_{20}\delta_1\delta_2\left(1 + \frac{\Delta\omega_1^2}{\delta_1^2}\right) / B_1 B_2$$

† Note that here another effect is neglected, i.e. resonator length modulation due to boundary oscillations at the pump frequency (Adler & Breazeale, 1970; Eller, 1973); since, however, the pumping mode number is apparently not less than 3, the main role is that of the volume nonlinearity accounted for here.

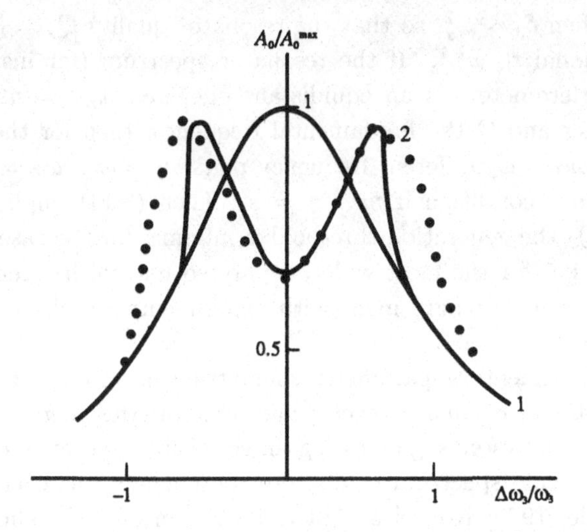

Figure 6.3. The resonance curve of a parametric generator at the pump frequency at an amplitude below the parametric generation threshold (curve 1) and above it (curve 2). The dots represent experimental results.

$$a_1^2 = \frac{\omega_{20}\omega_{30}\delta_2\delta_3}{B_1 B_2} \times$$
$$\times \left\{ -1 + \frac{\Delta\omega_1 \Delta\omega_3}{\delta_1\delta_3} \pm \left[\frac{A_0^2 \omega_0 B_1 B_2}{4\delta_1\delta_2\omega_{10}\omega_{20}} - \left(\frac{\Delta\omega_1}{\delta_1} + \frac{\Delta\omega_3}{\delta_3} \right)^2 \right]^{1/2} \right\},$$
$$a_1^2/a_2^2 = \omega_2\delta_2/\omega_1\delta_1. \tag{3.15}$$

It should be emphasized that, clearly, $\Delta\omega_1/\delta_1 = \Delta\omega_2/\delta_2$. It is of interest that a_3 does not depend on A_0 at all, i.e. parametric energy transformation completely prevents amplitude growth at the pump frequency with growth of the excitation amplitude. At low detunings the values of $\bar{A}$ and A_{th} coincide (soft regime), whereas when

$$\left(\frac{\Delta\omega_3}{\delta_3} \right)^2 > \frac{\delta_1 + \delta_2}{\delta_3}$$

it appears that $\bar{A} < A_{\mathrm{th}}$, and then two stable states of equilibrium (3.15) occur (hard regime). Within this domain the dependences of a_1, a_2 and a_3 on a_0 have a hysteretic character. It can be noted also that in the resonance curve of $a_3(\Delta\omega_3)$ a minimum is set up due to the energy absorbed in the parametrically excited modes (Figure 6.3).

Let us turn our attention to some special cases. Consider a liquid- or gas-filled resonator with rigid reflectors at its ends. If losses are due to viscosity then $\delta_3 \sim \omega_s^2$, so that the resonator quality $Q_s = (\delta_3/2\omega_s)^{-1}$ is proportional to ω_s^{-1}. If the resonator spectrum (for instance that of an interferometer) is an equidistant one, i.e. $\omega_{n0} = n\Omega$ where n is an integer and Ω the fundamental frequency, then for the pumping frequency $\omega_\Omega = \omega_s$ different frequency pairs $\omega_1 = \omega_m$, $\omega_2 = \omega_n$ satisfy the resonance condition if $m + n = s$. Then (3.14) implies that, as $A_{\text{th}} \sim Q_1 Q_2$ the generation threshold is minimal for the case of $m = 1$, $n = s - 1$, i.e. for the most widely separated natural frequencies. This conclusion is in fair agreement with experimental results obtained by Eller (1973).

Early experiments on parametric sound transformation were performed in plane parallel resonators such as acoustic interferometers filled with water. In such systems (possessing an equidistant set of resonance frequencies) a wide spectrum of discrete components was excited (Adler & Breazeale, 1970; Korpel & Adler, 1965). In order to eliminate unwanted components one can use an interferometer with selectively reflecting boundaries. It was probably this possibility that was realized by Yen (1975). Finally, parametric sound generation has been elaborately studied in waveguide resonators (Ostrovsky *et al*, 1972, 1978). In particular, a rectangular waveguide with acoustically soft lateral walls and rigid ends, filled with water, was used. The cross-sectional area was $3 \times 23\,\text{cm}^2$, i.e. the above condition $b^2/a^2 = 5/3$ was satisfied. The wave frequencies were $f_1 = 42\,\text{kHz}$ (1, 1 mode) and $f_2 = 84\,\text{kHz}$ (1, 3 mode). The threshold (corresponding to a Mach number of about 5×10^{-3}) being exceeded, a parametric generation regime was set up, and good agreement with the above formulae was observed (see Figure 6.3, where a resonance curve is depicted with a typical trough).

Another example is that of a resonator in the form of a metal rod bent into a ring. For comparatively low-number modes of such a rod the dispersion is small and resonance conditions are approximately satisfied; then mode qualities within this range are about the same. An experiment was carried out with an aluminium resonator 635 mm long, and with diameter 12 mm, its quality factor being $Q \simeq 500$, with a threshold Mach number $M \sim 5 \times 10^{-7}$. At $f_3 = 10\,\text{kHz}$, although dispersion was small, a modest number of frequencies was generated. In other cases, however (at large f_3), extremely broad spectra were generated as well. It is of interest that for an ultrasonic pump ($f_3 \simeq 20\,\text{kHz}$) a rather loud sound was produced at a lower frequency, i.e. a direct transformation of ultrasonic energy into audible sound was realized.

It can also be noted that for a solid resonator with a fair agreement of all "kinematic" parameters with theory, the experimental absolute value of the threshold amplitude was considerably less than the theoretical one (threshold strain $\varepsilon_{\mathrm{th}} \approx 1\text{--}3 \times 10^{-5}$ as compared with a predicted value of 10^{-4}). This is believed to be related to the presence of an anomalous metal nonlinearity, as reported in Chapter 1.

6.4 Acoustic solitons in a liquid with gas bubbles

Now let us consider processes of another kind which are also connected with nonlinearity and dispersion. We are referring to cases when dispersion is low at lower frequencies, while being rather pronounced at higher ones. Such systems are typically described by the Korteweg–de Vries equation (KdV) which has already been deduced in the first chapter for liquids with bubbles and for porous media. Here we discuss a most characteristic solution of the KdV equation, a soliton, in particular acoustic systems. Moreover, we shall try to take other effects, like small dissipation, into account, which allows us to relate solitons to a more general class of solutions in the form of shock waves with an oscillating front.

Let us, however, embark first on the KdV equation

$$\frac{\partial v}{\partial x} - \alpha v \frac{\partial v}{\partial y} - \beta \frac{\partial^3 v}{\partial y^3} = 0. \tag{4.1}$$

It is well known that this equation is completely integrable and, in particular, possesses a class of stationary solutions in the form of progressive waves propagating with constant velocity without waveform change. Here $v = v(\xi)$, $\xi = y + bx$, $y = t - x/c_0$, (b constant), and (4.1) becomes an ordinary differential equation. This equation has a set of periodic finite solutions known as cnoidal waves; they are described analytically by elliptic functions. Further, there exists a localized solitary solution, a soliton:

$$v = v_0/\cosh^2(\xi/\Delta) \tag{4.2}$$

where

$$\Delta = \sqrt{\frac{12\beta}{\alpha v_0}}, \qquad b = \frac{\alpha v_0}{3}. \tag{4.3}$$

The exact nonstationary solutions of (4.1) are also known, these being primarily the multi-soliton solutions: two or more solitons interact as they converge from infinity, and then eventually diverge retaining their initial parameters in the asymptotic large-time limit. This, among other properties, provides a basis for a close analogy between solitons and material particles. It is also known that any localized positive pulse of sufficiently large amplitude disintegrates asymptotically at large time, to form a finite number of solitons plus an oscillating "tail", i.e. radiation. Solitons have been observed in virtually all possible kinds of physical situations (waves on the surface of water, in plasma, and in electromagnetic lines, etc.). Let us cite three examples of acoustic solitons; the first refers to a liquid with bubbles, the second to solid waveguides of the thin rod type, while the third is concerned with granular media.

Solitons in a liquid with bubbles

The KdV equation for the case when the characteristic wave frequency is much less than the resonance frequencies of the bubbles was derived in Chapter 1. It was also demonstrated there that the equation holds in the range of low (albeit not too low) frequencies when a bubble is compressed adiabatically rather than isothermally. If the losses can be neglected, then there exist solitons with the following parameters:

$$\Delta = \frac{1}{\omega_0}\sqrt{\frac{6}{Gp_s c}}, \quad b = \frac{Gp_s}{3}, \tag{4.4}$$

where $p_s = \rho_0 c v_0$ is the peak pressure in the soliton; the rest of the notation is the same as that used in (3.27) from Section 3 of Chapter 1.

The equation allowing for thermal losses has not been analysed in its full form. However, more simple phenomenological models have been considered; these were treated in the book by Nakoryakov *et al.* (1983). The simplest model deals with a viscous term which, along with the dispersive one, yields a combined Korteweg–de Vries–Burgers (KdVB) equation:

$$\frac{\partial v}{\partial x} - \alpha v \frac{\partial v}{\partial y} - \beta \frac{\partial^3 v}{\partial y^3} = \delta \frac{\partial^2 v}{\partial y^2}. \tag{4.5}$$

Passing to the variable $\xi = y + bx$, we again obtain a class of stationary solutions. On the phase plane a finite solution corresponds to a separatrix connecting the equilibrium points. This solution describes a shock wave as a transition between two constant velocity values; this transition can have oscillations due to dispersion. At low δ the first oscillations are quite close to solitons (solitons corresponding to a closed separatrix).

In the latter case a nonstationary solution can be developed as well, describing a slowly decaying soliton. Let us seek a solution in the form of a soliton (4.2) with a slowly changing parameter v_0, but with the same relations between Δ, b and v_0. The methods of constructing such solutions are well established nowadays. In the first-order approximation the relation for v_0 may be deduced from an energy balance equation. Indeed, let us multiply the reference equation (4.5) by v and integrate with respect to y over an infinite interval (actually, over the soliton localization domain). Then, substitution of the solution (4.2) yields (Ostrovsky, 1983)

$$\frac{dv_0}{dx} = -\frac{4}{45}\left(\frac{\alpha b}{\beta}\right)v_0^2 = -\gamma v_0^2, \quad \gamma = \frac{4}{45}\frac{\alpha b}{\beta}$$

and hence

$$v_0 = v_{00}(1 + \gamma v_{00} x)^{-1}. \tag{4.6}$$

This formula looks like (2.25) in Chapter 2, describing sawtooth wave damping; at large x the soliton amplitude decays as x^{-1} and ceases to depend on the initial value v_{00}. Some other nonstationary solutions of equation (4.5) of transient character were obtained by Gurevich & Pitaevsky (1987).

Current experimental data concerning this problem are representative enough, although the majority of them are concerned with strong shock waves. Due to the bubbles the shock formation distance turns out to be considerably less than that in "pure" water. The situations considered above are more closely approached by a series of experiments described by Gasenko *et al.* (1977), Nakoryakov *et al.* (1983). They employed two main parameters of the problem, depicting the role of losses and dispersion: $Re = v_0 \ell_0/\delta$ (Reynolds number) and $\sigma = \ell_0(\sqrt{\beta/v_0})^{-1}$ (Ursell parameter), ℓ_0 being a characteristic length of the perturbation, v_0 a typical wave amplitude. An analysis of the stationary solutions of (4.5) implies that for $\sigma/Re > \sqrt{2}$ these solutions have the form of shock waves with a monotonic profile, while for $\sigma/Re < \sqrt{2}$ oscillations arise on the shock front. At small σ and large Re a linear short-wave train arises, whereas at very large σ a finite number of solitons is created. Note that for a soliton $\sigma = 12$. These calculations are in good agreement with experimental results (Kuznetsov *et al.*, 1978) where the compression pulse propagation in a glycerol–water solution containing carbon dioxide bubbles was investigated. The photographs in Figure 6.4 show the initial pulse shape and its profiles at a distance of 1 m from the source for various medium parameters; they are summarized in Table 6.1, where the initial pulse amplitude p_0 and its length ℓ_0 are specified. All these cases refer to the domain $\sigma/Re < \sqrt{2}$ when, according to theory, dispersion effects are significant. At $\sigma > 14$ (Figure 6.4(c)–6.4(e)) the initial pulse

Table 6.1

Figure	σ	Re	σ/Re	p_0 MPa	ℓ_0 cm	Signal characteristics
5.7a	12.1	-	-	0.048	10	Radiated pulse
5.7b	-	-	0.05	0.035	90	Shock wave
5.7c	52.8	3520	0.015	0.0945	20	Multi-soliton perturbation
5.7d	30	1500	0.021	0.048	16	Two-soliton perturbation
5.7e	12.1	356	0.034	0.029	10	Single soliton
5.7f	3.4	48	0.071	0.0035	6	Wave packet

breaks down into solitons, while for $\sigma < 14$ (Figure 6.4f) a linear wave packet arises (see also Table 6.1 summarizing these results).

It seems of interest that, while for the case of carbon dioxide gas bubbles, solitons were quite distinct, for helium bubbles they were not observed at all. Estimates based on the results of Chapter 1 demonstrate (see also Kobelev & Ostrovsky, 1980) that here the thermal wavelength is large, so that the bubble compression process is close to isothermal and soliton solutions are impossible (see also Chapter 1).

6.5 Acoustic solitons in solids

Solitons in rods

Now let us address the longitudinal elastic wave propagation in a thin rod with radius a. A well-known fact is that the velocity of linear long waves in such a rod is equal to $c_\ell = \sqrt{E/\rho}$, where E is Young's modulus and ρ the density. The finite thickness of the rod brings about the sound velocity dispersion, which becomes more pronounced as the wavelength shortens. This can, in particular, result in soliton formation. These problems have been tackled in various studies (e.g. Ostrovsky & Sutin, 1977; Samsonov, 1988).

The equation describing finite-amplitude waves can be deduced on the basis of general equations discussed in Section 2 of Chapter 1. Let us write down an expression for the internal energy density:

$$e = \mu u_{ik}^2 + \frac{\lambda}{2} u_{ii}^2 + \frac{A}{3} u_{ik} u_{i\ell} u_{k\ell} + B u_{ik}^2 u_{\ell\ell} + \frac{C}{3} u_{\ell\ell}^3 \tag{5.1}$$

(notation here is the same as in Chapter 1).

In a longitudinal wave with wave number k satisfying the condition $ka \ll 1$, longitudinal displacements u_r exceed by far the transverse

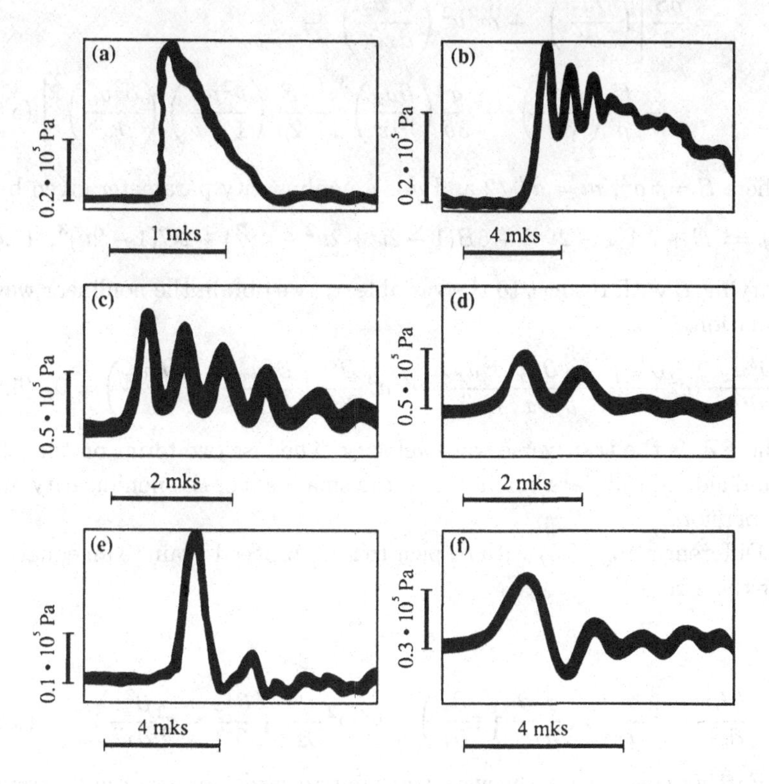

Figure 6.4. Experimental data on the propagation of a single pulse in a liquid with gas bubbles (Kuznetsov *et al.*, 1978). (a) the initial wave profile; (b–f) wave profiles at distance 1 m from source (see Table 6.1).

(radial) displacements u_r, the latter depending on the radial coordinate r. At low ka one can use the Love approximation, according to which

$$u_r = -\nu r \frac{\partial u_x}{\partial x}, \tag{5.2}$$

where ν is the Poisson's ratio. Integrating the energy density over the rod cross-section and taking (5.2) into account, we obtain (in Lagrangian variables again) a one-dimensional Lagrangian (Lagrange function density per unit length)

$$L = \int \left[\frac{\rho}{2} \left(\frac{\partial u_x}{\partial t} \right)^2 - \varepsilon \right] \mathrm{d}s$$

$$= \frac{\rho S}{2}\left[\left(\frac{\partial u_x}{\partial t}\right)^2 + m^2\nu^2\left(\frac{\partial^2 u_x}{\partial x \partial t}\right)^2 - \right.$$

$$\left. - \frac{E}{\rho}\left(\frac{\partial u_x}{\partial x}\right)^2 - \frac{g}{3\rho}\left(\frac{\partial u_x}{\partial x}\right)^3 - \frac{E}{2\rho}\left(\frac{\nu^2 m^2}{1+\nu}\right)\left(\frac{\partial^2 u_x}{\partial x^2}\right)^2\right], \quad (5.3)$$

where $S = \pi a^2$, $m = a/\sqrt{2}$ and g is a nonlinearity parameter given by

$$g = 3E + 2A(1 - 2\nu^3) + 6B(1 - 2\nu + 2\nu^2 - 4\nu^3) + 2C(1 - 2\nu)^3. \quad (5.4)$$

Varying L with respect to the variable u_x, we obtain the nonlinear wave equation

$$\frac{\partial^2 u_x}{\partial t^2} - c_\ell^2 \frac{\partial^2 u_x}{\partial x^2} - \frac{g}{\rho}\frac{\partial u_x}{\partial x}\frac{\partial^2 u_x}{\partial x^2} - \nu^2 m^2 \frac{\partial^2}{\partial x^2}\left(\frac{\partial^2 u_x}{\partial t^2} - c_t^2 \frac{\partial^2 u_x}{\partial x^2}\right) = 0, \quad (5.5)$$

where c_t is the transverse wave velocity. The last two terms on the left-hand side of (5.5) are small due to the smallness of the nonlinearity and dispersion.

Differentiating (5.5) with respect to x we proceed again to an equation for the strain

$$\varepsilon = \frac{\partial u_x}{\partial x},$$

$$\frac{\partial^2 \varepsilon}{\partial t^2} - c_\ell^2 \frac{\partial^2 \varepsilon}{\partial x^2} = \frac{g}{\rho}\frac{\partial}{\partial x}\left(\varepsilon \frac{\partial \varepsilon}{\partial x}\right) + \nu^2 m^2 \frac{\partial^2}{\partial x^2}\left(\frac{\partial^2 \varepsilon}{\partial t^2} - c_t^2 \frac{\partial^2 \varepsilon}{\partial x^2}\right). \quad (5.6)$$

Further considering the wave travelling towards positive x and passing to variables $y = t - x/c_\ell$ and $x' = x$ (and omitting the prime in what follows) we derive a KdV equation for ε or, what is essentially the same, for the velocity $v = \partial u/\partial t = -c_\ell \varepsilon$:

$$\frac{\partial v}{\partial x} - \alpha v \frac{\partial v}{\partial y} - \beta \frac{\partial^3 v}{\partial y^3} = 0, \quad (5.7)$$

where

$$\alpha = \frac{g}{2\rho c_\ell^4}, \quad \beta = \frac{\nu^2 m^2}{c_\ell^3}\left(1 - \frac{c_t^2}{c_\ell^2}\right).$$

The solution (4.2) in the form of a soliton here acquires parameters

$$\Delta = \nu m \sqrt{24 \frac{\rho}{g} \frac{c_\ell}{v_0}\left(1 - \frac{c_t^2}{c_\ell^2}\right)}, \quad b = \frac{g v_0}{6\rho c_\ell^3}. \quad (5.8)$$

A longer compression pulse breaks down into solitons, which makes the displacement wave in the rod take on the form of “steps”.

Excitation and observation of solitons in such systems would be of keen interest, but the experiments seem not to be so simple. Consider, for example, waves in a steel wire of 1 mm diameter, such that $c_\ell \approx 3 \times 10^3\,\mathrm{m\,s^{-1}}$, and accept a peak deformation value ε_0 in a wave of 10^{-5} (this value corresponds approximately to the static yield strength of the material); then the velocity amplitude in a soliton is $v_0 = 0.5\,\mathrm{m\,s^{-1}}$, the soliton length being $\ell_0 \simeq 1\,\mathrm{cm}$ (Ostrovsky & Sutin, 1977). A practical method of exciting such solitons is associated, probably, with sound generation by a laser pulse. An investigation by Dreiden *et al* (1984) deals with excitation of a longitudinal sound pulse in a polystyrene rod of 1 cm diameter and 5.5 cm length by focusing a laser beam on its end. As a result a compression pulse with a pressure amplitude $p_0 = 13.5\,\mathrm{atm}$ and length $\ell_0 = 0.8\,\mathrm{cm}$ was excited; as was estimated by the authors, the central portion of a pulse propagates without deformation and has parameters approaching those of a soliton.

It has to be noted that the dispersion parameter sign in (5.7) can in principle be different, depending on the material properties, so that for $\beta > 0$ compression solitons exist, while for $\beta < 0$ rarefaction solitons should occur; however, the latter is possible only for materials with a shear wave velocity exceeding that of longitudinal waves.

Of a certain interest is the problem of soliton propagation in a rod with smoothly varying cross-section $S(x)$. In this case the Lagrangian (5.3) depends on x, and an additional term

$$\frac{E}{2\rho S}\frac{\partial S}{\partial x}\frac{\partial u_x}{\partial x}$$

arises in equation (5.5). For the KdV equation (5.7) this yields

$$\frac{\partial v}{\partial z} - \alpha v \frac{\partial v}{\partial y} - \beta \frac{\partial^3 v}{\partial y^3} + \frac{v}{2S}\frac{\partial S}{\partial z} = 0. \tag{5.9}$$

If the latter term is of the same order as the others it will radically change the solution character. We draw under discussion only two limiting cases. Should the inhomogeneity, albeit smooth, be strong enough to make this term dominate over the two previous ones (i.e. the inhomogeneity is "stronger" than nonlinearity and dispersion) the propagation occurs in a linear way, the wave profile suffers practically no change, and the wave amplitude will be proportional to $S^{-1/2}$.

If, however, the last term in (5.9) is small compared with the others, its effect can be accounted for by perturbation theory. In particular, provided the boundary condition is in conformity with a soliton then the inhomogeneity leads, to the first approximation, to a smooth change of the soliton parameters. A similar problem of soliton decay has already been treated in the previous section with respect to a liquid with gas bubbles. Again multiplying (5.9) by v and integrating over y readily furnishes

$$v = v_0 \left(\frac{S}{S_0}\right)^{-2/3}. \tag{5.10}$$

Therefore, the soliton amplitude has a stronger dependence on S than that of a linear wave: although the wave energy in both cases is conserved, the soliton duration depends on its amplitude such that a soliton compresses with growth of amplitude, while broadening with its decrease.

In the same way one can allow for smooth inhomogeneity of the other parameters of a rod. This problem was investigated in studies by Molotkov and Vakulenko (1980), Samsonov (1984), and Samsonov and Sokurinskaya (1987).

An analogous consideration can also be effected for a thin plate (Ostrovsky & Sutin, 1977). Within this plate longitudinal waves also are described by equation (5.6), where now (d being the plate thickness)

$$c_\ell^2 = \frac{E}{\rho}(1-\nu^2)^{-1}, \quad \nu^2 m^2 \to d^2\nu^2/12(1-\nu^2),$$

$$g = \frac{3E}{\rho(1-\nu^2)} + A\left[1 - \left(\frac{\nu}{1-\nu}\right)^3\right]$$
$$+ 3B\left[1 - \frac{\nu}{1-\nu} + \left(\frac{\nu}{1-\nu}\right)^2 + 2\left(\frac{\nu}{1-\nu}\right)^3\right]$$
$$+ C + \left(\frac{1-2\nu}{1-\nu}\right). \tag{5.11}$$

The fact that similar equation parameters for waves in an unbounded medium, in a rod and in a plate are all different provides a theoretical possibility for measuring all the three third-order constants A, B, C for a given material.

Waves in granular media

Let us briefly discuss another interesting model, of a nonlinear medium made up of a set of touching elastic spheres (granules). The problem of the elastic interaction of two solids (a contact problem in the theory of elasticity) was solved by Hertz back in the last century; its description can be found in the book by Landau and Lifshitz (1986). If no account is taken of material elastic nonlinearity, then the geometric parameters of the two solids and their interaction force (compressive force) F are related through

$$F = \frac{2E}{3(1-\sigma^2)} \left(\frac{R_1 R_2}{R_1 + R_2} \right)^{1/2} [(R_1 + R_2) - (x_2 - x_1)]^{3/2}, \quad (5.12)$$

where E is Young's modulus for the sphere material, R_1, R_2, are their radii and x_1, x_2 are their centre coordinates (for definiteness $x_2 > x_1$). It is obvious that for undeformed spheres $x_2 - x_1 = R_1 + R_2$. The Hertz law holds good under the condition that the contact surface size is much less than the particle radius, and the characteristic time of the process is much longer than the sphere's natural radial oscillation period $T \approx 2.5\, R/c_l$, where c_l is the sound velocity in the sphere material.

It is noteworthy that a relation of the same type, i.e. $F \sim h^{3/2}$ where h is the relative displacement, is retained for nonspherical bodies as well, so that qualitatively all that follows refers to them also.

Let us consider, following Nesterenko (1983), a one-dimensional chain formed by contacting spheres of the same radius R, or else formed by their three-dimensional packing if the propagation takes place along the x-axis running through the sphere's contact points. Let a constant compressive force F_0 be applied at the chain ends, creating an initial convergence of neighbouring particles amounting to h_0. Then the displacement u_i of the i-th particle from equilibrium, taking (5.12) into account, will satisfy the equation

$$\ddot{u}_i = A(h_0 - u_i + u_{i-1})^{3/2} - A(h_0 - u_{i+1} + u_i)^{3/2}, \quad (5.13)$$

where $A = E(2R)^{1/2}/3(1-\nu^2)M$, M being the particle mass.

We shall analyse the case of small nonlinearity, when $|u_{i\pm1} - u_i| \ll h_0$, i.e. wave displacements are small compared with the reference displacement. Then expanding the differences in (5.13) and retaining terms of the first and the second order in amplitude, we obtain

$$\ddot{u}_i = \alpha(u_{i+1} - 2u_i + u_{i-1}) + \beta(u_{i+1} - 2u_i + u_{i-1})(u_{i-1} - u_{i+1}), \quad (5.14)$$

where $\alpha = (3/2)Ah_0^{1/2}$, $\beta = (3/8)Ah_0^{-1/2}$.

Similar equations are typical of a rather wide class of nonlinear systems with discrete parameters. Here, above all, the well-known Fermi–Pasta–Ulam problem of a "nonlinear string" must be mentioned, which served as a powerful stimulus for investigation of the problems of stochastization and reversibility in nonlinear distributed systems. Numerical calculations were carried out to demonstrate that in such discrete systems solitons can exist. Simultaneously, nonlinear effects in electromagnetic nonlinear systems (discrete transmission lines) were studied. Our task here, however, is not to analyse the peculiarities of discrete systems but to proceed instead to their distributed, continuum analogues. To accomplish this, let us consider long-wave perturbations, whose characteristic scale λ is large compared with the distance $2R = a$ between particle centres. Then, expanding the differences in (5.14) with respect to the x-coordinate, the following nonlinear wave equation results:

$$\frac{\partial^2 u}{\partial t^2} = c_\ell^2 \frac{\partial^2 u}{\partial x^2} + 2c_\ell \gamma \frac{\partial^4 u}{\partial x^4} - \ell \frac{\partial u}{\partial x}\frac{\partial^2 u}{\partial x^2}, \tag{5.15}$$

where

$$c_\ell^2 = a^2 A h_0^{1/2}, \quad \gamma = c_\ell R^2/6, \quad \ell = c_\ell^2 R/h_0.$$

This equation is of the same type as, for example, (5.5). It yields, in precisely the same manner, a KdV equation of type (5.7), the solutions of which include, in particular, solitons.

As an effective sound velocity here, a quantity c_ℓ is employed that can have a value much less than that of the longitudinal wave velocity in the material of the granules, i.e. the chain may serve as a kind of "delay line". Moreover, as $c_\ell \sim \sqrt{h_0}$ and $\varepsilon \sim h_0^{-1}$, the wave velocity drops while the nonlinearity of the system increases with precompression decrease. Because of this, the case $h_0 = 0$ is of special interest, when (5.14) loses its applicability because the nonlinearity is not inherently small (though relative deformation of an individual particle may still remain small). The long-wave approximation being still valid, (5.13) implies that

$$\begin{aligned}\frac{\partial^2 \varepsilon}{\partial t^2} =& c_1^2 \left[\frac{3}{2}(-\varepsilon)^{1/2}\frac{\partial \varepsilon}{\partial x} + \frac{a^3}{8}(-\varepsilon)^{1/2}\frac{\partial^4 \varepsilon}{\partial x^4} \right. \\ & \left. - a^2 \frac{\partial \varepsilon}{\partial x}\frac{\partial^2 \varepsilon}{\partial x^2} \Big/ (1-\varepsilon)^{1/2} - \frac{a^2}{64}\left(\frac{\partial \varepsilon}{\partial x}\right)^3 \Big/ (-\varepsilon)^{1/2} \right], \\ c_1^2 =& \frac{E}{3\rho_0(1-\sigma^2)}. \end{aligned} \tag{5.16}$$

Here $\varepsilon = \partial u/\partial x$ is the strain, which is assumed negative. The fact is that here nonlinearity is not small, and u involves the initial displacement due to the compressive force as well, so that the condition $\varepsilon < 0$ simply means that the rarefaction deformation in a wave does not exceed in absolute value the static one – otherwise granules in the "rarefaction" phase will diverge, losing contact. Naturally, at low wave amplitudes (5.16) is transformed into (5.15).

Searching for the stationary solutions of (5.16), depending on the variable $\eta = x - Vt$, with V constant, we obtain for the variable $z = (-\varepsilon)^{5/4}$

$$\frac{a^2}{10}\frac{\mathrm{d}^2 z}{\mathrm{d}\eta^2} + z - \frac{V^2}{c_1^2} z^{3/5} + C z^{-1/5} = 0, \tag{5.17}$$

where C is an integration constant. In the most interesting case, $C > 0$, there exist finite periodic solutions, as well as a solitary solution of a "soliton-on-pedestal" type (the "pedestal" z_1 is defined by the initial compression). Provided z_2 (the peak value) and z_1 are close to each other, KdV-type solitons arise. If, however, $z_1 \ll z_2$, then in the domain where $z_2 > z \gg z_1$, the solution approaches that obtained for $C = 0$, namely

$$-\varepsilon \approx \left(\frac{5V^2}{4c_1^2}\right)^2 \cos^4\left[\frac{\sqrt{10}}{5a}(x - Vt)\right], \tag{5.18}$$

so that the soliton velocity is related to the peak deformation value through

$$\frac{V}{c_1} \approx 0.9|\varepsilon_{\max}|^{1/4}. \tag{5.19}$$

It is remarkable that in this limiting case the soliton length ceases to depend on its amplitude, namely

$$L_s \approx 1.8a \tag{5.20}$$

(here L_s is determined at a level of 0.5 with respect to the peak value). These dependences, clearly, are radically different from the "weakly nonlinear" ones when the KdV equation holds. Recall that this is connected with the significant nonlinearity of the static relation $\sigma(\varepsilon)$, i.e. of the equation of state of the medium. In fact, (5.12) implies (for equal R) that

$$\sigma = E|\varepsilon|^{3/2}/3(1 - \nu^2), \tag{5.21}$$

and the local sound velocity (dispersion not accounted for) can be written as

$$c_\ell^2 = \frac{1}{\rho_0}\frac{\mathrm{d}\sigma}{\mathrm{d}\varepsilon} = \frac{E|\varepsilon|^{1/2}}{2\rho_0(1 - \nu^2)}. \tag{5.22}$$

which is in agreement with (5.16).

In an unloaded chain, i.e. at $\varepsilon = 0$, this quantity vanishes. Generally, an unloaded system of touching particles can be referred to as a system with anomalous nonlinearity in the same sense as that discussed in Chapter 1.

These conclusions were tested both numerically and experimentally (Nesterenko & Lasaridi, 1985). In the experiment, steel balls with diameter $a = 4.75\,\mathrm{mm}$ were employed (their number varied from 20 to 40 for different cases), being placed in a quartz tube. In this case $\rho_0 = 7.8 \times 10^3\,\mathrm{kg\,m^{-3}}$, $E = 2 \times 10^{11}\,\mathrm{N\,m^{-2}}$, $\nu = 0.29$. When excited by impact at one end the perturbation in the chain broke down into solitons whose parameters were in fair agreement with the above-cited results. Pulse durations amounted to 10–20 μs, while compression force amplitudes in them were estimated as being equal to 10–80 N. The particle velocities would not exceed 10 $\mathrm{m\,s^{-1}}$, i.e. the "Mach number" with respect to the sound velocity in steel (but not to the wave velocity) remained low, so that in a way, an "acoustic" situation can be spoken of here. Later on the acoustical aspects of such a model were studied both theoretically and experimentally (Belyaeva *et al.*, 1992). As follows from (5.15), the quadratic nonlinearity parameter is equal to $(2\varepsilon_0)^{-1}$, where ε_0 is a static "prestrain". By measurement of the second harmonic amplitude this parameter was found to achieve values of 10^3 and more, which is interesting in view of possible diagnostics of soils.

6.6 Wave interaction in a liquid with resonance bubbles

Very special are dispersive media made up of resonant oscillators, such as liquids with gas bubbles. Here we shall tackle certain resonance effects.

As was suggested in Chapter 1, wave propagation in a gas–liquid mixture is described by equations (3.22), (3.18) and (3.14) of Chapter 1,

$$\frac{1}{c_0^2}\frac{\partial^2 p'}{\partial t^2} - \frac{\partial^2 p'}{\partial x^2} = \rho_0 \frac{\partial^2 z'}{\partial t^2}, \tag{6.1}$$

$$z' = \int V'(R_0) n(R_0)\,\mathrm{d}R_0, \tag{6.2}$$

$$\ddot{V}' + \omega_0^2 V' - 3(\gamma+1)\beta\omega_0^2 (V')^2 - \beta(2V'\ddot{V}' + (\dot{V}')^2) + f\dot{V}' = \mathcal{X} p' \tag{6.3}$$

(notation here is that of Section 3 of Chapter 1).

Dispersion equation

Consider first a linear approximation. Neglecting nonlinear terms in (6.3) and searching for a solution in the form p', $V' \sim \exp \mathrm{i}(\omega t - kx)$ we recover a relation between V' and p' from (6.3),

$$V' = \frac{4\pi R_0 p'}{\rho_0(\omega_0^2 - \omega^2 + \mathrm{i}\omega f)}, \tag{6.4}$$

and then (6.1) and (6.2) yield the dispersion equation

$$k^2 = \frac{\omega^2}{c_0^2} + 4\pi \int \frac{n(R_0) R_0 \, \mathrm{d}R_0}{\xi + \mathrm{i}/Q}, \tag{6.5}$$

where $\xi = (\omega_0/\omega)^2 - 1$, and $Q = \omega/f$ is the bubble quality factor at $\omega = \omega_0$. From this it is clear that the presence of bubbles brings about wave phase velocity dispersion as well as wave decay. Most interesting is the case of low losses, when the imaginary part in (6.5) is small. Then the wave number k can be set equal to $(\omega/c) - \mathrm{i}a$, where

$$c(\omega) = c_0 \left[1 + \frac{4\pi c_0^2}{\omega^2} \int \frac{\xi n(R_0) R_0 \, \mathrm{d}R_0}{(\xi^2 + Q^{-2})}\right]^{-1/2}, \tag{6.6a}$$

$$a(\omega) \simeq \frac{2\pi c(\omega)}{\omega} \int \frac{n(R_0) R_0 \, \mathrm{d}R_0}{Q(\xi^2 + Q^{-2})}. \tag{6.6b}$$

Let us consider two limiting cases.

(a) Bubbles of equal radius $\bar{R}_0$, i.e. $n(R_0) = N\delta(R_0 - \bar{R}_0)$ where N is the total number of bubbles per unit volume. If losses are ignored (i.e. $Q \to \infty$) the dispersion equation (6.5) will acquire the form

$$k^2 = \omega^2 \left(\frac{1}{c_0^2} + \frac{4\pi \bar{R}_0 N}{\omega_0^2 - \omega^2}\right), \tag{6.7}$$

where the resonance frequency ω_0 corresponds to radius $\bar{R}_0$.

This relation is similar to the dispersion curve for a model of a dielectric as a system of identical oscillators (Lorentz model) in electrodynamics. It involves two branches, the low- and high-frequency ones, separated by a "nontransmission zone" where real values of ω correspond to imaginary ones of k. Here a field entering the medium is exponentially attenuated due to its reflection by bubbles. This zone is situated between ω_0 and

$$\omega_1 = \omega_0 \left(1 + \frac{4\pi}{\omega_0^2} \bar{R}_0 N c_0^2\right)^{-1/2}.$$

At low frequencies the sound velocity is close to the value $c_m < c_0$ recovered from (3.27) of Chapter 1, while at high frequencies it approaches c_0.

Now we shall account for small losses in a bubble. Here, according to (6.6*b*), a decay factor arises, namely

$$a = \frac{2\pi c_0 \bar{R}_0 N}{\omega Q(\xi^{-2} + Q^{-2})}, \tag{6.8}$$

where $\xi = \omega_0^2/\omega^2 - 1$.

These relations were experimentally verified by Fox *et al.* (1955).

(b) Bubbles with widely distributed values of radius. In real experiments the bubbles' radii are seldom grouped near particular values; more frequently the distribution function $n(r_0)$ is broad in that the resonance frequency band for all the bubbles exceeds the resonance curve width of an individual bubble, i.e.

$$\frac{\mathrm{d}n}{\mathrm{d}\omega_0}\frac{\omega_0}{Q} \ll n \quad \text{or} \quad \frac{\bar{R}_0}{Qn}\frac{\mathrm{d}n}{\mathrm{d}\bar{R}_0} \ll 1. \tag{6.9}$$

Under these conditions, when integrating (6.6) at each ω it is possible to perform the integration in the resonance vicinity only, setting $n(r_0)$ constant, $\omega_0 = \omega$. As a result we obtain for the decrement (see, for example, Clay & Medwin, 1977)

$$a = 2\pi^2 R_\omega^2 n(R_\omega), \tag{6.10a}$$

$R_\omega = R_0(\omega)$ being the bubble resonance radius at frequency ω.

It is noteworthy that this quantity does not depend on the individual bubble quality factor. The fact is that the energy consumption of an individual bubble increases with Q while the number of bubbles interacting with a harmonic field decreases (due to resonance curve narrowing). In this sense we see an analogy with a "collisionless" Landau damping in plasma (Ryutov, 1975).

For air bubbles in water under atmospheric pressure we have

$$a = 725 n(R_\omega) R_\omega^3. \tag{6.10b}$$

This formula is frequently used when measuring bubble concentrations in seawater as well as in laboratory tanks: the relation $n(r_0)$ is obtainable by measuring a at various frequencies.

Another interesting fact is that for a broad distribution function $n(R_0)$, the contribution of the bubbles to the real part of k, i.e. to the sound velocity variation, is sharply reduced. Indeed, assuming in (6.6*a*) that $\xi \to 0$ we obtain $k = \omega/c_0$, i.e. $c = c_0$. And only in the next approximation, assuming

$$n(r_0) = n(\bar{R}_0) + \frac{\mathrm{d}n}{\mathrm{d}\bar{R}_0}(R_0 - \bar{R}_0)$$

one can see that

$$\Delta c \sim \left.\frac{\mathrm{d}n}{\mathrm{d}R_0}\right|_{R_0=\bar{R}_0}$$

(similar to the corresponding situation for plasma waves). This conclusion, about the practical absence of wave phase velocity dispersion for such a case, has been confirmed experimentally (Kobelev *et al.*, 1979). It should be noted, however, that both the losses and nonlinearity parameter of such a medium (see below) still depend significantly on frequency.

Nonlinear wave interactions

We have already discussed long nonlinear waves in a gas–liquid mixture when the wave spectrum falls within the frequency range $\omega \ll \omega_0$. Here we set aside this restriction and consider the excitation of harmonics and combination frequencies in a liquid with bubbles, taking resonance effects into account.

In an early study by Zabolotskaya and Soluyan (1967) (see also the book by Rudenko & Soluyan, 1975) subharmonic parametric amplification has been treated for bubbles of equal size. The signal frequency was supposed equal to $\omega_0/3$ while that of pumping amounted to $2\omega_0/3$, so that the sum frequency is a resonant one, which promotes heavy absorption of the field at this frequency thus preventing its generation. In addition, nonresonant small-size bubbles were taken into account, yielding additional detuning. Calculations imply that for reasonable medium and pump parameters significant parametric amplification of a signal can be observed.

In this section, however, we shall pay attention mainly to the more realistic case of a wide distribution of bubbles over radii (Kobelev & Ostrovsky, 1980). Consider first the case of second harmonic generation in a plane wave. Represent the pressure perturbation as

$$p' = p_1(\mathbf{r})\exp(\mathrm{i}\omega t) + p_2(\mathbf{r})\exp(2\mathrm{i}\omega t) + \text{c.c.} \tag{6.11}$$

where index 1 denotes a given amplitude of the first harmonic while 2 stands for a small perturbation at frequency 2ω. Substituting (6.11) into the equation for bubble oscillations (6.3), and neglecting the perturbation p_2 in nonlinear terms, leads to a volume perturbation amplitude V_2 at frequency 2ω given by

$$V_2(\omega_0^2 - 4\omega^2 + 2\mathrm{i}\omega f) - \mathcal{X}p_2 = \beta V_1^2[3(\gamma+1)\omega_0^2 + \omega^2] \tag{6.12}$$

where V_1 is the linear response amplitude at frequency ω_1.

Substituting the relation between V_2 and p_2 into (6.2) and then into

(6.1) yields a wave equation for p_2 which in three dimensions (when p'_{xx} is replaced by $\nabla^2 p'$) has the form

$$\nabla^2 p_2 + (k_{2\omega}^2 - \mathrm{i}a_{2\omega})p_2 = k_{2\omega}^2 \varepsilon_{2\omega} p_1^2, \tag{6.13}$$

where $k_{2\omega} = 2\omega/c_0$ (here perturbation of the phase velocity is disregarded) and $a_{2\omega}$ is derived from (6.10) for frequency 2ω. Equation (6.13) has the same form as that for a nondispersive medium (see Chapter 5) but the nonlinearity parameter $\varepsilon_{2\omega}$ is defined by the expression

$$\varepsilon_{2\omega} = 3\gamma\pi c_0^4 \int_0^\infty \frac{[\omega_0^2(\gamma+1) - \omega^2] n(r_0)\,\mathrm{d}r_0}{r_0[\omega_0^2 - 4\omega^2(1 - \mathrm{i}Q_{2\omega}^{-1})][\omega_0^2 - \omega^2(1 - \mathrm{i}Q_{2\omega}^{-1})]}. \tag{6.14}$$

Under the same assumptions as those used when deducing formula (6.10), i.e. considering n as constant over the resonance curve width, and assuming that $n(r_0)$ must decay quickly enough as $r \to 0, \infty$ to provide convergence of the integral, we obtain a rather simple expression,

$$\varepsilon_{2\omega} = \frac{\pi c_0^4}{3\omega^4} n(r_{2\omega}). \tag{6.15}$$

Therefore, to this approximation the nonlinearity parameter depends on the concentration of bubbles with resonance frequency 2ω alone, while the bubbles resonant at frequency ω make no contribution whatsoever. At the same time the oscillation amplitude of an individual bubble with radius r_ω at frequency 2ω is far in excess of that for a bubble with radius $r_{2\omega}$ at the same frequency. This can be attributed to the fact that the oscillation phases of bubbles having radii in the vicinity of r_ω but on opposite sides of r_ω differ by π, while at frequency 2ω this difference amounts to 2π. Therefore, the former cancel each other's fields and do not contribute to the total field, whereas the latter make a cophased contribution. Note that for bubbles with equal radii the major contribution would have been that of bubbles with radius r_ω (Zabolotskaya & Soluyan, 1972).

Let us consider now the generation of a wave of difference frequency $\Omega = \omega_1 - \omega_2$ by the given (pump) waves at two close frequencies ω_1 and ω_2. In the same way as above, the wave equation may be derived in the form

$$\nabla^2 p_\Omega + \frac{\Omega^2}{c_0^2} p_\Omega = -\frac{\varepsilon_\Omega \Omega^2}{\rho_0 c_0^4} p_1 p_2^*, \tag{6.16}$$

where now

$$\varepsilon_\Omega = \frac{2\pi c_0^4 \sqrt{\rho_0/3}}{\gamma p_0} \tag{6.17}$$

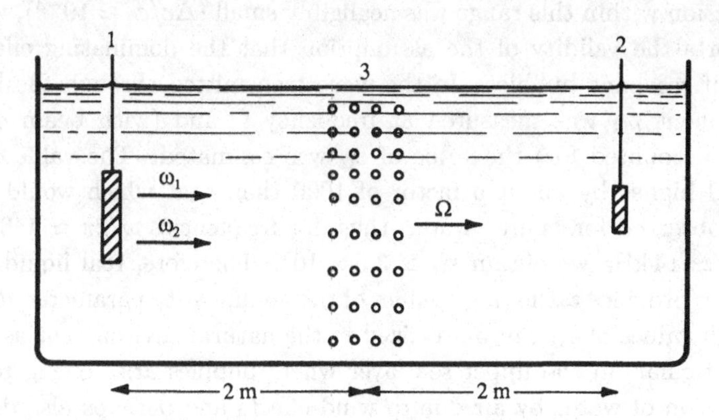

Figure 6.5. Difference frequency wave generation in water containing bubbles. (1) transducer; (2) receiver; (3) bubble layer.

$$\times \int \frac{\omega_0[3(1+\gamma)\omega_0^2 - (\omega_1^2+\omega_2^2-\omega_1\omega_2)]n(r_0)\,dr_0}{(\omega_0^2-\omega_1^2+\mathrm{i}\omega_1^2/Q_1)(\omega_0^2-\omega_2^2-\mathrm{i}\omega_2^2/Q_2)(\omega_0^2-\Omega^2+\mathrm{i}\Omega^2/Q_3)}.$$

Owing to the proximity of ω_1 and ω_2 one can set $n(\omega_1) \simeq n(\omega_2) \simeq n(r_\omega)$ and $Q_1 = Q_2 = Q$. Then, assuming again a wide distribution of bubble radii, one can obtain from (6.17)

$$\varepsilon_\Omega = \mathrm{i}\pi^2 c_0^4 \left[\frac{(3\gamma+2)n(r_\omega)}{\omega^3(\Omega-\mathrm{i}\omega/Q)} - \frac{n(r_\Omega)}{\Omega^2\omega^2}\right]. \qquad (6.18)$$

To further simplify this expression, suppose that $\Omega/\omega \ll Q^{-1}$ i.e. Ω is less than the bubble resonance curve width. Further, in real conditions the concentration usually drops with radius (typically $n \sim r_0^{-\alpha}$ where $\alpha \simeq 3$–4). Then, neglecting the second term in (6.18) we have the nonlinearity parameter equation:

$$\varepsilon_\Omega \simeq 3.93 \times 10^{-2} n(r_\omega) Q \lambda^4 \qquad (6.19)$$

(λ is the wavelength at frequency ω).

Comparing (6.15) and (6.19) we see that the ratio $\varepsilon_\Omega/\varepsilon_{2\omega}$ is of the order of $50\,Qn(r_\omega)/n(r_{2\omega}) \gg 1$. This is attributed to an already-mentioned misphasing of the bubble oscillations under the frequency doubling while for small difference frequency generation this misphasing is not observed. On the other hand, for bubbles of equal radii, $\varepsilon_{2\omega} \gg \varepsilon_\omega$.

Relation (6.19) was verified in an experiment (Kobelev & Ostrovsky, 1980), where a sound beam crossed a bubble layer created electrolytically. The bubble concentration was measured by damping, in accordance with (6.10a), within the range 90–290 kHz. The sound velocity dispersion within this range was negligibly small ($\Delta c/c_0 \approx 10^{-3}$), which supports the validity of the assumption that the dominating effect is that of resonant bubbles. In the wave transmitted through the layer, component p_Ω was measured at frequency Ω and (with beam divergence accounted for) the value of ε_Ω was estimated. The value of p_Ω proved higher by about a factor of 1000 than that which would have been observed for "pure" water; thus, for frequencies $\omega/2\pi \simeq 140\,\text{kHz}$, $\Omega/2\pi \simeq 14\,\text{kHz}$ we obtain $\varepsilon_\Omega \simeq 3.5 \times 10^3$. Therefore, real liquid with bubbles provides rather high values of the nonlinearity parameter. Large enough values of ε_Ω can be realized in the natural environment as well, in particular, in the upper sea layer where bubbles arise owing to the saturation of water by air due to wind effects and perhaps also due to the influence of sea organisms. The quantity ε_Ω evaluated from (6.19) in the majority of cases exceeds the value typical of water by a factor of 10 or even 100.

In this connection the possibility of enhancing the effectiveness of a parametric array by placing bubbles in the interaction region is worth discussion. This idea has been put forward more than once (see, for example, Zabolotskaya & Soluyan, 1967), but its realization has proved rather difficult because of substantial losses associated with resonant bubbles: in fact, the ratio ε_Ω/a, in accordance with (6.10) and (6.19), does not depend on bubble concentration at all! A more reliable way to avoid this difficulty is to use a bubble layer (a "curtain") situated in the far-field zone of the primary wave beam and oriented across its axis instead of "bubbling" the whole bulk of the beam (Kustov *et al.*, 1982). In this case the directivity of the parametric array will be determined by the pump-irradiated area of the layer; thus, one obtains an aperture antenna rather than a travelling-wave one. This permits one to keep a relatively narrow directional pattern, while keeping losses at an acceptably low level.

Experiments were performed in a $4 \times 5 \times 5\,\text{m}^3$ tank (Figure 6.5). A 10 cm diameter pump transducer radiated a biharmonic primary beam in which one frequency was fixed at 130 kHz while the other could vary from 135 to 160 kHz so that $\Omega = 5\text{–}30\,\text{kHz}$. Pump pulses had a duration of 40 ms. At a distance 2 m ahead of the transmitter, a 14 cm thickness hydrogen bubble layer was created with the help of a wire grid. A signal hydrophone was located at a distance of 2 m beyond the layer. Measurements of bubble concentrations within the layer (from the resonance attenuation at different frequencies, as was mentioned above) had shown a wide distribution over radii, so that resonance bubbles play the crucial role and the formula (6.19) for the nonlinearity parameter is applicable.

In the experiment, the level of the signal amplitude and the directivity pattern were both registered. We shall not discuss the results here in any detail, but we mention that for example at $\Omega = 20$ kHz the axial pressure amplitude was four times larger than it could be without bubbles. This gain may seem not very impressive, but one should remember that in the latter case the interaction length equals about 150 m (losses in water are small), while the bubble layer reduces it to 14 cm (just the layer thickness) and the total length of the antenna (distance from the transmitter to the layer) is 2 m. Anyway, after removing the bubbles, the secondary field could not be registered by the hydrophone at all. Thus, the "bubble curtain" method seems to be promising, at least for laboratory tank conditions where the limitations on the antenna length are obviously very severe.

6.7 Parametric sound transmitter in a waveguide

The field of a parametric sound transmitter acquires specific properties in a waveguide. Here, resonance conditions may be satisfied for non-collinearly propagating waves, so that the direction of a low-frequency radiation maximum does not coincide with that of the primary wave (pump) propagation.

Let us consider the behaviour of a parametric transmitter field in a plane waveguide when a discrete set of normal waves is excited (Zaitsev *et al.*, 1987). One wall of the waveguide ($z = 0$) will be assumed free, while the other ($z = H$) is rigid. A waveguide formed by a layer adjacent to a liquid halfspace (the Pekeris model), see Tolstoy & Clay (1964), may be analysed in the same way. At an angle β to the waveguide axis x, a highly directional biharmonic pump beam with frequencies ω_1 and ω_2, is propagated. If losses are sufficiently low, this beam may be subject to multiple reflections from the waveguide walls. The field generation at a resonance frequency is described by the equation (see Chapter 5)

$$\nabla^2 p_s' + k^2 p_s' = -\frac{\varepsilon\Omega}{\rho_0 c_0^4} p_1' p_2'^*. \tag{7.1}$$

Here $k = \Omega/c_0$, $\Omega = \omega_1 - \omega_2$ is the difference frequency, p_1', p_2' correspond to pressures in the primary waves, specified in the form

$$p_i' = p_i(\mathbf{r}_1)\exp(-\mathrm{i}k_1\xi - a\xi_1), \quad i = 1, 2, \tag{7.2}$$

where $\xi(x,z)$ is the ray coordinate, a being the absorption factor.

The low-frequency field p'_s satisfying the conditions on the waveguide walls will be sought in the form of an expansion in the waveguide's natural modes: $p'_s = \Sigma A_n(x,y)y_n(z)$ where $\{y_n\}$ is a system of orthogonal eigenfunctions. Thus, for the waveguide with one soft and one rigid boundary

$$f_n = \sin(\sigma_n z), \tag{7.3}$$

where $\sigma_n = \pi(2n-1)/2H$ is a vertical wave number, n being the mode number.

A solution to (7.1) can be formulated by means of the Green's function for a plane waveguide (Tolstoy & Clay, 1969),

$$p'_s = \frac{\mathrm{i}\varepsilon\Omega^2}{\rho_0 c_0^4}\frac{2}{H}\sum y_n(z)\int \mathrm{d}x'\mathrm{d}y'\mathrm{d}z' p_1(r)p_2^*(r')f_n(z')x$$
$$\times \exp\left(-\mathrm{i}k\xi' - 2a\xi'\right) H_0^{(2)}\left(\mathcal{X}_n\sqrt{(x-x')^2+(y-y')^2}\right). \tag{7.4}$$

Here $H_0^{(2)}$ is the Hankel function of the second kind; the horizontal wave number $\mathcal{X}_n$ satisfies the relation $k^2 = \mathcal{X}_n^2 + \sigma_n^2$. The integral is taken over the region occupied by the pump beam. We shall be seeking the low-frequency field in a far-field zone ($x \gg a^{-1}$) where its structure has been completely formed. In addition we shall assume that the pump beam satisfies the following conditions: the characteristic width $d(x)$ is much less than the vertical scale σ_n^{-1} of the mode radiated; in each beam cross-section the phase variation of virtual low-frequency sources is considerably less than π.

This approximation is similar to the conditions permitting one to describe the field of a parametric transmitter with a weakly diverging pump beam in free space within the framework of the Westervelt model.

Provided these conditions are satisfied, the integral in (7.4) in can be reduced to the form

$$p'_s = \frac{\varepsilon\Omega^2}{4c_0^4}\sqrt{\frac{2}{\pi}}\exp\left(\frac{\mathrm{i}\pi}{4}\right)\frac{1}{H}W\sum f_n(z')\int\int f_n(z)\frac{\exp(-\mathrm{i}\mathcal{X}_n x')}{\sqrt{\mathcal{X}_n x}}$$
$$\times\, \delta(\xi(x',z'))\exp\left(-\mathrm{i}k\xi' - 2a\xi' + \mathrm{i}\mathcal{X}_n x'\cos\theta\right)\mathrm{d}x'\mathrm{d}z', \tag{7.5}$$

where $W = \int\int(p_1 p_2^*/\rho c_0)\,\mathrm{d}s$ is the power of the pump beam, θ is the angle measured from the transmitter axis in a horizontal plane, and $\delta(\xi)$ is the delta function. A further analysis of (7.5) will be performed for three characteristic cases: horizontal orientation of a transmitter; small slope, when the pump beam reflection from the waveguide boundaries is negligible; and an arbitrary slope of the beam axis when reflection cannot be disregarded.

For a horizontally oriented beam (7.5) yields an expression for a difference frequency field

$$p_s' = -\frac{\varepsilon\Omega^2}{4c_0^3}\sqrt{\frac{2}{\pi}}\exp\left(\frac{\mathrm{i}\pi}{4}\right)\frac{W}{H}\sum f_n(z)f_n(z_0)\frac{\exp(-\mathrm{i}\mathcal{X}_n x)}{\sqrt{\mathcal{X}_n x}}(\mathrm{i}\Delta_n + 2a)^{-1}, \tag{7.6}$$

where $\Delta_n = k - \mathcal{X}_n \cos\theta$ is the deviation from synchronism between a nonlinear source and the n-th mode. From (7.6) it is clear that a diagram maximum is observed for an angle $\theta = 0$. For weak damping ($a \ll \Delta_n$), and $\theta = 0$, (7.6) coincides with an expression for the field on the axis obtained by Karabutova and Novikov (1968).

The expression for the directivity diagram half-width of each mode (at the half-power level) can also be readily derived:

$$\Delta\theta_n = \left[\sqrt{2\gamma_n^4 + (4a/k)^2} - \gamma_n^2\right]^{1/2}, \tag{7.7}$$

where $\gamma_n = \arctan(\sigma_n/\mathcal{X}_n)$ is the slope angle of the Brillouin wave corresponding to the n-th mode (here $\gamma_n \ll 1$ is assumed). Equation (7.17) implies that the diagram width grows with the mode number (i.e. with angle γ_n). For a sufficiently short antenna ($a \gg \gamma_n^2 k$), when desynchronism has not yet shown up, (7.7) yields the familiar expression for the directivity diagram half-width in free space: $\Delta\theta = 2(a/k)^{1/2}$ (see (2.8) of Chapter 5). Mode amplitudes in the direction $\theta = 0$ are proportional to $[\gamma_n^2/2 + (2a/k)^2]^{-1/2}$ and decrease with mode number, because for higher modes the angle γ_n increases and desynchronism becomes more pronounced.

Consider now the case of an inclined primary beam whose axis slope angle is supposed to be so small ($\beta \leq Ha$) that the pump beam decays before reflection from the waveguide boundary. Representing the eigenfunction (7.2) as a pair of plane Brillouin waves and integrating (7.5) we obtain the following formula for the low-frequency field:

$$p_s' = -\frac{\varepsilon\Omega^2}{4c_0^3}\sqrt{\frac{2}{\pi}}\exp\left(\frac{\mathrm{i}\pi}{4}\right)\frac{2}{H}W\sum f_n(z)\frac{\exp(-\mathrm{i}\mathcal{X}_n x)}{\sqrt{\mathcal{X}_n x}}$$
$$\times\left(\frac{\exp[\mathrm{i}(\sigma_n z_0 - k\sin\beta z_0)]}{\mathrm{i}\Delta_n(\gamma_n) + 2a/\cos\beta} - \frac{\exp[-\mathrm{i}(\sigma_n z_0 + k\sin\beta z_0)]}{\mathrm{i}\Delta_n(-\gamma_n) + 2a/\cos\beta}\right) \tag{7.8}$$

where

$$\Delta_n = k(\cos\beta - \cos\gamma_n\cos\theta) + k(\sin\beta - \sin\gamma_n)\tan\beta.$$

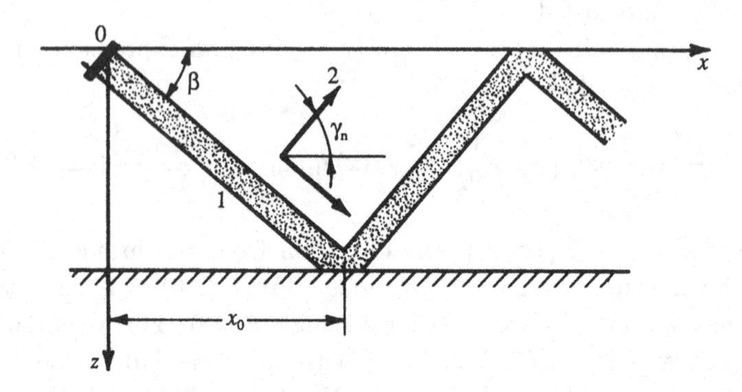

Figure 6.6. Example of the parametric array in waveguide geometry in the case of multiple pump-beam reflections. (1) pump-beam; (2) Brillouin wave directions.

Thus, the value of the n-th mode excitation factor is defined by a pump interaction with each of the two Brillouin waves, the more efficient interaction being that with the wave directed at a sharper angle to the pump beam, since the value Δ_n is less for this wave.

The behaviour of the diagram in a horizontal plane is similar to the case considered earlier, its maximum being at $\theta = 0$ as before.

The most efficient excitation of a mode numbered n is observed at $\Delta_n = 0$, which corresponds to the case when the primary beam axis slope angle coincides with that of a Brillouin wave ($\beta = \pm\gamma_n$). Here other modes have only a weaker excitation, such that the excitation factor of the n-th mode decreases by a factor of $\sqrt{2}$ at $\Delta\beta = |\beta - \gamma_n| = 2(a/k)^{1/2}$, which, by the way, corresponds to the directivity diagram half-width in free space. Hence at small β one can selectively excite low number modes which are typically most interesting from a practical viewpoint.

To provide the possibility of selective excitation of a higher mode, one should allow for pump beam reflections from the boundaries (Figure 6.6). In the case of multiple re-reflections, the integral with respect to x in (7.5) must be broken down into a sum of integrals over elementary rectilinear portions of length x_0. The final result of parametric interaction is defined by interference of the secondary fields from all such portions, such that phase relations between each portion's contributions to the total field are determined by the boundary conditions on the waveguide walls.

This interaction may be calculated most easily for a waveguide with two rigid boundaries when phase shifts under boundary reflection are absent. Here $f_n = \cos(n\pi z/H)$ and integral (7.5) has the form

$$p'_s = \frac{\varepsilon\Omega^2}{4c_0^3}\sqrt{\frac{2}{\pi}}\exp\left(\frac{\mathrm{i}\pi}{4}\right)\frac{2}{H}W\sum f_n(z)\frac{\exp(-\mathrm{i}\mathcal{X}_n x)}{\sqrt{\mathcal{X}_n x}}D_n(\theta,\beta), \quad (7.9)$$

where

$$D_n^{-1} = \mathrm{i}[k(\cos\beta - \cos\gamma_n\cos\theta) + k(\sin\beta - \sin\gamma_n)\tan\beta] + 2a/\cos\beta.$$

Evidently, in this case a parametric interaction for all modes is most efficient when directed along the pump beam axis, just as in the case of free space. The excited mode amplitudes depend, however, on the vertical angle β, and attain a maximum at strict synchronism with a corresponding Brillouin wave: $\beta = \gamma_n$. The angular domain corresponding to this excitation is $\Delta\beta = 2(a/k)^{1/2}$, that is, it is defined by the total effective length of the antenna a^{-1}. Therefore, for small enough beam dissipation, giving $\Delta\beta < \gamma_{n+1} - \gamma_n$, the possibility arises to excite selectively individual modes, the higher ones included.

For a waveguide with one free boundary things evolve in another way. Here $f_n \sim \sin\sigma_n z$ and the secondary field is expressed through

$$p'_s = -\frac{\varepsilon\Omega^2}{4c_0^3}\sqrt{\frac{2}{\pi}}\exp\left(\frac{\mathrm{i}\pi}{4}\right)\frac{2}{H}W\sum f_n(z)\frac{\exp(-\mathrm{i}\mathcal{X}_n x)}{\sqrt{\mathcal{X}_n x}}D_n(\theta,\beta). \quad (7.10)$$

The diagram factor D_n now acquires a more complex form:

$$D_n = F_1(\gamma_n)F_2(\gamma_n) - F_1(-\gamma_n)F_2(-\gamma_n),$$
$$F_1(\gamma_n) = \frac{1 - \exp(-\mathrm{i}\Delta_n x_0 - 2ax_0/\cos\beta)}{\mathrm{i}\Delta_n(\gamma_n) + 2a/\cos\beta},$$
$$F_2(\gamma_n) = [1 + \exp(-\mathrm{i}\Delta_n x_0 - 2ax_0/\cos\beta)]^{-1}, \quad (7.11)$$

where $x = H/\tan\beta$ and

$$\Delta_n = k(\cos\beta - \cos\gamma_n\cos\theta) + k(\sin\beta - \sin\gamma_n)\tan\beta.$$

In (7.11) the factor F_1 characterizes the radiation field of an individual portion of length x_0 where the pump beam has a constant slope, while the factor F_2 is determined by the field interference from different portions. It should be noted that the maxima of these factors do not coincide. This is easily understood; indeed, with weak decay, when a large number of reflections takes place over the antenna length ($a^{-1} \gg x_0$), we have $F_1 \approx \sin(\Delta_n x_0/2)/(\Delta_n/2)$. In this case, the angular dependence of F_1 is smoother than that of F_2 so that a parametric radiation maximum is defined here mainly by the form of the function F_2. Analysing it for low angles β, γ_n, $\theta \ll 1$ one can obtain the following expression for the principal maxima of the mode directivity diagram in a horizontal plane: $\theta_{\max} \simeq \pm[2\pi\beta/(kH) - (\beta - \gamma_n)^2]^{1/2}$. It is evident that at a certain given slope angle of the beam, different modes are formed whose radiation maxima are observed for various angles. Let us emphasize that the radiation maximum of the mode n coincides with $\theta = 0$ provided that $2\pi\beta/(kH) - (\beta - \gamma_n)^2 = 0$. In this case a desynchronism between the interacting waves due to nonlinearity is compensated by the phase shift at reflection from the waveguide free boundary.

Parametric transmitter operation in a waveguide has been studied experimentally in laboratory tanks by Bjørnø *et al.* (1979). We consider here some results of a model experiment (Zaitsev *et al.*, 1989). The waveguide was formed by a water layer 40–180 mm thick. A receiving hydrophone was placed at a distance of about 4 mm from the transmitter. The highly directional biharmonic pump beam had a centre frequency of 3.1 MHz, while the difference frequency could vary within the range 200 ± 50 kHz. An analysis of the parametric radiation mode structure was performed by a method of pulse selection based on the spatial separation of individual mode pulses due to their group velocity discrepancies. Figure 6.7 depicts predicted and experimental dependencies of the first mode field level on the waveguide depth, for a horizontally oriented transmitter. A wide dispersion of the calculated values is caused by inaccurate measurements of the pump beam power, which is a parameter of the theoretical model. Note that at low waveguide depths the field level in it may be even less than in free space, which is associated with a severe desynchronism between the pump beam and the first mode. The synchronism increases with depth so that the field grows, to attain a maximum exceeding its level in free space at a certain depth. At a still larger depth the waveguide's focusing influence becomes insignificant, and the radiation proceeds practically as in an unbounded medium. A certain rise of the predicted field level over the experimental one can be attributed to the above-mentioned effect of finiteness of the real pump beam width, which is most important at small waveguide depths.

The use of the impulse method for normal mode selection allows one to investigate the individual mode field level dependence on the primary beam orientation in a vertical plane. Figure 6.8 depicts, for example, experimental and calculated curves for the first mode. In this case a given mode excitation condition requires that the transmitter axis orientation should differ both from the waveguide axis direction and from the wave vector direction of the corresponding Brillouin wave, whereas

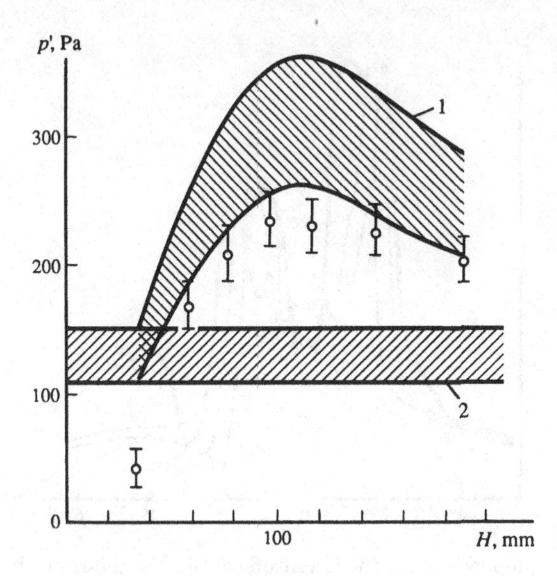

Figure 6.7. Dependences of the first-mode field level on the waveguide depth H. Region 1, waveguide theory, equation (7.10); region 2, free-space theory; open circles, experiment (Zaitsev *et al.*, 1987).

the measured values of the optimal angle are in fair agreement with the calculated ones. Some experimental results in investigations of a parametric transmitter sound field in a shallow sea are presented in a paper by Donskoy *et al* (1993).

References

Adler, L. & Breazeale, M.A. (1970). Generation of fractional harmonics in resonant ultrasonic wave system, *JASA* **48,** 1077–83.

Akhmanov, S.A. & Khokhlov, R.V. (1964). *Problems of Nonlinear Optics* (Viniti, Moscow).

Andreev, V.G., Vasil'yeva, O.A., Lapshin, E.A. & Rudenko, O.V. (1985). Processes of second harmonic generation and parametric amplification in a medium with selective absorption, *Akust. Zh.* **31,** 12–16.

Belyaeva, I.Y., Ostrovsky, L.A., & Timanin (1991). Experiment on harmonic gerneration in grainy media, *Acoust. Lett.* **15,** 221–24.

Bjørnø, L., Folsberg, J. & Pederson, L. (1979). Parametric arrays in shallow water, *Journal de Physique* **40** *supplement 11,* CB–71.

Bloembergen, N.S. (1975). *Nonlinear Optics* (W.A. Benjamin, Inc., New York-Amsterdam).

Clay, C.S. & Medwin, H. (1977). *Acoustical oceanography: principles and applications* (J. Wiley & Sons, New York).

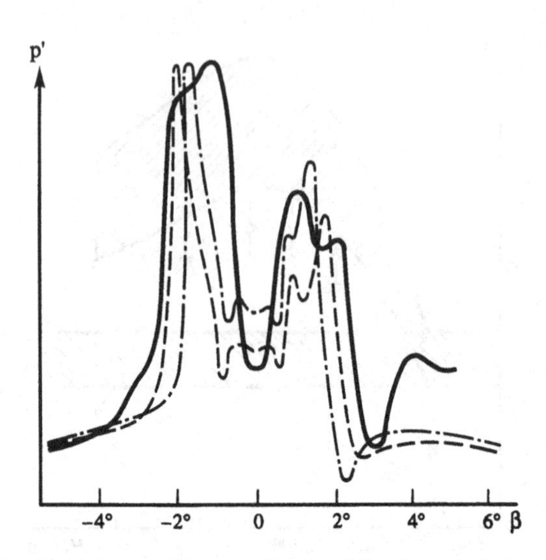

Figure 6.8. The dependence of the amplitude of the first mode on the orientation of the primary beam. p' is in arbitrary units. Solid curve, experiment; broken curves, theoretical predictions for two frequencies $f = f_0 \pm 15$ kHz, where $f_0 =$ 3.1 kHz is the main frequency.

Donskoy, D.M., Zaitsev, V.Y., Naugol'nykh, K.A. & Sutin, A.M. (1933). Experimental investigations of a powerful parametric transmitter field in a shallow sea, *Akust. Zh.* **39, 2,** 266–73.

Dreiden, G.V., Ostrovsky, Yu.U., Samsonov, A.M., Semenova, I.V. & Sokurinskaya, E.V. (1984). Formation and propagation of train solitons in non-linearly elastic solids, *Sov. Phys. Techn. Phys. ZTF* **33,** 1237–41.

Eller, A.J. (1973). Fractional-harmonic frequency pairs in nonlinear systems, *JASA* **53,** 758–65.

Fox, F. & Curley, S. (1955). Phase velocity and absorption measurements in water containing air bubbles, *JASA* **27,** 534–9.

Gasenko, V.G., Nakoryakov, V.E. & Shreiber, I.R. (1977). Burgers–Korteweg–de Vries approximation in wave dynamics of gas–fluid systems, in *Nonlinear Processes in Two-phase Media,* pp. 17–31, ed. S. Kutateladze (Novosibirsk. Inst. Teplofiziki SO AN SSSR.,).

Gol'dberg, Z.A. (1972). Evaluation of efficiency of parametric amplification of acoustic waves in liquids, *Prikl. & Teor. Fizika, Kazakh. Univ. Alma-Ata* **3,** 38–43.

Gurevich, A.V. & Pitaevsky, L.P. (1987). Average description of waves in Korteweg–de Vries–Burgers equation, *Zh ETF* **93 (9),** 871–82.

Hamilton, M.F. & Ten Cate, J.A. (1987). Sum and frequency generation due to noncollinear wave interaction in rectangular duct, *JASA* **87,** 1705.

Kaplan, A.E., Kravtsov, Y.A. & Rylov, V.B. (1966). *Parametric generator and frequency divider* (Moscow, Sov. Radio.).

Karabutova, N.E. & Novikov, B.K. (1986). Work of parametric radiator of

sound in plane waveguide, *Akust. Zh.* **32,** 65–70.
Keller, J.B. & Miltman, M.H. (1971). Finite-amplitude sound-wave propagation in a waveguide, *JASA* **49,** 329–33.
Khokhlov, R.V. (1961). Theory of shock radio-waves in nonlinear lines, *Radiotekhnika i Elektronika* **6,** 917.
Kobelev, Yu.A. & Ostrovsky, L.A. (1980). Models of gas-liquid mixture as a nonlinear dispersive medium, p. 143, in *Nonlinear Acoustics,* ed. V. Zverev & L.Ostrovsky (IPF. AN SSSR., Gorky).
Kobelev, Y.A., Ostrovsky, L.A. & Sutin, A.M. (1979). Effect of self-transparency for acoustic waves in liquid with gas bubbles, *Lett. to ZhETF* **30,** 423.
Korpel, A. & Adler, R. (1965). Parametric phenomena observed on ultrasonic waves in water, *Appl. Phys. Lett.* **7,** 106–7.
Kuznetsov, V.V., Nakoryakov, V.E., Pokusaev, B.G. & Shreiber, I.R. (1978). Propagation of perturbation in a gas-liquid mixture, *J. Fluid Mech.* **85,** 85–96.
Landau, L.D. & Lifshitz, E.M. (1987). *Theory of Elasticity* (Moscow, Nauka).
Lazaridi, A.I. & Nesterenko, V.F. (1985). Observation of waves of a new type in one-dimensional grained medium, *Zh. Prikl. Mekh. i Tekhn. Fiz.* **3,** 115–18.
Moffett, M.B. & Mellen, R.H. (1981). On absorption as a means of saturation suppression, *J. Sound Vibr.* **76,** 295.
Molotkov, L.D., & Vakulenko, S.A. (1980). Nonlinear longitudinal waves in an inhomogeneous rod, *Zap. Nautchn. Seminara LAMI* **99,** 64–73.
Nesterenko, V.F. (1983). Propagation of nonlinear compression pulses in grained media, *Zh. Prikl. Mekh. i Tekhn. Fiziki.* **5,** 136–48.
Nakoryakov, V.E., Sobolev, V.V. & Shreiber, I.R. (1983). *Propagation of Waves in Gas- and Vapour-Liquid Media* (Inst. Teplofiziki, Novosibirsk).
Naugol'nykh, K.A. & Ostrovsky, L.A. (1980). Parametric transmitter of sound, in *Nonlinear Acoustics,* ed. V. Zverev, L. Ostrovsky (IPF AN SSSR, Gorky).
Ostrovsky, L.A. & Papilova, I.A. (1973). On nonlinear interaction and parametric amplification of waves in acoustic waveguides, *Akust. Zh.* **19,** 67–75.
Ostrovsky, L.A., Papilova, I.A. & Sutin, A.M. (1972). Parametric generation of ultrasound, *Lett. to ZhETF* **15 (8),** 456–8.
Ostrovsky, L.A., Soustova, I.A. & Sutin, A.M. (1978). Nonlinear and parametric phenomena in dispersive acoustic systems, *Acustica* **5,** 298–306.
Ostrovsky, L.A. & Sutin, A.M. (1977). Nonlinear elastic waves in rods, *Prikladnaya matematika i mechanika* **3,** 531–7.
Rudenko, O.V. & Soluyan, S.I. (1977). *Theoretical foundations of nonlinear acoustics* (Plenum, New York).
Rudenko, O.V. (1983). To the problem of artificial nonlinear media with resonant absorber, *Akust. Zh.* **20,** 398.
Ryutov, D.D. (1975). Analogue of Landau decay in a problem of sound wave propagation in liquid with gas bubbles, *Lett. to ZhETF* **22,** 446–9.
Samsonov, A.M. (1984). Soliton evolution in a rod with variable cross-section, *Sov. Phys. Doklady* **29,** 586–7.
Samsonov, A.M. (1988). On the existence of longitudinal waves in an inhomogeneous nonlinearly-elastic rod, *Prikl. Mat. i Makh.* **51,** 483–8.

Samsonov, A.M. & Sokurinskaya, E.V. (1987). Solitary longitudinal waves in an inhomogeneous nonlinearly-elastic rod, *Prikl. Mat. i Mekh.* **51,** 483–8.
Tolstoy, I. & Clay, K.S. (1966). *Ocean Acoustics* (McGraw-Hill, New York).
Yen Nai-chyan (1975). Experimental investigation of subharmonic generation in an acoustic interferometer, *JASA* **57,** 1357–62.
Woodsum, H.C. (1981). Author's reply, *J. Sound Vib.* **76,** 297.
Zabolotskaya, E.A. & Soluyan, S.I. (1967). On a possibility of acoustic wave amplification, *Akust. Zh.* **13,** 296.
Zabolotskaya, E.A. & Soluyan, S.I. (1972). Radiation of harmonics and combination frequencies by air bubbles, *Akust. Zh.* **18,** 472.
Zaitsev, V.Yu, Kurin, V.V. & Sutin, A.M. (1989). Model investigations of mode structure of the field of a parametric transmitter in acoustic waveguide, *Akust. Zh.* **35,** 266–71.
Zaitsev, Y., Ostrovsky, L.A. & Sutin, A.M. (1987). Mode structure of parametric radiation field in acoustic waveguide, *Akust. Zh.* **33,** 37–41.
Zarembo, L.K. & Serdoboloskaya, O.Yu. (1974). To the problem of parametric amplification and parametric generation of acoustic waves, *Akust. Zh.* **20,** 726–32.

7
Self-action and stimulated scattering of sound

7.1 Cubic nonlinearity in acoustics

In this chapter we shall discuss some effects associated with a cubic (or, generally, "odd") medium nonlinearity rather than a quadratic one. These include the effects of self-focusing, "self-transparency", stimulated scattering, wave front conjugation and others. In classical nonlinear acoustics the role of cubic nonlinearity is rarely substantial and until recently cubic terms have not had attention lavished upon them. The role of cubic corrections, violating oddness of the field in a sawtooth wave, can only be mentioned (Rudenko & Soluyan, 1975). In the 1980s the study of self-action and stimulated scattering effects in both theoretical and experimental acoustics was given a much more pronounced impetus; it was stimulated also by advances in nonlinear optics and plasma physics.

The question arises: can "cubic" effects emerge before discontinuities in a wave appear, bringing about its rapid decay? At first glance, this is impossible for a nondispersive medium, as cubic terms in the governing equations are much less in magnitude than quadratic ones that are responsible for nonlinear profile distortions. However, the presence of dispersion or specific nonlinearity mechanisms (e.g. thermal ones) may change this situation drastically. For example, as was demonstrated in Chapter 1, there exist media with "anomalous" nonlinearity, such as solids with complex structures, in which "cubic" effects can even dominate over quadratic ones. The number of experimental demonstrations of the effects mentioned above is continually growing nowadays in acoustics. Moreover, the related problems are of considerable heuristic interest.

Many of the phenomena tackled in this chapter are connected with the effects of nonlinearity on the average velocity of an acoustic wave. To a quadratic approximation in the field amplitude, such an effect is not felt, becoming pronounced only at the third order in the wave amplitude. Indeed, in the governing equations for the sound pressure p', let the terms proportional to p'^3 be present. Substituting a nearly harmonic field into them of the form

$$p' = \frac{1}{2}\left[A(\mathbf{r},t)\exp(i\omega t) + A^*(\mathbf{r},t)\exp(-i\omega t)\right] \tag{1.1}$$

where the amplitude dependence on t is considered slow, we shall obtain terms of the type $A^3\exp(3i\omega t)$ (the third harmonic) and $|A|^2 A\exp(i\omega t)$, i.e. a nonlinear response at the fundamental frequency. If it is of a reactive nature (i.e. is not connected with energy losses), one can attribute to it corrections of the order of $|A|^2$ to the wave propagation velocity.

It should be immediately noted that in "common" homogeneous media, the value of the cubic acoustical nonlinearity is usually extremely small. Having in mind, e.g., gas or liquid, let us expand the equation of state $p(\rho)$ in a series, retaining the cubic term

$$p' = c_0^2\rho' + \frac{1}{2}\frac{\partial c_0^2}{\partial \rho}\rho'^2 + \frac{1}{6}\frac{\partial c_0^2}{\partial \rho^2}\rho'^3, \tag{1.2a}$$

or, for an ideal gas,

$$p' = c_0^2\left(\rho' + \frac{\gamma-1}{2\rho_0}\rho'^2 + \frac{(\gamma-1)(\gamma-2)}{6\rho_0^2}\rho'^3\right). \tag{1.2b}$$

Because $M = (\rho'/\rho_0)_{\max}$ is essentially the acoustic Mach number, then the ratio between cubic and linear terms in (1.2*b*) will be of the order of $(\gamma-1)(\gamma-2)M^2/6$, but M will seldom be in excess of 10^{-3} in acoustics.

It should be emphasized that such effects can be associated with quadratic nonlinearity as well, but, as previously, at the third infinitesimal order. Should the equations include terms with p'^2, the harmonic field will produce a disturbance in the form of a second harmonic ($\sim A^2\exp(2i\omega t)$) and a constant component ($\sim |A|^2$). Now substituting the sum of the basic harmonic and the disturbances mentioned into quadratic terms, one obtains terms proportional to $|A|^2 A\exp(i\omega t)$.

In addition, if we speak about an "averaged" nonlinearity manifested during many periods of the field oscillations, other mechanisms can be involved. The most universal seems to be the thermal one: if wave energy is absorbed in the medium, the temperature T of the latter is subject to variation such that the thermodynamic equation of state $p = p(\rho, T)$ will be sensitive to thermal effects.

The set of equations describing sound propagation in a viscous heat-conducting medium has the form (Landau & Lifshitz, 1986)

$$\frac{\partial \rho'}{\partial t} + \rho_0 \nabla \cdot \mathbf{v} = 0, \tag{1.3a}$$

$$\frac{\partial \mathbf{v}}{\partial t} + \frac{\nabla p'}{\rho_0} = \nu \nabla^2 \mathbf{v} + \frac{1}{\rho_0}\left(\zeta + \frac{\eta}{3}\right) \nabla\nabla \cdot \mathbf{v}, \tag{1.3b}$$

$$\rho_0 c_p \frac{\partial T'}{\partial t} - \kappa \nabla^2 T' = \eta \frac{\partial v_i}{\partial x_k}\left(\frac{\partial v_i}{\partial x_k} + \frac{\partial v_k}{\partial \xi} - \frac{2}{3}\delta_{ik}\frac{\partial v_\ell}{\partial x_\ell}\right) + \zeta(\nabla \cdot \mathbf{v})^2. \tag{1.3c}$$

Here T' is the perturbation of the medium temperature imposed by the wave, κ is the thermal conductivity factor, c_p the specific heat at constant pressure. In equations (1.3) we neglect hydrodynamic nonlinearity, accounting only for thermal effects. Setting the sound speed c to be a function of the period-averaged temperature $T_a = T_0 + \overline{T}$, where $\overline{T}$ is the average temperature perturbation due to wave energy dissipation, the equation of state may be written in the form $\rho' = p'/c_0^2(\overline{T})$, or (Zabolotskaya & Khokhlov, 1976)

$$\rho' = \frac{1}{c_0^2}p' - \frac{2}{c_0^3}\frac{\mathrm{d}c_0}{\mathrm{d}T}p'\overline{T}, \tag{1.4}$$

where $c_0 = c(T_0)$, and

$$\frac{\mathrm{d}c_0}{\mathrm{d}T} = \left(\frac{\partial c}{\partial T}\right)_{T=T_0}.$$

These equations imply that thermal nonlinearity is, in essence, connected with dissipation, in that the thermal variations in (1.3c) are governed by energy dissipation due to viscosity and thermal conductivity. Moreover, the mechanism of such nonlinearity is inertial in time and nonlocal in space, so that in the general case it cannot be reduced to a local relation of the type (1.2).

Let us assume that variations of all the required quantities are those of an "almost plane" wave, i.e. a wave beam or pulse,

$$p' = \frac{1}{2}\left\{p(\mathbf{r},t)\exp[\mathrm{i}(\omega t - kx)] + \text{c.c.}\right\}, \tag{1.5}$$

$$v = \frac{1}{2}\left\{v(\mathbf{r},t)\exp[\mathrm{i}(\omega t - kx)] + \text{c.c.}\right\}, \tag{1.6}$$

$$T' = \frac{1}{2}\left\{T(\mathbf{r},t)\exp[\mathrm{i}(\omega t - kx)] + \text{c.c.}\right\} + \overline{T}(\mathbf{r},t) \tag{1.7}.$$

In (1.7) a slowly changing temperature component $\overline{T}$ is added, which is responsible for the thermal "self-action" of the waves on $v = v_x$. The equation for $\overline{T}$ is deduced, through averaging of (1.3c), as

$$c_p \rho_0 \frac{\partial \overline{T}}{\partial t} - k \nabla \overline{T} = \frac{b \rho_0 \kappa^2}{2} |v|^2, \tag{1.8}$$

where $b = \xi + 4\eta/3$ as above.

Hence, one can easily estimate the characteristic relaxation time of the thermal process as $\tau_T \approx \ell^2/\mathcal{X}$, where $\mathcal{X} = \frac{\kappa}{\rho_0 c_p}$ is the thermal diffusivity, and ℓ is the characteristic scale of the perturbation (say, the beam width) (Bunkin & Lyakhov, 1984). If, for instance, $\ell = 0.5$ cm, then for benzene and glycerol one obtains $\tau_T \approx 260$–270 s. Should the acoustic pulse duration exceed this value, one may speak about a steady-state process and neglect the term $\partial \overline{T}/\partial t$ in (1.8). Then, supposing that $\nabla^2 \overline{T} \approx \overline{T}/\ell^2$, one can estimate the value of $\overline{T}$ as

$$\overline{T} \sim \frac{\ell^2 \kappa^2 b \rho_0 |v|^2}{2k}. \tag{1.9}$$

Therefore, the ratio between the nonlinear and linear terms in (1.4) will be of the order

$$m = (\ell \kappa)^2 b \rho_0 |v|^2 \frac{\mathrm{d}c_0}{\mathrm{d}T} / k c_0. \tag{1.10}$$

The values of $c_0^{-1} \mathrm{d}c_0/\mathrm{d}T$ for various liquids seem not to differ radically, being of the order of $10^{-3}\,\mathrm{K}^{-1}$. For water this value amounts to $1.7 \times 10^{-3}\,\mathrm{K}^{-1}$ while $\kappa = 0.6\mathrm{W\,m^{-1}\,K^{-1}}$. Substituting the values for all parameters into (1.10) yields, for example for $\ell = 0.5$ cm, ultrasound frequency 5 MHz and pressure amplitude 1 atmosphere, the value $m \simeq 8.6 \times 10^4 M^2$, where $M = v/c_0$ is the acoustic Mach number. This value is rather significant, though being hardly achievable because of the pronounced inertial character of the thermal mechanism.

It can be noted for the sake of comparison that the corresponding value of hydrodynamic nonlinearity (cf. (1.2b)) is only of order M^2, which is much smaller than the above-mentioned value. It is noteworthy also that the value of $\mathrm{d}c_0/\mathrm{d}T$ for water is positive, being negative for the majority of other liquids.

Another extreme case is concerned with short pulsed disturbances having characteristic time $\tau \ll \tau_T$. In this case one can disregard the term with $\nabla^2 T$ in (1.8) and obtain instead of (1.10)

$$m \sim \frac{\tau k^2 b |v|^2 (\mathrm{d}c_0/\mathrm{d}T)}{\mathcal{X} \rho_0 c_0 c_p}. \tag{1.11}$$

This is less than the maximum achievable value of (1.10) by a factor of τ/τ_T.

One should bear in mind that even though the nonlinearity parameter is proportional to the viscous coefficient b, the role of dissipation is a dual one, i.e. by increasing local nonlinearity it shortens the region in which nonlinear effects show up, since for large b the value of $|v|^2$ decays rapidly with distance.

Apart from thermal mechanisms, similar effects can be provoked by the averaged flows associated with sound absorption. To take them into account one should introduce into (1.6) the term $\bar{v}$ describing these flows, for which the following hydrodynamic equation holds (Bunkin & Lyakhov, 1984):

$$\frac{\partial \bar{v}}{\partial t} - \nu \nabla^2 \bar{v} = -\frac{1}{4}\left(\frac{\partial}{\partial x} + \frac{1}{c_0}\frac{\partial}{\partial t}\right)|\bar{v}|^2, \tag{1.12}$$

where ν is the shear viscosity.

In the general case "detection" of pressure oscillations is also possible (a component $\bar{p}$ emerges), but this effect is insignificant as a rule.

Accounting for two self-action mechanisms, the thermal one and that connected with the flow, the equation for the pressure amplitude in a beam can be derived by substituting expressions (1.5)–(1.7) into the acoustic equations (1.3). The result is the parabolic equation (see Chapter 4)

$$\left(c_0\frac{\partial}{\partial x} + \frac{\partial}{\partial t} + \frac{\mathrm{i}c_0^2}{2\omega}\nabla_\perp^2 + c_0 a\right)p = \mathrm{i}\omega p\left(\frac{\bar{v}}{c_0} + \frac{\gamma_p}{2}\bar{T}\right), \tag{1.13}$$

where $a = b\omega^2/2\rho_0 c_0^3$, $\gamma_p = 2/c_0(\mathrm{d}c_0/\mathrm{d}T)_p$, and

$$\nabla_\perp^2 = \frac{\partial^2}{\partial y^2} + \frac{\partial^2}{\partial z^2}.$$

Equation (1.13) together with equation (1.8) (where in the right-hand terms the relation $v = p/\rho c_0$ is used) and (1.12) form a closed system. In what follows, only temperature effects will be considered.

As was mentioned above, the effects that are of interest here are supposed to show up in a region where the wave remains quasi-harmonic, i.e. their evolution length L is less than the distance $L_N \sim (\varepsilon k M)^{-1}$ at which a discontinuity arises in a wave due to ordinary quadratic nonlinearity. More elaborate estimates for various effects can be found in the review by Bunkin and Lyakhov (1984).

7.2 Self-focusing of sound

Self-focusing effects are known to be related to the dependence of the wave velocity c, averaged over a period, on its intensity I. If $\mathrm{d}c/\mathrm{d}I < 0$ then for a wave beam with intensity maximum on its axis, peripheral rays show a tendency to refract towards the axis where the sound velocity is less (while refractive index n is higher) and as the intensity grows with focusing, the refraction manifestation becomes still more pronounced, giving rise to a "self-focusing" instability. If $\mathrm{d}c/\mathrm{d}I > 0$, we can speak about "self-defocusing" although, as we see later, this process is not associated with instability.

In acoustics (as in optics earlier) the self-focusing effect was initially discussed by Askaryan (1966) and later was treated more thoroughly by Zabolotskaya and Khokhlov (1976). In the 1980s the first experiments on acoustic self-focusing were carried out.

Let us discuss here a few simple models of the process. Neglecting quadratic nonlinearity we shall first proceed from the wave equation

$$\frac{\partial^2 \rho'}{\partial t^2} = \nabla^2 p', \tag{2.1}$$

and suppose that $\nabla p' = c^2 \nabla \rho'$ where c^2 is a slowly varying function of coordinates and time. Then (2.1) takes on the form

$$\frac{\partial^2 \rho'}{\partial t^2} = \nabla \cdot (c^2 \nabla \rho'). \tag{2.2}$$

This equation is the same as in linear geometrical acoustics where $c^2(\mathbf{r})$ is considered to be a known function. Here c^2 will be regarded as a function of the field intensity I or, equivalently, of the field amplitude squared, $|A|^2$. Traditionally, we restrict ourselves to the case of small nonlinearity, when $c^2 = c_0^2(1 + \beta|A|^2)$ and c_0, β are constants. Then equation (2.2) acquires the form in terms of p'

$$\frac{\partial^2 p'}{\partial t^2} = c_0^2 \nabla^2 p' + c_0^2 \beta \nabla \cdot (|A|^2 \nabla p'). \tag{2.3}$$

Now consider a wave beam with a narrow angular spectrum,

$$p' = A(\mathbf{r}) \exp[\mathrm{i}(\omega t - kx)] + \text{c.c.}, \tag{2.4}$$

where $A(\mathbf{r})$ is again a slowly varying amplitude. Substituting this expression into (2.3), we neglect small terms of higher order, i.e. the terms of type $A_{xx}|A|_x^2$ and $\nabla A \cdot \nabla |A|^2$. Then one obtains the parabolic equation (cf. 1.13):

$$\nabla_\perp^2 A - 2ik\frac{\partial A}{\partial x} - \beta k^2 |A|^2 A = 0. \tag{2.5}$$

Most often two cases are considered, namely that of a two-dimensional beam when there is no field dependence on the coordinate z, and the more realistic one of an axisymmetric beam. In the latter case it is convenient to pass to cylindrical coordinates $A = A(r, x)$, and then instead of (2.5) we have

$$\frac{1}{r}\frac{\partial}{\partial r}\left(r\frac{\partial A}{\partial r}\right) - 2ik\frac{\partial A}{\partial x} - \beta k^2 |A|^2 A = 0, \tag{2.6}$$

where now r is the distance from the beam axis.

Let us commence with a plane two-dimensional beam, assuming $A = A(x, y)$. Substitution of $A = s\exp(i\varphi)$, where s and φ are real functions, yields, after separating real and imaginary parts in (2.5),

$$2\frac{\partial s}{\partial y}\frac{\partial \varphi}{\partial y} + s\frac{\partial^2 \varphi}{\partial y^2} - 2k\frac{\partial s}{\partial x} = 0, \tag{2.7a}$$

$$\frac{\partial^2 s}{\partial y^2} - s\left(\frac{\partial \varphi}{\partial y}\right)^2 + 2ks\frac{\partial \varphi}{\partial x} - \beta k^2 s^3 = 0. \tag{2.7b}$$

The former equation embodies energy conservation; it can be written as

$$k\frac{\partial s^2}{\partial x} = \frac{\partial}{\partial y}\left(s^2\frac{\partial \varphi}{\partial y}\right). \tag{2.8}$$

Concerning the latter equation, one can say that if the term s_{yy} is neglected, which is possible under smooth enough variations of amplitude and phase, equations (2.7) will correspond to the geometrical acoustics approximation. In this case it is convenient to introduce the transverse component of the wave number $q = \varphi_y$ as a new variable, which yields (after dividing the second equation by s and differentiating with respect to y)

$$k\frac{\partial I}{\partial x} = \frac{\partial}{\partial y}(qI), \tag{2.9a}$$

$$k\frac{\partial q}{\partial x} = q\frac{\partial q}{\partial y} + \frac{\beta k^2}{2}\frac{\partial I}{\partial y}, \tag{2.9b}$$

where $I = s^2$ is proportional to the field intensity.

Transverse instability of a beam

Let us analyse the plane wave stability with respect to small perturbations of its spatial structure. Assuming $I = I_0 + I'$, $q = q'$, where I, $q \sim \exp i(\alpha x - \delta y)$ and linearizing the system (2.9), we have

$$\alpha = \pm\delta(\beta I/2)^{1/2}. \tag{2.10}$$

If $\beta > 0$ then (for real δ) the value of α will be real as well. If, however, $\beta < 0$ then α becomes imaginary, which means exponential growth of disturbances in space, corresponding to the initial stage of self-focusing (Bespalov & Talanov, 1967).

Similar effects are certainly observed in the three-dimensional case as well, when the disturbances depend on two transverse coordinates y, z.

In order to take diffraction into account we have to consider (2.7) in its full form. It can be also written in variables s and $q = \partial\varphi/\partial y$ as

$$k\frac{\partial s^2}{\partial x} = \frac{\partial}{\partial y}(s^2 q), \tag{2.11a}$$

$$2k\frac{\partial q}{\partial x} - 2q\frac{\partial q}{\partial y} - 2\beta k^2 s\frac{\partial s}{\partial y} + \frac{\partial}{\partial y}\left(s^{-1}\frac{\partial^2 s}{\partial y^2}\right) = 0. \tag{2.11b}$$

Linearizing this system and searching for a harmonic solution for disturbances we obtain, instead of (2.10),

$$\alpha = \pm\frac{\delta}{2}\sqrt{\beta I + \delta^2/k^2}. \tag{2.12}$$

Therefore, it is clear that even for $\beta < 0$ only those sufficiently large-scale disturbances can grow for which $\delta < k\sqrt{-\beta I}$, while for shorter "ripples" at the front, smoothing of the disturbances due to diffraction impedes the cumulative self-refraction.

As a more realistic model we shall discuss an axisymmetric beam and account only for the thermal mechanism of nonlinearity. Thus from (1.13) and (1.8) of Section 1, one can obtain equations for the pressure and temperature amplitudes:

$$2\mathrm{i}k\frac{\partial p}{\partial x} = \frac{1}{r}\frac{\partial}{\partial r}\left(r\frac{\partial p}{\partial r}\right) - \frac{\mathrm{i}k^3 b}{\rho_0 c}p - \frac{2k^2}{c_0}\frac{\mathrm{d}c_0}{\mathrm{d}T}Tp, \tag{2.13}$$

$$c_p\rho_0\frac{\partial T}{\partial t} = \frac{\mathcal{X}}{r}\frac{\partial}{\partial r}\left(r\frac{\partial T}{\partial r}\right) + \frac{k^2 b}{2\rho_0 c_0^2}|p|^2, \tag{2.14}$$

where r is a transverse coordinate (slow thermal diffusion along the beam is disregarded) and $T = \bar{T}$.

If we again set $p = s\exp(i\varphi)$, $q = \partial\varphi/\partial r$ and confine ourselves to a geometric acoustics approximation, equation (2.13) will be reduced to a pair of equations analogous to (2.9),

$$k\frac{\partial I}{\partial x} + q\frac{\partial I}{\partial r} + I\left(\frac{\partial q}{\partial r} + \frac{q}{r}\right) + kaI = 0, \tag{2.15a}$$

$$k\frac{\partial q}{\partial x} + q\frac{\partial q}{\partial r} = -\frac{k^2}{c}\frac{\mathrm{d}c_0}{\mathrm{d}T}\frac{\partial T}{\partial r}, \tag{2.15b}$$

$a = b\omega^2/2\rho_0 c_0^3$ being an attenuation factor depending on both the medium viscosity and heat conductivity, $I = s^2$.

Problems of this type have been addressed in optics more than once (Akhmanov *et al.*, 1968). Let us consider only the simplest examples (Zabolotskaya & Khokhlov, 1976). We shall assume the beam to be Gaussian and the process to be quasistationary (i.e. the terms with $\partial T/\partial t$ in (2.14) will be ignored). Then the system (2.14)–(2.15) will be satisfied by a self-similar solution in the form of a Gaussian beam:

$$s = \frac{s_0}{f(x)}\exp\left(-\frac{x}{2R_0} - \frac{r^2}{R_0^2 f^2}\right), \tag{2.16}$$

where $f(x)$ is a dimensionless beam width, R_0 being the beam radius for $x = 0$.

By integrating equation (2.14) and accounting for (2.16) in the stationary case, the temperature gradient can be found:

$$r\frac{\partial T}{\partial r} = \frac{a\exp(-ax)}{2\pi^2 xc}G_0\left[\exp\left(-\frac{2r^2}{R_0^2 f^2} - 1\right)\right], \tag{2.17}$$

where $G_0 = G(x = 0)$, and

$$G = \frac{\pi}{\rho c_0}\int_0^\infty s^2 r\,\mathrm{d}r$$

is the total beam power.

If one considers the field near the beam axis ("aberrationless approximation") all functions can be expanded in powers of r, and equation (2.15a) yields the ray equation

$$\frac{\mathrm{d}^2 f}{\mathrm{d}x^2} = \frac{a\exp(-ax)}{R_T f}, \tag{2.18}$$

where

$$R_T = \frac{2\kappa c}{I_0 \mathrm{d}c_0/\mathrm{d}T} \tag{2.19}$$

is a parameter characterizing nonlinear refraction (now $I_0 = G_0/\pi R_0^2$).

Omitting the details of integrating (2.18), we shall estimate the effective focusing length X_T as that at which f undergoes a change of order unity. If one assumes that $aX_T \ll 1$, then, according to (2.18), $X_T^2 \sim R_T/a$, or

$$X_T = \left(\mathcal{X} c_0 \Big/ a \frac{\mathrm{d}c_0}{\mathrm{d}T} I_0 \right)^{1/2} . \tag{2.20}$$

As was mentioned above, for the majority of liquids, except for water, the sound velocity diminishes with temperature. Because of this a sound beam in water is defocused, while in other liquids self-focusing can be observed. As for the focusing distance X_T, say, for the case of acetone where $c_0^{-1}\mathrm{d}c_0/\mathrm{d}T = -3.8 \times 10^{-1}\,\mathrm{K}^{-1}$, then if the sound has frequency 100 kHz and intensity $I_0 \simeq 0.3\,\mathrm{W\,cm^{-2}}$ the value of X_T comes to 1.1×10^4 cm, i.e. about $10^4\lambda$. If the frequency is as much as 1 MHz, then for the same intensity we find $X_T \simeq 1$ m. This is considerably less than the sound decay length $L_d \approx a^{-1}$.

A frequently used parameter is a critical power quantity G_{cr} for which the nonlinear refraction compensates completely for the diffraction divergence (thus, to find G_{cr} one has to solve the problem taking diffraction into account). This quantity one can assess from (2.12) as that corresponding to a zero value of the increment α. For a sound beam with width $\ell = 12$ cm at frequency 1 MHz in acetone we obtain $G_{\mathrm{cr}} = 0.1$ W.

For water the sound beam diverges due to nonlinearity, even without diffraction. In this case $c_0^{-1}\mathrm{d}c_0/\mathrm{d}T = 1.6 \times 10^{-3}\,\mathrm{K}^{-1}$ and given the same conditions as before (frequency 100 kHz, intensity $0.3\,\mathrm{W\,cm^{-2}}$) we have $X_T = 2.8$ m for the characteristic nonlinear length.

In the cases considered the distance of discontinuity formation exceeds the value of X_T, which supports the validity of the approximation used.

Note that for gases, the relation $c(T)$ is defined by the state equation $c^2 = \gamma RT/\mu$ (μ being the molecular weight and R the gas constant). For gases $\mathrm{d}c/\mathrm{d}T > 0$, so that the sound beam defocuses.

It should be pointed out, however, that the analysis performed above (in conformity with the results obtained by Zabolotskaya & Khokhlov (1976)) fails to allow for some important aspects, in particular convection effects and generation of averaged flows (bringing about defocusing). This allowance yields pessimistic conclusions as to stationary self-focusing (Bunkin & Lyakhov, 1984), claiming it to be possible only in liquids with viscosity higher than that of glycerol. More promising estimates are obtained for short pulses, where the self-focusing proves possible for viscous liquids of glycerol type at frequencies of 1–10 MHz and pulse durations equal to 1–10 sec. One of the versions observed experimentally is the self-focusing at an interface deformed in response to radiation sound pressure. Such an experiment was carried out for frequency 1.5 MHz and beam intensity $I_0 = 1\,\mathrm{W\,cm^{-2}}$ (beam radius being $d = 0.5\,\mathrm{cm}$). The water surface was bent upward a few millimetres (Bunkin *et al.*, 1986). Thermal self-focusing was also observed (Assman *et al.*, 1985, 1986; Andreev *et al.*, 1985). Thus, in the work by Assman *et al.*, propagation of an ultrasound beam in benzene (viscosity $\eta \approx 6.4 \times 10^{-3}$Ps) and glycerol ($\eta = 14$Ps) was observed. The radius of a sound beam with frequency $f = 2\,\mathrm{MHz}$ was equal to $d = 0.75\,\mathrm{cm}$, and for experimental impulses having power up to 50 W and duration of the order of one second the process was essentially nonstationary; the self-focusing length was measured to be 12 cm and 8 cm for benzene and glycerol, respectively. In this case the distance of discontinuity formation due to quadratic nonlinearity was considerably greater (15–20 cm). During experiments significant narrowing of the beam (by about a factor of 2) and even its disintegration into several beams ("filaments") due to self-focusing effects were observed.

As for liquids with low viscosity, the length of thermal self-focusing in them is rather large but it can be shortened substantially simply by discontinuity formation. Indeed, the strong absorption of discontinuous waves can bring about medium heating and eventually, notwithstanding the rapid decay, pronounced self-refraction can develop (Karabutov *et al.*, 1988).

7.3 One-dimensional modulated waves

Let us consider the evolution of amplitude and frequency-modulated waves in a medium with dispersion and cubic nonlinearity. Both dispersion and nonlinearity lead to distortions of the envelope shape, i.e. amplitude and frequency variations. This interplay of various factors yields numerous effects such as harmonic wave instability, formation of localized structures (shock waves and envelope solitons), etc. These processes have been studied since the 1960s in optics, plasma physics and other fields of physics (Ostrovsky, 1963, 1966; Whitham, 1965, 1974). In acoustics, however, due to the absence of dispersion and weak cubic nonlinearity in traditional media, they were paid little attention, until recently.

Nevertheless, it is quite possible (for instance, when sound propagates in a waveguide filled with a medium having anomalous nonlinearity) that the processes of acoustic wave self-action become important.

It will be assumed here that the equation of the adiabat in the medium has the form

$$\rho' = \frac{p'}{c_0^2} + gp'^3, \quad g = \text{constant}. \tag{3.1}$$

The last term is supposed to be smaller than the linear one while exceeding other nonlinear terms entering the equations of acoustics. Then the propagation of sound will be described through the following model equation:

$$\frac{1}{c_0^2}\frac{\partial^2 p'}{\partial t^2} - \nabla^2 p' = g\frac{\partial^2 p'^3}{\partial t^2}. \tag{3.2}$$

The solution of this equation for a sound wave in a waveguide with rectangular cross-section is sought in the form

$$p' = A(x,\tau)f(y,z)\exp i(kx - \omega t) + \text{c.c.} \tag{3.3}$$

and where $\tau = t - x/c_g$, c_g is the group velocity at frequency ω, and the variation $A(x,\tau)$ is considered slow. The function $f(x,y)$ defining the transverse structure of the waveguide mode satisfies, as before, the boundary conditions imposed at the waveguide walls. Substituting (3.3) into (3.2) yields, after calculating the right-hand side and selecting the main-frequency terms, a nonlinear Schrödinger equation (Karpman, 1973; Whitham, 1974)

$$2ik\frac{\partial A}{\partial x} + \left(k\frac{\partial^2 k}{\partial \omega^2}\right)\frac{\partial^2 A}{\partial \tau^2} + \lambda|A|^2 A = 0, \tag{3.4}$$

where

$$\lambda = 3g\omega^2\left(\int |f|^4\,\mathrm{d}y\mathrm{d}z\right)\Big/\left(\int |f|^2\,\mathrm{d}y\mathrm{d}z\right),$$

which is exhaustively studied in modern mathematical physics.† It has, in particular, an exact multi-soliton solution derived using the known inverse scattering method. Here we restrict ourselves to a simpler analysis (Ostrovsky, 1966, 1968). Setting $A = s\exp(i\varphi)$ again and separating real and imaginary terms, we obtain from (3.4)

$$2k\frac{\partial s}{\partial x} + n\left(2\frac{\partial s}{\partial \tau}\frac{\partial \varphi}{\partial \tau} + s\frac{\partial^2 \varphi}{\partial \tau^2}\right) = 0 \tag{3.5}$$

† A more thorough derivation of equation of the type (3.4) for acoustics waveguides was performed by Zabolotskaya & Schwartsburg (1988), who took into account also the quadratic terms in (3.1) dominating in "normal" media.

$$-2ks\frac{\partial\varphi}{\partial x}+n\left[\frac{\partial^2 s}{\partial\tau^2}-s\left(\frac{\partial\varphi}{\partial\tau}\right)^2\right]+\lambda s^3=0, \tag{3.6}$$

where $n = kk_{\omega\omega}$ is the dispersion parameter. In a way similar to the one used in the previous section these equations can be transformed into a more explicit "hydrodynamic" form by introducing the frequency deviation $\delta = \varphi_\tau$ and differentiating equation (3.6) (divided first by s) with respect to τ. Thus

$$k\frac{\partial s^2}{\partial x}+n\frac{\partial}{\partial\tau}(\delta s^2)=0, \tag{3.7}$$

$$2\left(k\frac{\partial\delta}{\partial x}+n\delta\frac{\partial\delta}{\partial\tau}\right)+3\lambda\frac{\partial s^2}{\partial\tau}+n\frac{\partial}{\partial\tau}\left(\frac{1}{s}\frac{\partial^2 s}{\partial\tau^2}\right)=0. \tag{3.8}$$

Note that the former equation (containing no nonlinearity parameter) embodies the condition of energy conservation.

Most tractable is the case of sufficiently smooth perturbations, when the last term in (3.8), containing the high derivatives of s, can be ignored. The remaining set of equations for s^2 and δ proves quasilinear, furnishing an analogue of the geometrical-acoustics approximation. If it is linearized near constant values of s^2 and δ, i.e. we substitute $I = I_0 + I'$, $\delta = \delta_0 + \delta'$ where $I_0 = s_0^2$, the characteristic equations are promptly formed as

$$\frac{d\tau}{dx}=\frac{n}{k}\left(\delta_0\pm 3\sqrt{\mu I_0}\right),\quad \frac{d\delta}{dI}=\pm\left(\frac{3\nu k}{nI_0}\right)^{1/2}=\pm\left(\frac{\mu}{I_0}\right)^{1/2} \tag{3.9}$$

where $\mu = 3\lambda/2n$.

Hence, an important conclusion (the Lighthill criterion, Lighthill (1965)) can be reached: if $\mu > 0$ then the set (3.7), (3.8) will be hyperbolic within the long-wave limit, while for $\mu < 0$ it will prove elliptic. Therefore, the result depends qualitatively on the sign ratio for the dispersion parameter n and on the nonlinearity parameter ν.

In the hyperbolic case the variables s and δ exhibit "hydrodynamic" behaviour. In particular, two Riemann waves exist that have the form (in reference variables x, t)

$$\begin{aligned} I &= s^2 = F\left[x - c_g\left(1\mp 3\frac{n}{k}\sqrt{\mu I}\right)t\right],\\ \delta &= \delta_0 \pm 2\sqrt{\mu I}. \end{aligned} \tag{3.10}$$

One of these waves can be referred to as "fast" while the other is "slow", in comparison with the linear velocity c_g.

As usual in a Riemann wave, a multivaluedness of the amplitude and frequency arise at a finite distance; after that the term with high derivative of s cannot be discarded in (3.8). One can certainly introduce a "jump postulate" and consider discontinuities, i.e. envelope shock waves (both fast and slow) such that integrating (3.7) and (3.8) at a jump yields the boundary conditions as an analogue of the shock adiabat. However, for such shocks really to exist losses are inevitable, as before, though they are of the "envelope viscosity" type connected with variations in the amplitude, rather than with instantaneous field values. These properties are typical of the losses of relaxation type, e.g. thermal nonlinearity.

In the elliptic case the characteristics (3.9) are complex, but some analytical solutions are known as well. Especially interesting here is the problem of the stability of a harmonic wave with s, δ = constant (incidentally, provided δ is constant one may set $\delta = 0$ without loss of generality, since the choice of the constant carrier frequency ω is arbitrary). It follows from (3.9) that if we set $s = s_0 + s'$ and represent perturbations in the form $\exp[i(\Omega\tau - Kx)]$, then (for $\delta = 0$)

$$K = \pm 3\frac{n\Omega}{k}\sqrt{\mu I_0}. \tag{3.11}$$

For $\mu < 0$ this implies that K is imaginary and that in the general case perturbations grow exponentially. Therefore, self-modulation arises, i.e. the harmonic wave is unstable with respect to modulating disturbances. Obviously this corresponds to the ellipticity of the envelope equations in their "hydrodynamic" limit.

Let us return to (3.7) and (3.8) in full form. First of all, the stability of a harmonic wave may easily be investigated in this general case as well. Then instead of (3.11) we obtain

$$K = \pm 3\frac{n\Omega}{k}\left[\mu I_0\left(1 + \frac{\Omega^2}{4\mu I_0}\right)\right]^{1/2}. \tag{3.12}$$

Thus, even in the elliptic case instability manifests itself, in fact, only at comparatively low modulation frequencies, $\Omega < 2\sqrt{|\mu I_0|}$.

It seems useful to note an analogy between the results of this and previous sections: a transverse instability of the self-focusing type corresponds in a sense to the longitudinal self-modulation instability considered here.

Let us discuss now stationary solutions of the system (3.7), (3.8) in the form of a stationary progressive wave: with s, δ depending on $\tau - bx = \xi$, where b is proportional to a nonlinear correction to the group velocity, we arrive at the second-order equation

$$\frac{d^2s}{d\xi^2} - \left(2\mu s^2 - C_1\right) s + \frac{s\delta}{n}\left(n\delta - 2kb\right) = 0, \tag{3.13a}$$

where

$$n\delta = bk - \frac{C_2}{s^2} \tag{3.13b}$$

(C_1 and C_2 being integration constants).

First, the case $C_2 = 0$ will be considered. Then, according to (3.13b), the frequency deviation δ is constant and one can, as previously, assume without loss of generality that $\delta = 0$; then $b = 0$ which just means that the wave propagates with the linear group velocity appropriate to frequency ω. For $\delta = 0$ the (3.13) solution type depends again on the sign of the parameter μ. For $\mu > 0$ (stable case), s changes its sign in all finite solutions; there exists, among them, a solitary wave (in optics a "dark soliton", cf. Hasegawa & Tappert (1974)) in the form of a "dip" in a harmonic wave,

$$s = s_0 \tanh\left(s_0\sqrt{\mu}\xi\right). \tag{3.14}$$

If, however, $\mu < 0$ (unstable case) there exists a solution in the form of a localized pulse, i.e. an envelope soliton (Ostrovsky 1966; Karpman, 1973),

$$s = \frac{s_0}{\cosh(s_0\sqrt{\mu}\xi)}. \tag{3.15}$$

The solutions represented by (3.14) and (3.15) are depicted in Figure 7.1.

If $C_2 \neq 0$, then for $\mu > 0$ equation (3.13) has two nonzero equilibrium points (the straight line $s = 0$ on the phase plane is singular, as is evident from (3.13)). In this case "dark solitons" are also possible and, moreover, if relaxation losses are added envelope shock waves can really exist.

Now let us discuss, in brief, specific features of the acoustic problem as applied to the waveguide case (Zabolotskaya & Schwartsburg, 1988). Making use of the dispersion equation for a waveguide mode in the form $k_{mn}^2 = \omega^2/c_0^2 - \gamma_{mn}^2$ where γ_{mn} is the transverse wave number and m, n are integers, one can easily observe that the dispersion parameter $n_{mn} = \gamma_{mn}^2/c_0^2k^2 < 0$, because of which a modulation instability develops for $\nu > 0$ or $g > 0$, which corresponds to sound velocity growth with wave

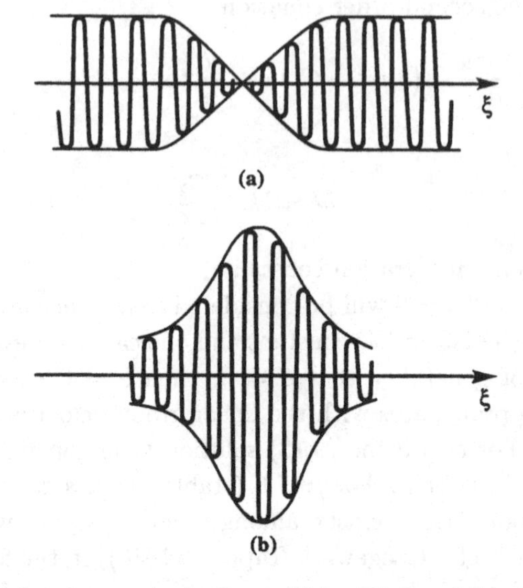

Figure 7.1. Modulated waves in the form of (a) a "dip" and (b) a localized pulse.

amplitude. The characteristic scale of the nonlinear waves (3.14) and (3.15) is of order

$$L_s \approx \frac{c_g}{\sqrt{\mu I}} \approx \frac{\gamma_{mn} c_0}{3\omega^2 \sqrt{gI}} \left(\frac{\int |f|^2 \, \mathrm{d}y\mathrm{d}z}{\int |f|^4 \, \mathrm{d}y\mathrm{d}z} \right)^{1/2}. \tag{3.16}$$

For a rectangular waveguide with free boundaries the eigenfunctions are

$$f = \sin \frac{m\pi}{d} y \sin \frac{n\pi}{\ell} z$$

(where d, ℓ are cross-section dimensions) and the expression in parentheses is equal to 16/9; here,

$$\gamma_{mn}^2 = \pi^2 \left(\frac{m^2}{d^2} + \frac{n^2}{\ell^2} \right).$$

According to (3.1), as a measure of relative nonlinearity one can take $\beta = g c_0^2 I$ which results, e.g. for a square waveguide cross-section ($d = \ell$), in

$$L_s \approx \frac{\pi c_0^2 \sqrt{m^2 + n^2}}{\omega^2 d \sqrt{\beta/2}}.$$

Let us cite a single example, of a waveguide in the form of a square cross-sectioned rubber rod with a width $d = 5\,\mathrm{cm}$. Rubber behaviour is, to a large extent, the same as that of water, where $c_0 \approx 1.5 \times 10^3\,\mathrm{m\,s^{-1}}$ and the shear modulus is small. Therefore, the excitation of shear waves can be ignored here. In this case for a frequency of 10 kHz we have $L_s \simeq 6/\sqrt{\beta}$, and for $\beta = 10^{-3}$ (which seems to be possible in porous rubber) this comes to $L \approx 2\,\mathrm{m}$ which corresponds to about 13 wavelengths. For water, real values of β will not exceed 10^{-6} and under similar conditions $L_s \approx 60\,\mathrm{m}$; thus, observation of envelope waves becomes difficult. It should be noted, however, that in a liquid with bubbles, related effects can be associated with the averaged bubble drift (see below).

7.4 Stimulated scattering of sound

Now let us proceed to the effects of so-called stimulated scattering. Their peculiarity, as compared with common three-wave resonant interactions, is associated with the participation in the interaction of a specific oscillation mode that is typically of low frequency and has its own dispersion properties. In optics one can observe stimulated Raman scattering (when this mode is represented by molecular oscillations), Mandelstam–Brillouin scattering by acoustic (hypersonic) waves, scattering by thermal waves, concentration scattering by density fluctuations, etc. Stimulated scattering processes are also possible in acoustics.

For stimulated scattering the possibility of wave amplification is typical because of the internal feedback, even without any reflection at the boundaries. It is important here that, while in a "quadratic" medium efficient interaction is provided by satisfying resonance conditions for the frequencies and wave numbers, for the "cubic" case these conditions can be fulfilled automatically. Indeed, let us consider a triad of waves with frequencies ω_1, ω_3, and $\omega_2 = \omega_1 \mp \omega_3$, where ω_3 is the frequency of the aforementioned "special" mode of the medium, and

$$p' = \frac{1}{2}\Big\{p_1(x)\exp[\mathrm{i}(k_1x - \omega_1t)] + p_2(x)\exp[\mathrm{i}(k_2x - \omega_2t)] \\ + p_3(x)\exp[\mathrm{i}(k_3x - \omega_3t)] + \mathrm{c.c.}\Big\}, \tag{4.1}$$

where p_1, p_2, p_3 are pressure amplitudes which depend slowly on x. Wave 1 (pumping) will be assumed as given and the others are small, so that $p_2, p_3 \ll p_1$. Then quadratic terms ($\sim p^2$) in the reference equations will make contributions at the difference frequency ω_2 proportional to $p_1p_3^*$ (or $p_1^*p_3$ for the sum frequency ω_3), while being proportional to $p_1p_2^*$ (or $p_1^*p_2$) at ω_3. Cubic terms will yield, for the same frequencies, terms of type $p_1 p_1^* p_2$ and $p_1 p_1^* p_3$. The same may result from quadratic terms, but at the next order in p_1, if "quadratic" corrections to p are defined using the perturbation method and equations for amplitudes are written taking these corrections into consideration. Therefore, the equations for amplitudes p_2 and p_3 have the following form:

$$2\mathrm{i}k_2 \frac{\mathrm{d}p_2}{\mathrm{d}x} = G_1 p_1 p_3^* \exp \mathrm{i}\Delta kx + \mathrm{i}G_3 |p_1|^2 p_2 - \mathrm{i}\delta_2 p_2, \tag{4.2a}$$

$$2\mathrm{i}k_3 \frac{\mathrm{d}p_3}{\mathrm{d}x} = G_2 p_1 p_2^* \exp \mathrm{i}\Delta kx + \mathrm{i}G_4 |p_1|^2 p_3 - \mathrm{i}\delta_3 p_3, \tag{4.2b}$$

where G_i are complex interaction coefficients determined from the particular medium model and the δ_i are decay factors, Δk being the same mismatch of the wave numbers that was described in the previous chapter. Should the quadratic nonlinearity dominate, the second terms on the right-hand sides of (4.2) will be less in absolute value than the first ones, though still capable of playing the main role as they are always in resonance, whereas the factor $\exp(\mathrm{i}\Delta k\, x)$ would not produce a cumulative effect unless synchronism conditions are satisfied. In what follows, for the sake of simplicity we shall treat exactly such cases, neglecting terms with G_1, G_2. Then equations (4.2*a*) and (4.2*b*) become independent: coupling between the pump wave p_1 and the "scattered" wave p_2 is realized directly without a "special" wave p_3. The coefficients G_3 and G_4 can be obtained both directly from the cubic terms in the basic equations, and from the nonresonance quadratic terms in the second approximation. These coefficients are, in general, interrelated. In the processes of interest, they have an imaginary part (in what follows, we denote only their imaginary parts by $G_{3,4}$). Then, according to (4.2*a*), the scattered field amplitude changes exponentially ($p_2 \propto \exp gx$) with rate constant

$$g = \frac{G_3 |p_1|^2 - \delta_2}{2k_2}. \tag{4.3}$$

For $G_3 > 0$, on exceeding the pump intensity threshold value, that is, $|p_1|^2 > \delta_2/G_3$, amplification of the scattered wave is observed.

Let us demonstrate this with some particular examples.

Stimulated sound scattering in a liquid with gas bubbles

This problem was considered by Zabolotskaya (1977, 1984). Let all the bubbles have one and the same radius R_0 and quality factor Q near the resonant frequency ω_0. This case is similar to Raman scattering where light interacts with molecular vibrations having a specified resonant frequency. This problem (in the one-dimensional formulation) is reduced to the solution of the wave equation (see Chapter 1)

$$\frac{\partial^2 p'}{\partial x^2} - \frac{1}{c_0^2}\frac{\partial^2 p'}{\partial t^2} = -\rho_0 N \frac{\partial^2 V'}{\partial t^2}, \tag{4.4}$$

together with the equation for the volume perturbation of a single bubble taking cubic nonlinearity into consideration. The latter equation was deduced in the papers by Zabolotskaya and Vorobyev (1974) and Zabolotskaya and Soluyan (1976) in the form

$$\ddot{V}' + \omega_0^2 V' + f\dot{V}' - \alpha V'^2 - \beta[2V'\ddot{V}' + (\dot{V}')^2] + \mu_1 V'^3 + \mu_2[V'^2\ddot{V}' + V'(\dot{V}')^2] = \mathcal{X} p' \tag{4.5}$$

Obviously, (4.5) is a generalization of equation (3.14) of Chapter 1; it contains additional cubic terms. The notation here is the same but with the following new parameters:

$$\alpha = \omega_0^2(\gamma+1)/2V_0, \quad \mu_1 = (\gamma+1)(\gamma+2)\omega_0^2/6V_0^2, \quad \mu_2 = 2/9V_0^2$$

(V_0 is the volume of a single bubble).

Let us now assume the frequency ω_3 to be close to ω_0; then for the other two frequencies losses in the bubbles are of nonresonant character and may be ignored. Substituting (4.1) into the right-hand side of (4.5) and then into (4.4) we arrive at a system of the form (4.2), where

$$\begin{aligned} \mathrm{i}G_3 &= \rho N \omega_3^2 \chi^2 q^2 \left[(\omega_0^2 - \omega_1^2)(\omega_0^2 - \omega_2^2)(\omega_0^2 - \omega_3^2 - \mathrm{i}\omega_3 f)\right]^{-1}, \\ \mathrm{i}G_4 &= \rho N \omega_3^2 \chi^2 q^2 \left[(\omega_0^2 - \omega_3^2 + \mathrm{i}\omega_3 f)(\omega_0^2 - \omega_1^2)(\omega_0^2 - \omega_2^2)\right]^{-1}, \\ q &= 3(1+\gamma)\omega_0^2\beta - \beta(\omega_1^2 + \omega_2^2 - \omega_1\omega_2). \end{aligned} \tag{4.6}$$

Then the maximum growth rate (for $\omega = \omega_0$), given by (4.3), is equal (Zabolotskaya, 1984) to

$$g = \frac{G_3|p_1|^2}{2k_2} = \frac{N\omega_1^2 q^2 |p_1|^2 \rho \chi^3}{2(\omega_0^2 - \omega_1^2)(\omega_0^2 - \omega_2^2)\omega_0 f k_2}. \tag{4.7}$$

It is instructive to recall that here $\omega_2 = \omega_1 - \omega_0 < \omega_1$, i.e. the scattering goes into the lower frequency (in optical terms, the Stokes component). For the anti-Stokes component (at frequency $\omega_1 + \omega_0$) amplification would not take place and the wave would always decay.

Zabolotskaya (1977) offered the following numerical estimate. For a concentration $NV_0 = 10^{-5}$ of bubbles having resonant frequency $\omega_0 = 2\pi \times 20\,\mathrm{kHz}$, and a pump frequency $\omega_1 = 1.5\omega_0$, and hence the scattered (Stokes) wave frequency $\omega_2 = \omega_1 - \omega_0 = 0.5\omega_0$, equation (4.7) yields (assuming $f = 0.1\omega_0$) the threshold pump intensity in water $I_1 \simeq 10^{-2}\,\mathrm{W\,cm^{-2}}$ (therefore, the acoustic Mach number is of the order of 10^{-5}). If the pump intensity I_1 is far in excess of the threshold, then (4.7) yields the following growth rate estimate: $g \simeq 0.3I_1\,\mathrm{cm^{-1}}$, provided I_1 is expressed in terms of $\mathrm{W\,cm^{-2}}$. It is clear from this that for $I_1 \sim 1\,\mathrm{W\,cm^{-2}}$ the distance at which the scattered wave increases significantly appears to be of the order of a few centimetres. However, the wavelength of the scattered field in water is $\lambda \simeq 15\,\mathrm{cm}$, so that formula (4.7) can serve, at best, for rough estimates only.

Note also that, when describing the third-order effects, only that contribution from the cubic terms of the bubble pulsation equation suggested in the form of equation (4.5) was allowed for, while terms obtained from the two-fold use of the quadratic terms mentioned above were omitted in this estimate.

Some other cases of stimulated sound scattering

As was mentioned before, the scattering at resonant elements, the bubbles, is similar to Raman scattering in optics. Scattering by various types of waves that do not have pronounced resonances but change the sound propagation velocity is also possible. The role of scatterers can be played by vortex modes, thermal waves and even hydrodynamic modes, i.e. acoustic flows in a viscous liquid. All these lead to analogues of the Rayleigh type of scattering in optics.

Sound scattering by vortex waves was one of the problems considered rather early (Pushkina & Khokhlov, 1971). The estimates show that the threshold for this effect in liquid is too high, while in gases it seems to be more realistic. Thermal scattering in liquids is also difficult to observe: it requires high viscosity (for a considerable heat release) but low thermal conductivity (such that thermal variations will have no time to be smoothed), i.e. liquids with high Prandtl number are needed. The threshold pump intensity for this mechanism is equal (Bunkin *et al.*, 1986) to

$$I_{\text{th}} = 8\kappa\omega_1 \left(c \left| \frac{\partial \ln c^2}{\partial T} \right| \right)^{-1}, \tag{4.8}$$

where κ is thermal conductivity, $c(T)$ is the sound velocity, ω_1 is the pumping frequency.

Finally, cubic nonlinearity is created due to the sound "detection", i.e. acoustic flow formation (see above). The threshold of this type of scattering is equal to

$$I_{\text{th}} = 4\eta c^2 a, \tag{4.9}$$

where $\eta = \rho_0 \nu$ is the viscosity coefficient and a is the sound decay rate.

For these types of stimulated scattering, pulsed regimes are the most accessible, as heat conduction and viscosity lead to smoothing of sound velocity variations (within a time of order 10^{-3} seconds at frequency 1 MHz).

Also under discussion have been concentration mechanisms in chemical solutions, and suspended particles in liquids. In the latter case, the generation threshold amounts to only $1\,\text{W cm}^{-2}$ for a particle concentration of 1 per cent. Such an effect has already been observed experimentally. "Concentration" instabilities are also possible for liquids with bubbles (see below).

We shall not be concerned in detail with these interesting problems, since they do not have a substantial experimental basis so far. Nevertheless, the possibilities of applying the stimulated scattering of sound to the acoustic spectroscopy of liquids have already been analysed (Zabolotskaya, 1984).

7.5 Wave front reversal

Reversal (or conjugation) of the wave front (WFR) is the effect of reversal of the wave propagation direction at each point whilst all other characteristics are preserved intact (the frontal surface and amplitude distribution on it). Mathematically WFR is the reversal of the phase sign, or conjugation of the wave's complex amplitude. In particular, if a primary field is given in the form

$$p'(\mathbf{r}, t) = A(\mathbf{r}) \exp(\mathrm{i}\omega t) = |A(\mathbf{r})| \exp[\mathrm{i}(\varphi(\mathbf{r}) + \omega t)] \tag{5.1}$$

then the reversed wave will appear as

$$p'_c(\mathbf{r}, t) = KA^*(\mathbf{r}) \exp(-\mathrm{i}\omega t) = K|A(\mathbf{r})| \exp[\mathrm{i}(-\varphi(\mathbf{r}) + \omega t)], \tag{5.2}$$

where K is a complex constant factor (characterizing the net amplification or attenuation and the constant phase shift) of the wave.

If we proceed from physical aspects, then the concept of WFR can be somewhat generalized to cover not simply change of the wave propagation direction to the opposite one, but a possible turn of the whole field pattern through the same angle in space as well. Then, for example, the focusing process of a primarily diverging field (not necessarily in the reverse direction) can be referred to as WFR. In addition, one can speak about WFR with frequency transformation, provided the wave front structure satisfies the aforementioned conditions. Thus, the following transformation can be generally referred to as WFR:

$$A(\mathbf{r})\exp[\mathrm{i}(\omega_1 t - \mathbf{k}_1 \cdot \mathbf{r})] \to A^*(\mathbf{r})\exp[\mathrm{i}(\omega_2 t - \mathbf{k}_2 \cdot \mathbf{r})]. \qquad (5.3)$$

In particular, a diverging spherical wave focuses into the source point after reversal. As obvious examples of reversing systems one may cite plane or focusing mirrors, lenses, etc., working in a homogeneous medium.

Of more interest here are the instances of WFR in inhomogeneous media, including randomly inhomogeneous ones, which raises the possibility of recovery of the phase structure of a field "damaged" by inhomogeneities. As the latter are not known beforehand, an *adaptive* reversal is implied here, say, a receiving antenna at each element of which phase conjugation of the signal received with respect to the radiated one is effected. Adaptive antennae have been used in radio-engineering as well as in hydroacoustics which, however, is not directly related to nonlinear waves.

Back in the 1970s the methods of wave front "self-reversal" were realized in nonlinear optics, while conjugation was realized due to nonlinear interaction with an intense wave pump (Bespalov & Pasmanik, 1982). As is clear from the previous sections, in resonant interactions of both quadratic and cubic types conjugation is realized naturally, therefore making WFR or related effects possible in principle.

In recent years a series of similar WFR mechanisms in acoustics has been discussed (Bunkin *et al.*, 1981). As in optics, both bulk effects and their surface analogues are possible when nonlinear interaction occurs at an interface. Let us briefly consider some WFR mechanisms that can be encountered in acoustics.

Four-wave mechanisms of WFR

A pumping field gives rise to a spatial inhomogeneity of the average parameter (sound velocity) in a medium, playing the role of a "lattice" at which a signal is scattered. It is clear that such a lattice cannot be constructed by a travelling wave and, therefore, a standing wave is employed, created by a pair of opposed pump waves (Figure 7.2). This pattern is sometimes interpreted as a holographic one: waves p_1 and p_s "record" a hologram while the wave p_2 "reads" it out, giving rise to a reversed wave p_c.

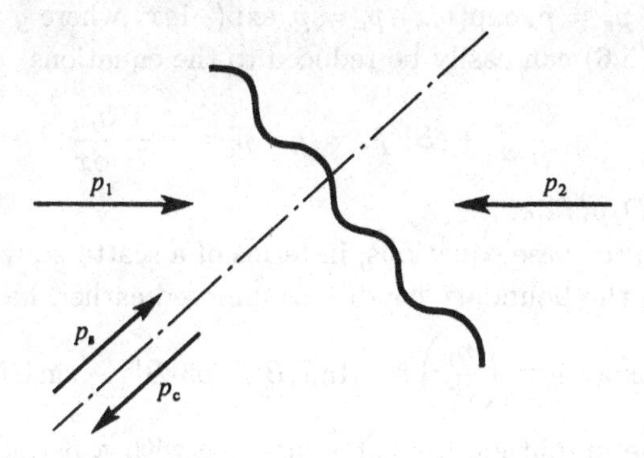

Figure 7.2. Four-wave mechanisms of WFR. Waves p_1 and p_s generate sound velocity perturbation (wavy line), which scatters the wave p_2, producing the reversed wave p_c. (Bunkin *et al*, 1981.)

The simplest version of such a reversal involves four plane waves of the same frequency,

$$p' = \frac{1}{2}\exp(\mathrm{i}\omega t)\left[p_1\exp(-\mathrm{i}\mathbf{k}_1\cdot\mathbf{r}) + p_2\exp(\mathrm{i}\mathbf{k}_1\cdot\mathbf{r}) + p_s\exp(-\mathrm{i}\mathbf{k}_2\cdot\mathbf{r}) + p_c\exp(\mathrm{i}\mathbf{k}_2\cdot\mathbf{r})\right] + \text{c.c.} \quad (5.4)$$

The equations of a medium with cubic nonlinearity contain terms of type p^3, which results in responses in the direction of the signal and of the scattered waves, namely

$$\begin{aligned}&\sim \exp\left[\mathrm{i}(\omega t - \mathbf{k}_2\cdot\mathbf{r})\right]\left[(|p_1|^2 + |p_2|^2)p_s\right]\\ &\sim \exp\left[\mathrm{i}(\omega t + \mathbf{k}_2\cdot\mathbf{r})\right]\left[(|p_1|^2 + |p_2|^2)p_c\right].\end{aligned} \quad (5.5)$$

Therefore, the equations for the amplitudes p_s and p_c for the given pump wave have the following structure (for simplicity we assume $|p_1| = |p_2|$; apart from this, in an isotropic medium $|\mathbf{k}_1| = |\mathbf{k}_2| = k$)

$$\begin{aligned}-2\mathrm{i}k\frac{dp_s}{dx} &= D_1|p_1|^2 p_s + D_2 p_1^2 p_c^*,\\ 2\mathrm{i}k\frac{dp_c}{dx} &= D_1|p_1|^2 p_c + D_2 p_1^2 p_s^*,\end{aligned} \quad (5.6)$$

where x is a coordinate in the direction of propagation of the signal, and D_1 and D_2 are interaction factors defined by a particular nonlinearity mechanism. The typical statement of the problem is as follows. At input, i.e. at a point $x = 0$, the incident signal wave $p_s(0) = p_{s0}$ is specified, while at the output point $x = L$ the scattered wave is absent, i.e. $p_c(L) = 0$. Given that these conditions are satisfied, after the replacement $p_s = \tilde{p}_s \exp(igx), p_c = \tilde{p}_c \exp(-igx)$ where $g = D_1|p_1|^2/2k$, the system (5.6) can easily be reduced to the equations

$$\frac{\partial^2 \tilde{p}_s}{\partial x^2} + |B|^2 \tilde{p}_s = 0, \quad \tilde{p}_c^* = -\frac{1}{B}\frac{\partial \tilde{p}_s}{\partial x}, \tag{5.7}$$

where $B = D_2 p_1^2/2ik$.

A solution to these equations, in terms of a scattered wave amplitude p_c satisfying the boundary conditions imposed earlier, has the form

$$p_c = \mathrm{i}\exp(-igx)\left(\frac{p_1}{p_1^*}\right) p_{s0} \left(\tan|B|L \cos|B|x - \sin|B|x\right), \tag{5.8}$$

i.e. the wave amplitude oscillates in space with a period proportional to the pump intensity. Therefore, the value of the reversal factor K entering formula (5.2) is equal to

$$|K| = \tan(|B|L). \tag{5.9}$$

It is important that the modulus and the phase of this factor do not depend on the direction of the vector **k** but on the pump intensity alone. This provides the preservation of the form of the front on reversal: an arbitrary wave can be expanded in plane waves such that the reversal process does not change the ratio between their amplitudes and phases. It can be indicated, though, that "indirect" dependence of $|K|$ on direction still exists since $|K|$ depends on the layer thickness L and, for a plane layer, to be definite, $L = L_0/\cos\psi$ where ψ is the angle of incidence of the signal on the layer and L_0 is the layer thickness. Because of this the transverse wave vector range is bounded. Let us also emphasize that for $|B|L = \pi/2$ the amplification factor tends to infinity, i.e. generation arises due to a mechanism of the stimulated scattered type, with feedback through a scattered wave.

The value of the parameter B, or to be more exact, of the factor D_2, is assessed, as was mentioned previously, via a particular nonlinearity mechanism. One such mechanism is realized in a liquid with bubbles. The equations for such a medium are given in the previous section (see (4.4) and (4.5)).

Substituting the sum (5.4) into (4.4), (4.5) for $f = 0$, we obtain equations of the type (5.7), with

$$D_1 = \frac{2\rho_0 N\omega^2\chi^2}{2\omega_0^2(\omega_0^2 - \omega^2)(\omega_0^2 - 4\omega^2)}\left[\left(\frac{3}{2}\mu_1 - m\omega^2\right)\omega_0^2(\omega_0^2 - 4\omega^2) - (\alpha - 3\beta\omega^2)^2\omega_0^2 - 2(\alpha - \beta\omega^2)^2(\omega_0 - 4\omega^2)\right]. \tag{5.10}$$

Near resonance one should allow for wave damping, setting $|\omega_0^2 - \omega^2| \simeq |\omega_0 f|$. Let us make an estimate of the distance $L_{\rm th}$ at which $BL = \pi/2$, i.e. essential amplification of the reversed wave will occur. Assume the bubble radius to be equal to 1.6×10^{-2} cm (i.e. the resonance frequency for the air bubbles will be approximately 300 kHz), volume content equal to $NV_0 \sim 10^{-5}$ and intensity of each of the pump waves equal to $0.1\,\mathrm{W\,cm^{-2}}$. Then for a frequency $\omega/2\pi = 12\,\mathrm{kHz}$ the "threshold" distance $L_{\rm th}$ amounts to 120 cm. Signal dissipation at this distance (for a bubble quality factor of 10) is still insignificant.

Parametric WFR

Another mechanism is associated with the case when the signal and the reversed wave have the same frequency ω, and the pump changing the sound velocity c has frequency 2ω. Therefore, in this case parametric resonance occurs. Let us assume that c varies homogeneously in space, i.e. $c = c(t)$. Then $p' = c^2(t)\rho'$, and to a linear approximation we have

$$\rho_0 \frac{\partial \mathbf{v}}{\partial t} + \nabla p' = 0, \quad \frac{\partial}{\partial t}\left(\frac{p'}{c^2}\right) + \rho_0 \nabla \cdot \mathbf{v} = 0,$$

or

$$\frac{\partial^2}{\partial t^2}\left(\frac{p'}{c^2}\right) = \nabla^2 p'. \tag{5.11}$$

If $c^2 = c_0^2(1+\alpha_m(t))$ where α_m is a small variable parameter, then (5.11) yields

$$\frac{1}{c_0^2}\frac{\partial^2 p'}{\partial t^2} - \nabla^2 p' = \frac{1}{c_0^2}\frac{\partial^2 \alpha_m p'}{\partial t^2}. \tag{5.12}$$

In accordance with what was said above, let us write the pressure as the sum of a signal and a reversed wave,

$$p' = p_s(x)\exp[\mathrm{i}(\omega t - \mathbf{k}\cdot\mathbf{r})] + p_c(x)\exp[\mathrm{i}(\omega t + \mathbf{k}\cdot\mathbf{r})] + \text{c.c.}, \tag{5.13}$$

(where x is again a coordinate along vector $\mathbf{k}$), and the variable parameter in the form

$$\alpha_m = \alpha_0 \cos 2\omega t = \frac{\alpha_0}{2}\left[\exp(2\mathrm{i}\omega t) + \exp(-2\mathrm{i}\omega t)\right]. \tag{5.14}$$

Substituting these expressions into (5.12) and separating the terms with $\exp \mathrm{i}(\omega t + \mathbf{k}\cdot\mathbf{r})$ yields the following equations for p_s and p_c:

$$-2\mathrm{i}\frac{\partial p_s}{\partial x} = \frac{k\alpha_0}{2}p_c^*, \quad 2\mathrm{i}\frac{\partial p_c}{\partial x} = \frac{k\alpha_0}{2}p_s^*, \tag{5.15}$$

which lead to equations (5.7), where now $|B| = \alpha_0 k/4$. Thus, here WFR is also observed, with the amplification factor (5.9).

It is useful to recall that in the "four-wave" WFR considered above the scattering occurs at a spatial lattice created by variation of the period-averaged sound velocity under the action of a standing pump wave; to realize this mechanism cubic nonlinearity is required. In the case under consideration, however, periodic variation of the instantaneous values of the sound velocity causes the effect. These variations can be of different origin, as has already been noted. Thus, for example, in papers by Bunkin *et al.* (1986) and Brysev *et al.* (1984) a scheme with electric pumping is suggested. If an elastic medium is placed in a capacitor creating, thereby, a variable homogeneous electric field in it, a sound velocity modulation may result, leading eventually to WFR. These processes seem to be more readily realized in solid piezoelectrics, where such effects as parametric amplification, pulsed echo, etc., have been observed for a long time. However, our interest here is primarily in purely acoustic effects when an acoustic field serves as the pump. In this case a specific difficulty is encountered: homogeneous pumping in an unbounded medium at frequency 2ω cannot be created by an acoustic wave.

If in (5.13) we assume p_s and p_c to be (slow) functions of time rather than of coordinates, the following formulae will be obtained instead of (5.15),

$$2\mathrm{i}\frac{\partial p_s}{\partial t} = \frac{\omega\alpha_0}{2}p_c^*, \quad 2\mathrm{i}\frac{\partial p_c}{\partial t} = \frac{\omega\alpha_0}{2}p_s^*. \tag{5.16}$$

Solutions of this system increase exponentially in time, i.e. parametric instability takes place. However, in a system with losses this instability may not be realized if α_0 is less than a threshold value.

WFR in a bubble layer

The example of a WFR mechanism in a liquid with bubbles considered above (see (5.10)) was based on the effects of cubic nonlinearity. Meanwhile, the effect of WFR can also be observed during a mere three-wave interaction in a bubbly medium provided the latter appears as a sufficiently thin layer in a homogeneous liquid (Kustov *et al.*, 1985).

Let a plane pump wave with frequency ω_1 be normally incident (from the left) on a plane bubble layer:

$$p_1' = p_1(x)\exp[\mathrm{i}(\omega_1 t - k_1 x) - a_1 x], \tag{5.17}$$

where coordinate x proceeds from the left-hand boundary of the layer,

a_1 being the absorption factor at frequency ω_1. In addition, a signal wave with inhomogeneous distribution of amplitude and phase along a transverse coordinate $\mathbf{r}$ is incident from the left,

$$p_2' = p_2(x)\exp[\mathrm{i}(\omega_2 t - k_2 x + \varphi(x)) - a_2 x], \tag{5.18}$$

where a_2 is the corresponding decay factor. These waves, when interacting at the layer, give rise to a scattered wave at the combination frequency $\omega_3 = \omega_1 - \omega_2$, which radiates into both the left- and the right-hand half-spaces. Its field outside the layer is defined by integrating over the secondary sources in the layer,

$$p_3' = \frac{1}{4\pi}\int \varepsilon_3 \omega_3^2 p_1' p'^*_2 (\rho_0 c_0^4 r')^{-1} \exp(-\mathrm{i}k_3 r' - a_3 r')\,\mathrm{d}\mathbf{r}', \tag{5.19}$$

where ε_3 is the nonlinearity parameter at frequency ω_3 and r' is the distance to the sources.

In the case considered, when p_1 and p_2 are defined by formulae (5.17) and (5.18), we have

$$p_1' p'^*_2 = p_1 p_2^* \exp[\mathrm{i}(\omega_3 t - \varphi) - \mathrm{i}(k_1 - k_2)x' - (a_1 + a_2)x']. \tag{5.20}$$

Let the primary waves decay strongly, so that $a_{1,2} > k_{1,2}$. Then, variation of the source phase along the x-axis is insignificant and radiation from the layer is similar to that from a plane aperture. The distribution of the amplitudes and phases of the secondary sources over the layer is defined by the factor $p_2^* \exp[\mathrm{i}(\omega_3 t - \varphi(\mathbf{r}))]$ in (5.20). It is evident that the phase distribution over coordinate $\mathbf{r}$ has the sign opposite to the signal phase, this conjugation meaning simply the front reversal. The direction θ_3 of the scattered waves is defined here by a synchronism condition along the layer,

$$k_3 \sin\theta_3 = -k_2 \sin\theta_2, \tag{5.21}$$

where θ_2 is the angle of incidence of the signal wave on the layer. Neglecting dispersion within the layer (which is justified for a wide distribution of bubble radii, when resonant bubbles respond at each frequency) yields

$$\sin\theta_3 = \omega_2(\omega_2 - \omega_1)^{-1}\sin\theta_2.$$

In this case the requirement of the vector synchronism $\mathbf{k}_3 = \mathbf{k}_1 - \mathbf{k}_2$ is not met exactly, but for a sufficiently thin layer, $k_3\ell \ll 1$, misphasing in the x-direction is insignificant and the main role is played by the transverse synchronism expressed by (5.21). If the signal has the form of a spherical wave, then for small angles θ_2 and θ_3 equations (5.19) and (5.20) yield the following expressions for the field at frequency ω_3 scattered forwards (+) and backwards (−), respectively,

$$p_3^{\pm} = \frac{\pi \varepsilon_3 p_1 p_2^* B_{\pm}}{2\rho c^2 \beta_{\pm} R_F (1 - R_0/R_F)} \left[1 - \exp(-\beta_{\pm}\ell)\right]. \tag{5.22}$$

Here $\beta_{\pm} = a_1 + a_2 \mp a_3$, $B_+ = \exp(-a_3\ell)$, $B_- = 1$, ℓ is the layer thickness, R_0 is the distance measured from the observation point to the layer, $R_F = \omega_3 R_2/\omega_2$ is the focal length for scattered waves, R_2 being the distance from the signal source to the layer.

It is clear that the scattered waves are brought into focus. If here $a_3\ell \gg 1$, then backward scattering prevails, while in the opposite case, if also $k_3\ell \ll 1$, the layer radiates two converging beams of about the same amplitude.

If, in particular, $\omega_1 = 2\omega_2$, i.e. $\omega_3 = \omega_2$, one obtains $\theta_3 = \pi \pm \theta_2$. This is the case of "true" WFR when, after scattering, the signal wave converges to the reference point. Nevertheless, even in the more general case one can speak of WFR: a diverging wave turns into a converging one with frequency transformation.

The experiment of Kustov *et al.* (1985) was carried out in a laboratory basin. A layer of hydrogen bubbles with thickness $\ell = 10$ cm was created by electrolysis. The pump beam at a frequency $f_1 = 100$ kHz was generated by a round transmitter of diameter 10 cm placed at a distance 2.6 m from the layer; the pump amplitude near the layer amounted to $p_1 = 3.2 \times 10^3$ Pa. A spherical radiator located at 1 m from the layer generated a signal wave at frequency $f_2 = 60$ kHz. Durations of the pump and signal pulses were equal to 1 and 0.3 msec, respectively, and θ_3 was 10°. The direction of the reversed wave propagation was 15° with respect to the pump radiator axis, in accordance with condition (5.21). The field at the difference frequency of 40 kHz had the form of two converging beams propagating on both sides of the layer. Accordingly, the beam width decreases initially, as it goes away from the layer, to increase eventually after passing through the focal point (Figure 7.3). The theoretical curve plotted is based on (5.19), and experimental results are given by dots. It should be emphasized that the phenomenon of WFR is observed here only for moderate wave intensity, when the effects of the redistribution of the bubble concentration in the acoustic field (see below) have not yet interfered.

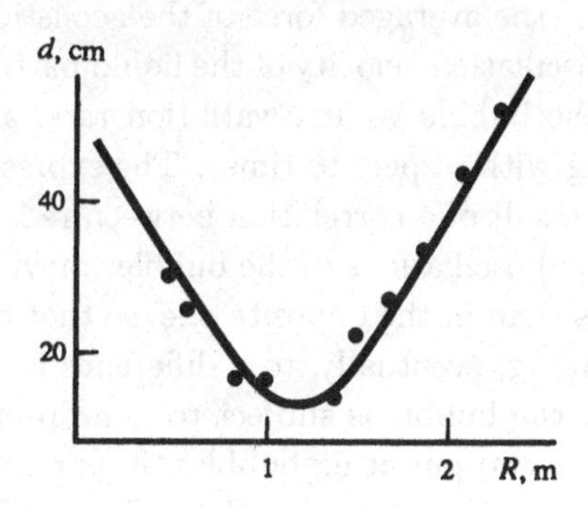

Figure 7.3. The dependence of beam width on distance, in water with bubbles. Solid curve – theory, dots – experiment.

7.6 Concentration nonlinearity in a liquid with gas bubbles

Various effects of self-action arise in a liquid with bubbles if their volume content and size distribution can change in space and time. In an acoustic field the bubbles, generally speaking, are not only subject to redistribution in space but they may appear (due to cavitation) or disappear (due to floating up, dissolution, or merging with other bubbles). All this frequently gives rise to rather strong nonlinearity (Sobolev, 1974). Here we shall briefly tackle these problems following the papers by Kobelev and Ostrovsky (1983, 1989).

Bubble motions in an external acoustic field are of two basic types: monopole pulsations and translational (dipole) oscillations. It is important that, due to the presence of the monopole pulsations, the velocity of the translational motions is not on average equal to zero: the bubble performs an average drift such that its displacements can be large, but only over a long enough time, as compared with the oscillation period (Sobolev, 1974). This averaged component of motion is in fact associated with the action of the radiation pressure of the acoustic field.

If the gas mass in a bubble is neglected, then the bubble's average velocity is essentially defined by a friction force and by the aforementioned averaged drift in the acoustic field due to radiation pressure as well as due to the influence of other bubbles.

The equation of the averaged bubble motion has the evident form

$$\frac{d\mathbf{U}}{dt} = \frac{1}{M}(\mathbf{F}_1 - \mathbf{F}_2), \tag{6.1}$$

where $\mathbf{U}$ is the average velocity, $M = (2/3)\rho\pi R^2$ is the associated mass of liquid, $\mathbf{F}_2 = 6\pi\rho R\mathbf{U}\nu$ is the Stokes force, ν is the viscosity coefficient, and $\mathbf{F}_1 = \rho\langle\dot{V}\mathbf{v}_a\rangle$ is the averaged force of the acoustic field action (here $\mathbf{v}_a$ is the acoustic oscillation velocity of the liquid particles in the absence of bubbles, $\dot{V}$ is the bubble volume variation rate, and angle brackets designate averaging with respect to time). The expression for $\mathbf{F}_1$ can be interpreted as the result of a correlation between radial (monopole) and translational (dipole) oscillations of the bubble: moving in one direction it has larger radius than in the opposite one, so that the field scattering pattern differs, leading, eventually, to a difference in radiation pressure values; this is why the bubble is subject to nonzero action as averaged over the period. For two pulsating bubbles, in particular, a well known expression for the Bjerknes force results: $F_1 = \rho\langle\dot{V}_1\dot{V}_2\rangle/4\pi r^2$ where r is the distance between the bubbles. Thus it is clear that bubbles oscillating in phase are attracted while those oscillating in antiphase are repelled. As another example, let us mention the behaviour of a single bubble in the field of an acoustic beam (Kobelev, 1983). It appears to depend on the field frequency: the phase relation between $\dot{V}$ and $\mathbf{v}_a$ is different for frequencies below and above resonance, so that small bubbles are drawn into the axial region of the beam while large ones are pushed out.

Such averaged effects are somewhat similar to those encountered in electrodynamics and plasma physics for charged particles, the difference, albeit important, being that here the main contribution to the averaged force is that of monopole rather than of dipole oscillations of the bubbles; therefore, the corresponding effects should be most vividly realized in acoustics.

Now let us consider the collective behaviour of bubbles in a quasi-harmonic field, and their reaction on the field. Let the field potential have the form

$$\varphi = \frac{1}{2}\psi(\mathbf{r}, t)\exp(i\omega t) + \text{c.c.} \tag{6.2}$$

such that the dependence $\psi(t)$ is slow enough. Then defining the bubble oscillation amplitude to a linear approximation (in this section we neglect the nonlinearity of individual bubbles) and substituting the result into the expression for $\mathbf{F}_1$, we obtain from (6.1) the following equation for the sound-initiated translational velocity of the bubble:

$$6(\mathbf{U} - \mathbf{U}_0)\nu = [-\alpha(\psi\nabla\psi^* + \psi^*\nabla\psi) + i\beta(\psi^*\nabla\psi - \psi\nabla\psi^*)], \tag{6.3}$$

where $\mathbf{U}_0$ is the drift velocity in the absence of sound and

$$\alpha = (1-\xi)^2\left[(1-\xi^2)^2+\delta^2\right]^{-1}, \quad \beta = \delta\left[(1-\xi^2)^2+\delta^2\right]^{-1},$$

$$\xi = \frac{\omega_0}{\omega}, \quad \delta = Q^{-1}.$$

Here we also disregard the bubble motion inertia (i.e. added mass), which is admissible in the majority of real cases.

Substituting (6.2) into the wave equation (3.22) of Chapter 1 and assuming the bubble concentration $N(R_0)$ to be slowly dependent on time, one obtains (Morse & Feshbach, 1958)

$$\nabla^2\psi + \left[k^2 - \int_0^\infty 4\pi R_0(\alpha + \mathrm{i}\beta)N(R,\mathbf{r},t)\,\mathrm{d}\mathbf{r}\right]\psi = 0, \tag{6.4}$$

where $k = \omega/c_0$. Here we omit the term ψ_{tt}/c_0^2, which is responsible for the envelope propagation in bubble-free liquid. This can be done provided that the bubble effect shows up at distances much less than the scale c_0T, where T is a characteristic modulation period.

To complete the system one needs to write down the equation for bubble concentration. This can be accomplished using the condition of number balance of bubbles within an element of phase volume $\mathrm{d}\mathbf{r}\mathrm{d}R$, taking into account the effects of "birth" and "death" of bubbles with given radius due to possible coagulation. This equation has the form (Kobelev & Ostrovsky, 1983; Maximov, 1985)

$$\frac{\partial N(R_0)}{\partial t} + \operatorname{div}(N(R_0)\mathbf{U}) = -N(R_0)\int_0^\infty N(R_0')W(R_0,R_0')\,\mathrm{d}R_0'$$
$$+\frac{1}{2}\int_0^{R_0} N(R''_0)N(R'_0)W(R'_0,R''_0)\frac{R_0^2}{R'^2_0}\,\mathrm{d}R'_0. \tag{6.5}$$

The right-hand side of (6.5) has the form of a "collision integral", composed of two terms of which the former describes the "departure" of bubbles out of the given element of phase space while the latter embodies the "arrival" of bubbles in it, due to the coagulation process. Here W is the probability of collision (i.e. coalescence) of a bubble pair, defined through the collision cross-section $\sigma(R_0, R'_0)$ by the formula (Kobelev, 1983)

$$w(R_0,R_0') = \sigma(R_0,R_0')|\mathbf{U}(R_0) - \mathbf{U}(R_0')| = 4\pi\ell\,\operatorname{sgn}\ell, \tag{6.6}$$

where

$$\ell = \left(\frac{R_0 + R'_0}{3\nu}\right)\left\{\frac{(\xi_1^2-1)(\xi_2^2-1)+\delta_1\delta_2}{[(\xi_1^2-1)^2+\delta_1^2][(\xi_2^2-1)^2+\delta_2^2]}\right\}|\psi|^2,$$

such that the radii R_0' and R_0'' are related through the condition of gas mass (volume) conservation (the resulting bubble has radius R_0 or R_0' in the first and second integral (6.5) respectively).

In what follows one-dimensional waves are treated first. In addition, to simplify the problem, we assume that initially the liquid contains bubbles of equal radii R_0, this being not too far from the radius at resonance, so that coagulation results in "departure" from the interaction region. Then the effect of coagulated bubbles will be neglected, under the assumption that their resonant frequency differs markedly from the field frequency. Then only the first term is retained on the right-hand side of (6.5), and equations (6.3)–(6.5) are reduced to

$$\frac{\partial^2\psi}{\partial x^2} + \left[k^2 - (\alpha + \mathrm{i}\beta)N\right]\psi = 0, \tag{6.7a}$$

$$\frac{\partial N}{\partial t} + \frac{\partial N\mathbf{U}}{\partial x} = -\left(\frac{2\beta}{3\delta\nu}\right)N^2|\psi|^2, \tag{6.7b}$$

$$6\pi\nu\mathbf{U} = -\alpha\frac{\partial|\psi|^2}{\partial x} + \mathrm{i}\beta\left(\psi^*\frac{\partial\psi}{\partial x} - \psi\frac{\partial\psi^*}{\partial x}\right). \tag{6.7c}$$

Now let us consider some particular effects described by equations (6.7) (Kobelev & Ostrovsky, 1983; Gorshkov & Kobelev, 1983).

Instability of a travelling wave

Homogeneous distribution of the bubbles in space in a travelling wave field is a possible equilibrium state which, however, proves unstable. Here, for the sake of simplicity, we shall neglect the wave decay, assuming $\beta = 0$. It should be noted that in this case (6.7*b*) corresponds to a "collisionless" approximation, in which bubble dimensions remain unchanged, the nonlinearity being associated with their redistribution in space.

Thus, for $\beta = 0$ the system (6.7) has an obvious solution in the form

$$\psi = \psi_0 \exp(-\mathrm{i}\eta x), \quad N = N_0(R_0' - R_0), \quad \mathbf{U} = 0, \tag{6.8}$$

where

$$\eta = \sqrt{k^2 - \alpha N_0}.$$

Let us analyse the stability of this solution. Set

$$\psi = \psi_0(1 + f(x,t))\exp(-\mathrm{i}\eta x), \quad N = N_0 + n(x,t),$$

where f, n are small perturbations such that their dependence on t is slow (on the scale of the field oscillation period). Substituting these formulae into (6.7) one obtains linear equations for ψ and n: in so doing one needs to take into consideration that n is real. Writing also $f = \psi_1 + \mathrm{i}\psi_2$, a real equation for, say, ψ_1 can be readily formulated:

$$\frac{\partial^3 \psi_1}{\partial t \partial x^2} + 4\eta^2 \frac{\partial \psi_1}{\partial t} - 2b \frac{\partial^2 \psi_1}{\partial x^2} = 0, \tag{6.9}$$

where $b = \alpha^2 N_0 |\psi_0|^2 / 6\pi\nu R_0$. Seeking a solution in the form $\psi_1 = \psi_{10} \exp[\mathrm{i}(\Omega t - Kx)] + \text{c.c.}$, the dispersion equation

$$\Omega = -2\mathrm{i} \frac{bK^2}{K^2 - 4\eta^2} \tag{6.10}$$

can easily be deduced.

The instability case, $\operatorname{Im}\Omega < 0$, is realized for all perturbations satisfying the condition $|K| > 2|\eta|$. For $|K|$ close to $2|\eta|$ one has to allow for the omitted dissipative terms in (6.7); according to (6.7a) $\eta^2 = K^2 - (\alpha + \mathrm{i}\beta)N_0$. Then instability occurs for the same inequality $|K| > 2|\eta|$, giving a maximum amplification rate ($\operatorname{Im}\Omega$) for $|K| = 2|\eta|$ of magnitude $b|\eta|/2\operatorname{Im}\eta$.

The following significant detail should be emphasized: because the quantity n is real, i.e. it has the form $\tilde{n} \exp \mathrm{i}(\Omega t - Kx) + \text{c.c.}$, then the perturbation must simultaneously contain components of the types $\exp(\pm \mathrm{i}Kx)$, i.e. waves travelling in opposite directions whose absolute values must be the same. In other words, the concentration perturbation must be "standing"; in the most unstable perturbation it has period $\pi/|\eta|$. In the field perturbations, waves of the form $\exp(-\mathrm{i}\eta x)$ (backward scattering) and $\exp(3\mathrm{i}\eta x)$ are present. The physical mechanism of this instability is now clear: when concentration inhomogeneities emerge a backward scattered wave arises, and a "standing" component appears in the field in which bubbles are grouped, further stimulating the scattering thereby, etc.

This result was extended to the three-dimensional case (Gorshkov & Kobelev, 1983) when the modulation wave propagates at an angle with respect to the direction of the carrier wave vector. In this case (taking losses into account)

$$\operatorname{Im}\Omega = 2b \frac{1 - (4\eta^2/K^2)\cos^2\theta}{\left[1 - (4\eta^2/K^2)\cos^2\theta\right]^2 + \left[4\cos^2\theta / L_0^2 K^2\right]}, \tag{6.11}$$

where $L_0 = \eta / 4\pi R_0 \beta N_0$ is the decay scale of the carrier wave. Therefore, it is clear that perturbations with wave vectors $K > 2|\eta| \cos\theta$ grow exponentially.

It should be noted that this type of instability, in the form of "self-scattering", underlies such effects as wave front reversal: if a fundamental wave diverges then perturbations due to instability will converge. Another possible effect is sound self-focusing associated with amplification of the transverse perturbations ($\theta = \pi/2$). The latter effect has recently been observed in a bubble layer (Belyaeva, 1993).

Real situations, however, are usually more complicated due to a wide distribution of bubble radii. In particular, radiation pressure leads to various average speeds of their convective motion.

If such a convective motion is significant, then one-dimensional instability vanishes or weakens markedly, while transverse instability (self-focusing) remains as pronounced as before.

Shock waves of envelopes

Interaction between the field and the bubbles leads also to the possible existence of stationary wave solutions in which the variables N and $\Psi = |\psi|^2$ depend on a single argument $x - ct$, where c is expressed by formula (3.27) of Chapter 1 (Gorshkov & Kobelev, 1983).

Let us again proceed from the system (6.7), assuming that

$$\psi = \psi_0(x, t) \exp(-ikx)$$

(ψ_0 being a slowly varying complex amplitude). Then equation (6.7*a*) acquires the form

$$(4\pi R_0)^{-1} \frac{\partial \psi_0}{\partial x} = \frac{\mathrm{i}\alpha - \beta}{2k} N \psi_0 \tag{6.12}$$

(here we have neglected the term $\partial^2 \psi_0 / \partial x^2$, which is of the second infinitesimal order). If ψ_0 is written as $s \exp(\mathrm{i}\varphi)$, expression (6.12) yields

$$\varphi(4\pi R_0)^{-1} = \frac{\alpha}{2k} \int N \,\mathrm{d}x, \tag{6.13a}$$

$$\frac{\partial \psi}{\partial x} + \frac{\beta}{k} N \psi = 0, \quad \psi = s^2. \tag{6.13b}$$

To the first approximation the function $\psi^* \psi_x - \psi \psi_x^*$ entering (6.7*c*) is equal to $2\mathrm{i}k|\psi_0|^2$, while the terms with ψ_{0x} can be ignored, so that the remaining equations (6.7) only contain $|\psi_0|^2 = \psi$, resulting in a closed set of equations (6.7*b*), (6.7*c*) and (6.13). Then this system can be reduced to dimensionless form:

$$\frac{\partial I}{\partial X} + nI = 0, \tag{6.14a}$$

$$(-c + W_1 I + 2W_2 nI)\frac{dn}{dX} + (1 - W_1 - nW_2)n^2 I = 0, \tag{6.14b}$$

where

$$n = \frac{N}{N_0}, \quad I = \frac{\psi}{\psi_0}, \quad X = \frac{x}{L_0}, \quad \tau = \frac{t}{T_0}, \quad L = \frac{k}{\beta N_0},$$

$$T_0 = \frac{3\delta\nu}{2\beta N_0 \psi_0}, \quad W_1 = \frac{\delta^2}{2[(1-\zeta)^2 + \delta^2]}, \quad W_0 = \frac{3\delta\nu V_0}{2k\psi_0},$$

$$W_2 = \frac{\delta(1-\xi^2)}{4kL_0[(1-\xi^2)^2 + \delta^2]}, \quad \zeta = X - V_0\tau, \quad V_0 = c - W_0$$

are dimensionless variables. The solutions to (6.14) in general are depicted in Figure 7.4 (a) where the phase trajectories of this set are given. Direct physical meaning can only be ascribed to the phase trajectories belonging to the upper right-hand quadrant (I, $n > 0$). The equations (6.14) are, however, invariant with respect to the transformations $(I, V_0) \to (-I, -V_0)$ and $(n, W_2, \zeta) \to (-n, -W_2, -\zeta)$. Because of this the phase trajectories from any quadrant can be brought to the first one by changing the signs of parameters W_2 and V_0. This means that Figure 7.4 (a) exhausts all possible solutions, with the exception of the cases $W_2 = 0$ (resonance bubbles) and $V_0 = 0$ (non-propagation case). The phase portraits for these two special cases are given in Figures 7.4 (b) and 7.4 (c).

Note that the curve $I = V_0/(W_1 + 2nW_2)$ (dashed line) is a singular one and there is a change of motion direction on it.

Of special interest here is the stationary shock wave represented by any of the trajectories coming out of the equilibrium point $n = 0$, $I = V_0/W_1$ and terminating at the axis $I = 0$ (which forms an equilibrium line). This wave is a transition front that completely removes bubbles of a given radius by inducing their coalescence. The velocity of this wave is determined only by the field intensity I_∞ as $\zeta \to -\infty$, namely, by $V_0 = W_1 I_\infty$, and V_0 does not depend on the equilibrium concentration n_∞. In dimensional variables this velocity is equal to

$$c = V_0 + k\delta\psi_0/3\nu[(1-\xi)^2 + \delta^2], \tag{6.15}$$

$$\xi = \omega/\omega_0.$$

The envelope shock waves exist for $V_0 > 0$ and any sign of W_2, that is for bubbles with radii both greater than and less than the resonance radius.

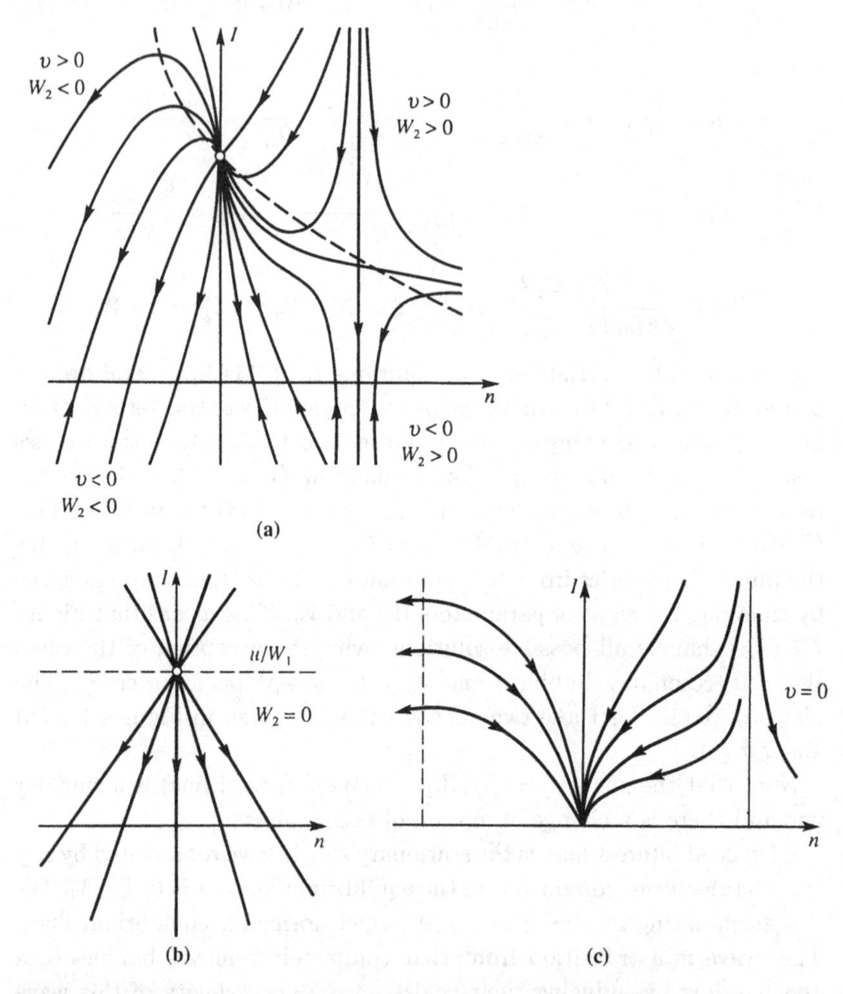

Figure 7.4. Phase plane trajectories for equation (6.14): (a) general case; (b) $W_2 = 0$ (resonance bubbles); (c) $V_0 = 0$ (non-propagation case).

In the resonance case $W_2 = 0$, the equations (6.14) can easily be integrated, leading to the following expressions for the variations of I, n in a shock wave:

$$I = \frac{I_\infty}{2}\left[1 - \tanh\left(\frac{n_\infty \zeta}{2}\right)\right], \quad n = \frac{n_\infty}{2}\left[1 + \tanh\left(\frac{n_\infty \zeta}{2}\right)\right]. \quad (6.16)$$

In nonresonance cases simple analytical solutions have not been obtained; it is worth noting, however, that for $W_2 < 0$ (small bubbles) shock waves exist only under the condition

$$\frac{\delta}{2kL_0} \le 1 - \xi^2 \le 0.$$

Since $\delta \ll 1$ and $kL_0 > 1$ (decay is supposed to be small over a wavelength) then $|1 - \xi^2|$ is small, which allows the shock wave to exist only near resonance; for $W_2 > 0$ (large bubbles) this restriction is lifted.

Effect of "nonlinear transparency" in acoustics

Let us now consider nonstationary processes connected with field penetration into a liquid with bubbles. In this case it is helpful to use equations similar to (6.14), but with back substitution of $\partial/\partial z$ for $\partial/\partial x$ and addition of the term n_τ to the left-hand side of (6.14*b*) (due to the nonstationarity of the process). Apart from this, in the case of nonresonant bubbles W_1, $W_2 \ll 1$ and the corresponding terms in (6.14*b*) can be ignored. As a result one obtains the following simple equations:

$$\frac{\partial I}{\partial X} + nI = 0, \quad \frac{\partial n}{\partial \tau} + n^2 I = 0. \tag{6.17}$$

These equations resemble those of nonlinear transparency in plasma physics and optics with, however, a significant difference: (6.17) contains n^2 rather than n, which leads to the absence of a stationary solution with finite n that is bounded as $X \to \infty$.

At the boundary $x = 0$ of a half-space with homogeneous initial concentration N_0, let a wave with amplitude ψ_0 be specified, i.e. $I(\tau, 0) = 1$, $n(0, z) = 1$. The appropriate solution can readily be obtained by writing $n = g^{-1}$, following which (6.17) yields

$$g_x + \ln g = 0,$$

from which equation it follows that

$$\int_{1+\tau}^{g} \frac{dy}{\ln y} = -X.$$

As a result the solution is expressed via the exponential integral function $\mathrm{Ei}(x)$,

$$X = \mathrm{Ei}(\ln(1+\tau)) - \mathrm{Ei}(\ln g), \tag{6.18}$$

after which I is obtainable from (6.17).

This result is depicted in Figure 7.6, where the dependence of I and n on z is given for various τ. The severe initial decay of the wave decreases, owing to coagulation, down to complete disappearance of the bubbles with the primary radius. The characteristic time of such medium "transparency" caused by the wave is of order $t_0 = 3\delta\nu/8\pi R_0 \alpha \psi_0 N_0$.

For a bubble layer of finite thickness one can also employ solution (6.18) if sound reflection at the layer boundary is disregarded. Let, for instance, a wave with frequency 30 kHz and pressure amplitude 3.2×10^3 Pa travel through a 20 cm layer of hydrogen bubbles. Assume the specific gas content to amount to $3 \cdot 10^{-5}$ and the bubble radius to be 0.7×10^{-2} cm. Then the previously introduced values of L_0 and t_0 are equal to 10 cm and 1 second, respectively, which yields $z = 2$ for the dimensionless layer thickness parameter. Therefore, the intensity of the sound after transmission through the layer builds up from 0.14 to 0.55 within a time interval of 4 seconds.

For resonant bubbles, their translational motion under the radiation pressure is significant, resulting in $W_2 = 0$, $W_1 = 1/2$, so that the equation for concentration has the form

$$n_\tau + \frac{1}{2} I n X + \frac{1}{2} n^2 I = 0 \tag{6.19}$$

(for simplicity we assume $W_0 = 0$, i.e. we neglect the liquid viscosity). The equation for I coincides here with the first equation of (6.17).

Here a difference is observed with respect to boundary conditions as well. If a wave is incident on the bubble layer boundary, then the boundary is displaced under the action of radiation pressure. Nevertheless this problem has an analytic solution (Kobelev & Ostrovsky, 1989). If the incident wave intensity is constant in time, this solution takes on the form

$$n = \left[1 + \frac{\tau}{2} \exp\left(\frac{\tau}{2} - X\right)\right]^{-1},$$

$$I = \left(1 + \frac{\tau}{2}\right) \exp\left(\frac{\tau}{2} - X\right) \left[1 + \frac{\tau}{2} \exp\left(\frac{\tau}{2} - X\right)\right]^{-1},$$

such that the boundary coordinate is $X = \tau/2$. A characteristic feature of this case is the establishment of a stationary envelope shock wave; in the case considered, this occurs for $X \approx 16$.

In the models described above, idealizations still abound. First of all there is the bubble "monoradius" idealization (in the resonance case, however, this makes a lot of sense, because the effect of the resonant bubbles is the most pronounced one). In addition, in real cases the bubbles float up, thereby continuously entering the beam region, eventually to leave it, etc. Nevertheless, the very existence of the nonlinear transparency effect and its basic qualitative peculiarities as prescribed by the theory were observed experimentally (Kobelev *et al.*, 1979). A wave of frequency 130 kHz travels through a bubble layer about 10 cm thick. In this case the relatively weak field (amplitude less than 10^3 Pa) decays by a factor of 100 (40 dB), while that of the stronger one ($p \simeq 3 \times 10^3$ Pa) does so by a factor of only 3 (10 dB). The relaxation time of the effect in this experiment was estimated as 1 second.

In order to clarify the mechanism behind the effect, the distribution of bubble radii was recorded for various intensities of the pump field; it was observed that the bubble concentration falls off with intensity, over a wide range of radii. In addition, experiments showed that this effect is of a nonresonant nature. All this testifies to the action of bubble coagulation under the influence of the acoustic field, due to averaged forces (Bjerknes forces). Direct evidence was also furnished: for larger bubbles, visible to the naked eye ($R > 40\,\mu$m), formation of bubble clusters and coalescence of the bubbles were observed visually.

However, detailed quantitative comparison between the theoretical and experimental results proved to be hard. Theory suggests, in particular, a strong dependence of the time of transition to a steady state on the bubble concentration ($t_0 \sim N^{-1}$), whereas experiment does not really reveal this dependence. This discrepancy is primarily associated with the assumption of the equality of bubble radii in the initial state. Scatter over the radii should weaken the dependence of t_0 on N; due to coagulation, bubbles would actually enter the resonance region, while in our theory coagulating bubbles are considered simply to "go out of play".

7.7 Cavitational self-focusing of sound

The first ideas about sound self-focusing in a bubble medium (Askaryan, 1966) were based on the cavitational effect: since the rate of bubble production depends on sound intensity, bubble concentration will become larger in the region of the field maximum, which brings about a self-refraction effect of the wave. For bubbles of small radius (smaller than the resonance one) the sound velocity drops with concentration, resulting in self-focusing: the rays bend from the wave beam periphery towards its axis. Later on the effect of self-focusing was studied theoretically in more detail (Sobolev, 1974; Iernetti *et al.*, 1980) and observed experimentally (Cutti *et al.*, 1980).

The main new problem here, as compared with the results of Section 2, is that of establishing a relation between the bubble concentration N

or volume content z (defining the sound velocity) and the sound field intensity I. Proceeding from experimental evidence, the authors cited above consider this relation to be linear:

$$N = a(I - I_{\text{th}}), \quad z = b(I - I_{\text{th}}), \tag{7.1}$$

where I_{th} is the intensity corresponding to the cavitation threshold, z is the gas volume content and a and b are constants. For relatively low frequency, when the bubble radii are less than the resonance radius, the sound velocity, as was mentioned, is defined only by the quantity z (see Chapter 1), namely by

$$c = c_0 \left(1 + \frac{\rho_0}{p_0} \frac{c_0^2 z}{\gamma}\right)^{-1/2} = c_0 \left(1 - \frac{c_0 b_1 (I - I_{\text{th}})}{2\gamma}\right). \tag{7.2}$$

Here γ is the adiabatic index; individual bubble nonlinearity is neglected, as in the previous section; $b_1 = c_0^2 b z \rho_0 / 2\gamma p_0$ and the difference between c and c_0 is supposed to be small.

In what follows one can proceed in the same way as in Section 2, writing the pressure perturbation in the form

$$p' = A(\mathbf{r}) \exp[i(\omega t - kx)]. \tag{7.3}$$

We shall proceed from the nonlinear parabolic equation (2.5),

$$-2ik\frac{\partial A}{\partial x} + \nabla_\perp^2 A + AF(|A|^2) = 0. \tag{7.4}$$

Here coordinate x is directed along the beam, $\nabla_\perp^2$ is the Laplacian with respect to transverse coordinates, and F is defined by the nonlinear part of the sound velocity; according to (7.2),

$$F = b_1(I - I_{\text{th}}).$$

One can easily see that with this expression, (7.4) yields a correction to the wave velocity corresponding to (7.2).

Substituting $A = s\exp(i\varphi)$ into (7.4), where s and φ are real variables, and setting $I = \alpha s^2$ where α is a real coefficient, yields, for an axially symmetric beam, the two equations

$$k\frac{\partial s^2}{\partial x} - \frac{\partial \varphi}{\partial r}\frac{\partial s^2}{\partial r} - s^2\left(\frac{\partial^2 \varphi}{\partial r^2} + \frac{1}{r}\frac{\partial \varphi}{\partial r}\right) = 0, \tag{7.5a}$$

$$\frac{2}{k}\frac{\partial \varphi}{\partial x} + \frac{1}{k^2}\left(\frac{\partial \varphi}{\partial r}\right)^2 = b_1(\alpha s^2 - I_{\text{th}}) + \frac{1}{k^2 s}\left(\frac{\partial^2 s}{\partial r^2} + \frac{1}{r}\frac{\partial s}{\partial r}\right), \tag{7.5b}$$

where r is again the transverse coordinate.

We shall seek an approximate solution to this set, describing the field near the beam axis by expanding s and φ in powers of r^2 (see also Section 2):

$$\varphi = \frac{r^2}{2}B(x) + G(x), \quad s^2 = \frac{s_0^2}{f}\left[1 - \frac{r^2}{(Df)^2}\right], \tag{7.6}$$

where B, G, f are functions of x and D is a constant.

It is clear that $B = R^{-1}$ where R is the wave front's radius of curvature and Df is the characteristic width of the beam. Substituting this solution into (7.5*b*), one can easily observe that $f_x = B(x)f$. The boundary conditions have the form

$$f(0) = 1, \quad f_x(0) = R_0^{-1},$$

i.e. the initial beam width is of the order of D while the initial radius of curvature is R_0.

From equation (7.5*a*), equating the factors for the same degrees of r we obtain the equation for f as

$$f^3\frac{\partial^2 f}{\partial x^2} + \frac{1}{D^2}2b_1 I_0 + \frac{1}{D^2} = 0,$$

($I_0 = \alpha s_0^2$), the solution to which, satisfying the boundary conditions, has the form

$$f^2 = \left(1 + \frac{x}{R_0}\right)^2 - gx^2, \tag{7.7}$$

where now

$$g = \frac{1}{D^2}\left(2b_1 I_0 + \frac{1}{k^2 D^2}\right).$$

Therefore, if $R_0^2 g > 1$, then the beam width diminishes with distance, at least for large x. The focal length X_f corresponding to condition $f = 0$ is equal to

$$X_f = \frac{R_0}{R_0\sqrt{g} - 1}.$$

In the paper by Cuiti *et al.* (1980) experimental evidence is given of the observed tendency of a cavitating beam towards self-focusing. A transmitter with radius 1.25 cm was operated in water at frequency 690 kHz. The beam was observed by a shadowgraph optical method. Observation was performed in the wave zone of the transmitter, at $x > 8$ cm. The gas volume content was about 8×10^{-7}. In the presence of cavitation, beam convergence was observed to the extent that the beam acquired complex structure. The authors believe that it was self-focusing that caused the cavitation efficiency for the focusing and plane transmitters to be approximately the same (due to nonlinear beam convergence in the latter case).

7.8 Phase locking of nonlinear oscillators in acoustics

In the conclusion of this chapter we shall briefly discuss a potentially important mechanism of acoustic field amplification or generation which, until recently, has only been studied in the "electromagnetic" sciences such as electronics and quantum radiophysics. Here we consider the collective behaviour of nonlinear resonance systems that oscillate incoherently with randomly distributed phases in the initial state, to be eventually partially synchronized due to a nonlinear mechanism of phase locking; as a result, a coherent field may be generated or amplified. This can be compared, for example, to the so-called "superradiation" effect created by a system of excited atoms (Zheleznyakov *et al.*, 1986) as well as to "masers on cyclotron resonance" in electronics (Gaponov *et al.*, 1967). Considering this process more extensively, we can say that the same principle of operation is typical of such quantum oscillators as masers and lasers.

Here we shall be concerned with a system of classical mechanical oscillators, i.e. acoustic monopoles, where effects of this kind are possible (Kobelev *et al.*, 1986). Note that such systems are in a sense more efficient than electromagnetic (dipole) ones, since a monopole has a stronger coupling (in comparative terms) with the field than a dipole. This known feature follows directly from comparing the efficiency of the radiation from the sphere resulting from volume variations (pulsations) with the radiation resulting from translational oscillations of the same amplitude.

In what follows one can keep in mind a gas bubble as a nonlinear oscillator. The presence of cubic nonlinearity is still important, since this leads to the dependence of the oscillator's average resonance frequency on amplitude.

A solution of the problem consists of, in general, three stages: analysis of the vibrations of one oscillator under the action of a weak coherent field, calculation of the collective contribution of the oscillators of a coherent field and, finally (if a wave is implied), solution of the field equations, and derivation of the complex dispersion relationship.

So, let us consider an oscillator with cubic nonlinearity described by the Duffing equation:

$$\ddot{v} + \mathcal{X}\dot{v} + \omega_0^2\left(v - \frac{\beta v^3}{3}\right) = F(t), \tag{8.1}$$

where F is the force associated with a coherent field and $\mathcal{X}$, β are decay and nonlinearity factors respectively. If this force is nearly harmonic, $2F = A(t)\exp(\mathrm{i}\omega t) + A^*(t)\exp(-\mathrm{i}\omega t)$, where $A(t)$ is a slowly varying function. Then, for a small detuning $\omega - \omega_0 = \varepsilon$, one can assume $2v = Y\exp(\mathrm{i}\omega\tau) + Y^*\exp(-\mathrm{i}\omega\tau)$, where Y is also slowly varying, $\tau = t - t_0$ and t_0 is the initial time defining the phase of a given oscillator. Then, as before, separating the first harmonic in v^3 and omitting terms with ε^2 and $\ddot{Y}$, we obtain

$$\frac{\partial Y}{\partial \tau} + \left[\frac{\mathcal{X}}{2} + \mathrm{i}\left(\varepsilon + \omega_0\beta|Y|^2\right)\right]Y = -\frac{\mathrm{i}}{2\omega_0}A\exp(\mathrm{i}\omega t_0). \tag{8.2}$$

Let us impose the initial condition $Y(0) = s_0\exp(-\mathrm{i}\varphi_0)$; then, as is evident, the free solution to (8.2) (corresponding to $A = 0$) has the form of decaying nonlinear oscillations,

$$Y_0 = s_0\exp(-\mathrm{i}\varphi_0 - \sigma(\tau)), \tag{8.3}$$

where

$$\sigma(\tau) = \frac{\mathcal{X}\tau}{2} + \mathrm{i}\left\{\varepsilon\tau + \frac{\eta_0}{\mathcal{X}}\left[1 - \exp(-\mathcal{X}\tau)\right]\right\} \tag{8.4}$$

with $\eta_0 = \omega_0\beta s_0^2$ as the normalized initial amplitude.

The forced solution (for $A \neq 0$) will be sought assuming A to be small, and consequently the forced response $y = Y - Y_0$ to the external field is small. Linearizing (8.2) with respect to y we obtain the equation

$$\frac{\partial y}{\partial \tau} + \left[\frac{\mathcal{X}}{2} + \mathrm{i}\left(\varepsilon + 2\mathrm{i}\omega_0\beta|Y_0|^2\right)\right]y + \mathrm{i}\omega_0\beta Y_0^2 y^* = -\frac{\mathrm{i}}{2\omega_0}A\exp(\mathrm{i}\omega t_0). \tag{8.5}$$

A fundamental system of solutions of equation (8.5) for $A = 0$ is composed, as is known, of derivatives of Y_0 with respect to parameters s_0 and φ_0 (this results directly from the way of deducing (8.5) from (8.2)). Thus

$$Y_1 = \frac{\partial Y_0}{\partial s_0}\left[1 - 2\mathrm{i}\left(\frac{\eta_0}{\mathcal{X}}\right)(1 - \exp(-\mathcal{X}\tau)\exp(-\mathcal{X}\tau)\right], \tag{8.6a}$$

$$Y_2 = \frac{\partial Y_0}{\partial \tau_0} = -\mathrm{i}y_0. \tag{8.6b}$$

Then one can readily obtain the general solution of (8.5) as

$$\begin{aligned}y =& C_1Y_1 + C_2Y_2 - \frac{\omega_0}{4s_0}y_1 \int_0^T \exp(\mathcal{X}\tau)\,[Y_2^* A\exp(\mathrm{i}\omega t_0] + \text{c.c.})\mathrm{d}\tau \\ & - Y_2 \int_0^T \exp(\mathcal{X}\tau)\,[Y_1^* A\exp(\mathrm{i}\omega t_0) + \text{c.c.}]\,\mathrm{d}\tau, \end{aligned} \tag{8.7}$$

C_1, C_2 being arbitrary constants.

The next stage is concerned with the calculation of the total coherent response of the system. Provided the oscillators are arbitrarily distributed over the resonant frequencies (for example, the bubbles over their radii) and over delay times t_0, then their response will be equal to

$$z' = \int y(t, t_0, \widetilde{\omega}_0) n(\widetilde{\omega}_0, t_0)\,\mathrm{d}\widetilde{\omega}_0 \mathrm{d}t_0, \tag{8.8}$$

where $\widetilde{\omega}$ is the current resonant frequency and $n(\omega_0, t_0)$ is a given distribution function such that the total number of oscillators is $N = \int n\,\mathrm{d}\omega_0 \mathrm{d}t_0$. In what follows we shall consider the bubbles' resonant frequency to be equal, so that $n(\widetilde{\omega}_0, t_0) = \tilde{n}(t_0)\delta(\widetilde{\omega}_0 - \omega_0)$.

Further on the problem must be rendered more specific by prescribing the function $\tilde{n}(t_0)$ and coherent field $A(t)$. Let us consider first the situation corresponding to that dealt with in electronics. Let each oscillator "live" for a finite period of time (from t_0 up to $t_0 + T$, such that $\omega_0 T \gg 1$); for instance, it enters and leaves the interaction domain instantaneously while the total number of oscillators N remains constant, so that $\tilde{n} = N/T$. Then

$$z' = \frac{N}{T} \int_0^T y(\tau)\,\mathrm{d}\tau, \tag{8.9}$$

$$y = Y - Y_0.$$

Substituting here the expression (8.7) for y yields $z' = A\mu\exp(\mathrm{i}\omega t) +$ c.c. where

$$\begin{aligned}\mu =& \frac{N}{T} \int_0^T (C_1Y_1 + C_2Y_2)\exp(\mathrm{i}\omega(t - \tau)) \\ & - \frac{\mathrm{i}}{2s_0 T}\frac{\partial}{\partial s_0}\left(\int_0^T Y_0 \int_0^{\tau'} \exp(\mathcal{X}\tau)\,\mathrm{d}\tau \mathrm{d}\tau' \right). \end{aligned} \tag{8.10}$$

Here the coherent field is supposed to be harmonic (A constant). For the initial conditions used here, $C_1 = C_2 = 0$.

Now one has to analyse the field equation. We shall be concerned with a plane sound wave satisfying the wave equation

$$p_{xx} - \frac{1}{c_0^2} p_{tt} = -\ell z'_{tt}, \tag{8.11}$$

where ℓ is the coupling coefficient between the field and the oscillators. The solution will be sought in the form of a travelling wave such that ω is real and $A = A_0 \exp(-ikx)$. Substituting here the expression for V gives us the dispersion equation

$$k^2 = k_0^2 \left(1 - \frac{\ell N}{T} \mu\right), \tag{8.12}$$

where $k_0 = \omega_0 / c_0$ is the wave number of sound in a medium containing no oscillators.

It is obvious that the fact of sound amplification or decay depends on the sign of the imaginary part of the wave number, or on the sign of $\mathrm{Im}\,\mu$, since

$$\mathrm{Im}\,k \simeq -\left(\frac{k_0 \ell N}{2T}\right) \mathrm{Im}\,\mu$$

(here it is accepted that $N\ell/T \ll 1$).

Expression (8.10) for μ (for C_1, $C_2 = 0$) may be reduced, taking into account (8.3) and (8.4), to

$$\mu = \frac{\mathrm{i}}{T} \frac{\partial}{\partial \eta_0} \left[\eta_0 \int_0^T \exp(-\sigma) \int_0^{\tau'} \exp(\sigma)\, \mathrm{d}\tau \mathrm{d}\tau' \right]. \tag{8.13}$$

In the case of nonzero decay the function $\sigma(\tau)$ is nonlinear, and in the general case equation (8.13) is not reduced to quadratures. Here we confine outselves to the case of small losses, when $(\mathcal{X}T)^2 \ll 1$, i.e. the amplitude of free oscillations within the time T would not be significantly changed. Here $\sigma = \lambda\tau$ where

$$\lambda = \frac{\mathcal{X}}{2} + \mathrm{i}\,(\varepsilon + \eta_0),$$

and (8.13) leads to

$$\mu = \mathrm{i} \frac{\partial}{\partial \eta_0} \left[\frac{\eta_0}{\lambda} \left(1 - \frac{1 - \exp(-\lambda T)}{\lambda T} \right) \right]. \tag{8.14}$$

If the decay is neglected altogether ($\mathcal{X} = 0$) then

$$\mathrm{Im}\,\mu = \frac{1}{q^2 T} \left[(1 - \cos qT) \frac{\varepsilon - \eta_0}{q} + \eta_0 T \sin qT \right], \tag{8.15}$$

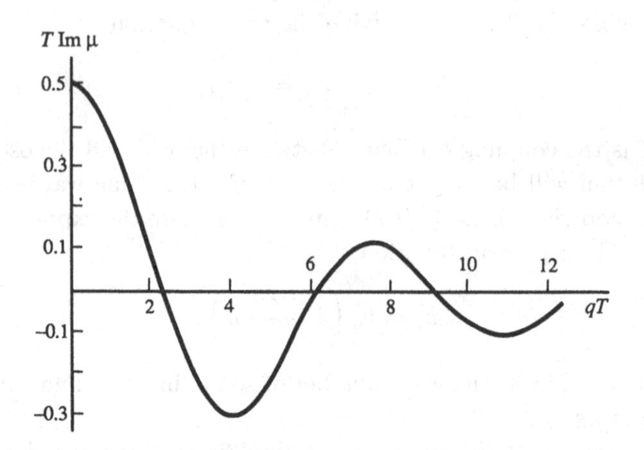

Figure 7.5. The dependence of the damping (amplification) rate Imμ on the total detuning q for a fixed oscillator amplitude.

where $q = \varepsilon + \eta_0$ is the total detuning with respect to the coherent field frequency.

The dependence of $\operatorname{Im}\mu$ on the total detuning q for a fixed amplitude η_0 of the oscillators is depicted in Figure 7.5. In the case of zero detuning the field will always decay ($\operatorname{Im}\mu = I/2$) due to the energy consumption for build-up of resonant oscillations, despite the absence (or smallness, to be more exact) of the losses in individual oscillators. Here we have obtained another analogue of the Landau damping in plasma (in the sense of independence of the decay coefficient on "individual" losses). As distinct from another known analogue (Ryutov, 1975), when the oscillators have different resonant frequencies (see Chapter 6), in the case considered here the oscillator dephasing is attained because of the scattering of starting times t_0.

The general dependence of $\operatorname{Im}\mu$ on q has an oscillatory nature, i.e. alternating zones of amplification and decay are observed. The maximum increment in the first zone is equal (for $qT \simeq 3\pi/2$) to

$$(\operatorname{Im}\mu)_{\max} \simeq \frac{\eta_0}{q^2} \simeq \frac{4\eta_0 T^2}{g\pi^2}, \tag{8.16}$$

while the quantity $\mathcal{X}$ determines the amplification threshold.

It should be noted that the available quality factor of acoustic systems (including bubbles) is rarely very high, and in realistic estimates one should as a rule account for the losses. The effect of losses for the problem under consideration was analysed in a study by Benditskaya *et al.*, (1988). Here we restrict our attention to a couple of notes. First the integral (8.13) can easily be evaluated for the second extreme case, when $\exp(-\mathcal{X}T/2) \ll 1$. Therefore

$$\operatorname{Im}\mu = \frac{2\mathcal{X}}{\mathcal{X}^2 + 4c^2}. \tag{8.17}$$

As could be expected, the field always decays under this condition. Let us note, however, that the maximum value of the decay rate is here much less than the value of $T/2$ obtained for $\mathcal{X} = 0$ and comparable values of the other parameters: resonance losses suppress the field much more efficiently than the losses in individual oscillators.

Of more importance is the fact that up to $\mathcal{X}T \simeq 1$ the instability growth rate remains of the same order of magnitude as for $\mathcal{X}T = 0$. This is evident even from equation (8.14), which is valid up to $\mathcal{X}T \approx 1$ (as only the terms of order $(\mathcal{X}T)^2$ are omitted in σ). The computer-aided solution of the problem shows that $\mu(\mathcal{X}T = 1)$ is in fact somewhat greater than the value $T/2$.

As to the particular estimates, say, for bubbles with quality factor $Q = \omega_0/\mathcal{X} \simeq 10{-}20$, the value of T must be such that, in optimal conditions, $\omega_0 T \approx 10{-}20$.

Meanwhile, the statement of the problem considered here seems to be more realistic for high-frequency systems; in this context, self-synchronizing of thermal phonon oscillations in solids deserves special consideration.

Other versions of oscillator self-synchronizing have been considered as well (Kobelev *et al.*, 1983). For instance, the oscillations can be maintained by periodic pulse "kicks", which leads under proper choice of the period to a more pronounced effect.

Finally, an interesting possibility is suggested by a closer analogue of the "superradiation" mentioned above, when the oscillators having an initially random phase generate a coherent pulse. This class of problems has also been considered for acoustic monopoles (Kobelev *et al.*, 1991; Ostrovsky & Soustova, 1991; Ilinsky & Zabolotskaya, 1992).

References

Akhmanov, S.A., Krindach, D.P., Migulin, A.V. *et al.* (1968). E–12 thermal self actions of laser beams, *IEEE J. Quant. Electronics* **4,** 568–77.

Andreev, V.G., Karabutov, A.A., Rudenko, O.V. & Sapozhnikov, O.A. (1985). Observation of sound self-focusing, *Pisma v ZhETF* **41,** 381–4.

Askarjan, G.A. (1966). Self-focusing and focusing ultra- and supersound, *Pisma v ZhETF* **4,** 144–7.

Assman, V.A., Bunkin, F.V., Vernin, A.V., Ljakhov, G.A. & Shipilov, K.F. (1985). Observation of thermal self-action of sound beam in liquid, *Pisma v ZhETF* **41 (4),** 148–50.

Assman, V.A., Bunkin, F.V., Vernin, A.V., Ljakhov, G.A. & Shipilov, K.F. (1986). Thermal self-action of sound beams in viscous fields, *Akust. Zh.* **32,** 138–40.

Belyaeva, I.A. (1993). Self-focusing effect in a liquid with gas bubbles, *Akust. Zh. 3J* **6,** 981–85.

Bendizkaya, L.M., Kobelev, Y.A., Ostrovsky, L.A. & Soustova, J.A. (1988). The energy loss in oscillator influence on the "classical laser" effect in acoustics, *Akust. Zh.* **34 (4),** 593–7.

Bespalov, V.J. & Talanov, V.J. (1966). On the thread-like structure of light beams in nonlinear liquids, *Pisma v ZhETF* **3,** 471–6.

Bespalov, V.J. & Pasmanik, G.A., eds. (1982). *Modern Trends in Investigations of Wave Front Reversal in Nonlinear Media* (IPF, Gorky).

Bespalov, V.J. & Pasmanik, G.A. (1986). *Nonlinear optics and adaptive systems* (Nauka, Moscow).

Brysev, A.P., Bunkin, F.V., Vlasov, D.V. & Kravtzov, Y.A. (1984). POFUS – parametric phase conjugative sound amplifier, pp. 19–30, in *Research on Hydrophysics (Proc. FIAN, V. 156),* ed. N. Basov (Nauka, Moscow).

Bunkin, F.V. & Liakhov, D.A. (1984). New problems of nonlinear acoustics of liquids, *Proc. FIAN* , 3–19.

Bunkin, F.V., Kravtzov, Y.A. & Liakhov, G.A. (1986). Acoustical analogy of nonlinear optical phenomenon, *UFN* **144,** 391–411.

Bunkin, F.V., Vlasov, D.V. & Kravtzov, J.A. (1981). On the problem of wave front reversal together with amplification of reversal wave, *Izv. AN SSSR Mechanics of Fluids and Gases* **8,** 1144–5.

Cuiti, P., Iernetti, G. & Sagoo, M.S. (1980). Optical visualization of nonlinear acoustical propagation in cavitating liquids, *Ultrasonics* **18,** 111–4.

Gaponov, A.V., Petelin, M.J. & Yulpatov, V.K. (1967). Induced radiation of excited classical oscillators and its utilisation in high frequency electronics, *Izv. Vuzov. Radiophysica.* **10,** 1414–53.

Gorshkov, K.A. & Kobelev, Y.A. (1983). The collective mechanism of self-action of sound influence on the propagation of acoustical waves in liquid with gas bubbles, Preprint *JPF AN SSSR Gorky* **84,** 24.

Hasegawa, A. & Tappert, F. (1973). Transmission of stationary nonlinear optical pulses in dispersive dielectric fibers, *Appl. Phys. Lett.* **23 (3),** 142–4.

Iernetti, G., Cuiti, P. & Sagoo, M.S. (1980). Different stages of acoustic self-focusing during cavitation, *Acustica* **46,** 228–9.

Ilinskii, Y.A. & Zabolotskaya, E.A. (1992). Cooperative radiation and scattering of acoustic waves by gas bubbles in liquids, *JASA* **92,** 2837–41.

Karabutov, A.A., Rudenko, O.V. & Sapozhnikov, O.A. (1988). The theory of thermal self-focusing including shock wave and acoustical streaming formation, *Akust. Zh.* **34,** 644–50.

Karpman, V.I. (1973). *Nonlinear Waves in Dispersive Media* (Nauka, Moscow).

Kobelev, Y.A. (1983). Nonlinear dipole oscillation of a spherical particle in a sound field, *Akust. Zh.* **29,** 783–9.

Kobelev, Y.A. & Ostrovsky, L.A. (1983). Collective self-action of sound in liquid with gas bubbles, *Pisma v ZhETF* **37,** 5–8.

Kobelev, Y.A. & Ostrovsky, L.A. (1989). Nonlinear acoustic phenomena due to a bubble drift in a gas liquid mixture, *JASA* **85,** 621–29.

Kobelev, Y.A., Ostrovsky, L.A. & Soustova, J.A. (1986). Autosynchronization of nonlinear oscillators in acoustics, *Izv. Vuzov. Radiophysics* **29 (3),** 1120–35.

Kobelev, Y.A., Ostrovsky, L.A. & Soustova, J.A. (1991). Nonlinear model of autophasing for classical oscillators, *Sov. Phys. JETP* **72,** 262–67.

Kobelev, Y.A., Ostrovsky, L.A. & Sutin, A.M. (1979). Self-transparency effects of sound waves in a liquid with gas bubbles, *Pisma v ZhETF* **30 (7),** 423–5.

Kustov, L.M., Nazarov, V.E. & Sutin, A.M. (1985). Acoustic wave front reversal on a bubble layer, *Akust. Zh.* **31,** 837–9.

Landau, L.D. & Lifshitz, E.M. (1986). *Hydrodynamics* (Nauka, Moscow).

Lighthill, M.J. (1965). Contributions to the theory of waves in nonlinear dispersive systems, *J. Inst. Math. Appl.* **1,** 837–9.

Maksimov, A.O. (1985). Distribution of bubbles coagulating in the sound field, *Akust. Zh.* **31,** 548–49.

Mikhailov, J.G. (1949). *Propagation of Ultrasound Waves in Liquids* (p. 152) (GTTI, Moscow).

Morse, P.M. & Feshbach, H. (1966). *Methods of Theoretical Physics* (J.L., Moscow).

Ostrovsky, L.A. (1963). On the theory of waves in nonstationary compressive media, *PM M.* **27 (5),** 924.

Ostrovsky, L.A. (1966). Propagation of the wave trains and space–time self-focusing in nonlinear medium, *ZhETF* **51 (10),** 1189–94.

Ostrovsky, L.A. (1968). Shock waves of envelopes, *ZhETF* **54 (4),** 1135–243.

Ostrovsky, L.A. & Soustova, J.A. (1991). Phase-locking effects in a system of nonlinear oscillators, *Chaos* **1,** 223–31.

Pushkina, N.J. & Khokhlov, R.V. (1971). On the stimulating scattering of sound on vorticity waves, *Akust. Zh.* **17,** 167–9.

Rudenko, O.V. & Soluyan, S.I. (1977). *Theoretical Foundations of Nonlinear Acoustics* (Plenum, New York).

Rutov, D.D. (1975). Landau damping analogy in a liquid with gas bubbles, *Pisma v ZhETF* **22 (1),** 46–8.

Sobolev, V.V. (1974). Propagation and self-focusing of sound in inhomogeneous gas–liquid medium, *Izv. AN SSSR, MJG* **1 (5),** 177–80.

Talanov, V.J. (1966). Self-similar wave beams in nonlinear dielectric, *Izv. Vuzov. Radio.* **9 (2),** 410–2.

Vorobiyev, E.M. & Zabolotskaya, E.A. (1974). Self-action of sound waves in the bubbles media, *Akust. Zh.* **20,** 623–4.

Whitham, G.B. (1965). A general approach to linear and nonlinear dispersive waves using a Lagrangian, *J. Fluid. Mech.* **22,** 273–83.

Whitham, G.B. (1974). *Linear and Nonlinear Waves* (Wiley–Interscience, New York).

Zabolotskaya, E.A. (1977). Two mechanisms of self-action of sound waves propagating in gas–liquid mixture, *Akust. Zh.* **23,** 591–5.

Zabolotskaya, E.A. (1984). Nonlinear acoustics combined methods of spectroscopy of gas bubbles in liquids, Research on hydrophysics, *Proc. FIAN* **156,** 31–41.

Zabolotskaya, E.A. & Khokhlov, R.V. (1976). Thermal self-action of sound waves, *Akust. Zh.* **22,** 28–31.
Zabolotskaya, E.A. & Soluyan, S.I. (1976). On the possibility of amplification of sound waves, *Akust. Zh.* **13,** 296–8.
Zabolotskaya, E.A. & Schwarzburg, A.B. (1968). The propagation of finite amplitude waves in a wave guide, *Akust. Zh.* **34 (5),** 892–956.
Zeleznyakov, V.V., Kocharovsky, V.V. & Kocharovsky, Vl.V. (1986). Cyclotron super radiation – classical analogy of Dicke super radiation, *Izv. Vuzov. Radiophyzika.* **29,** 1095–116.

Conclusion

It is quite possible that those who have had the patience to go through this book or at least to follow its main ideas will ask: "What is nonlinear acoustics after all?" Really, we must admit that such a clear and attractive definition as "small-but-finite-amplitude branch of fluid (and solid) mechanics" seems to be right in principle but is too narrow nowadays; however, we are unable to replace it by an another, equally convincing one. Moreover, the difference between the effects of NA and those of nonlinear optics or plasma physics which seemed early on to be easily expressible (in the first place by the difference in dispersion properties) is now getting more and more smooth...

At the same time this lack of definiteness in the definition of NA may be, as we believe, considered as an advantage because it reflects the broadening of the investigation field. In fact, any sufficiently wide physical concept may be described by a set of fundamental problems involved in it rather than by one or few firm statements. Thus, we hope that the contents of this book may give the reader at least a limited image of modern NA in a more or less understandable form.

But, still, it seems to be reasonable to discuss very briefly the main directions of NA development in order to try to predict what is to come. There are three aspects to be discussed (in the spirit of the preface to the book). The first is related to the position of NA as a part of nonlinear wave theory. Now we can easily see that practically all the models of this theory involving different kinds of nonlinearity, dispersion, wave geometry, etc. may prove to be of importance here and one can expect that the variety of theoretical problems will increase. The second aspect is the scope of specific acoustical problems, estimates and physical experiments. Here a tendency exists, as it seems to us, to move in frequency range from sound and ultrasound to both higher and lower frequencies (due to interpenetration of the methods and problems of NA with such disciplines as thermal physics, strength and structure of materials). And the third aspect is the problem of technical applications. Is NA really useful in these aspects? The question is not as simple as it may seem. For example, the enthusiasm of the 1960s about parametric arrays – the really new and unusual devices based clearly on NA principles which promised to solve a number of underwater acoustics problems – now appears to be declining. However, some special but interesting applications of parametric arrays, both in the ocean and other fields of acoustical location and in diagnostic methods are left and occupy a reliable position. Moreover, new possible applications such as the development of nonlinear methods of diagnosis of fluid and solid media inhomogeneities seem to attract more and more attention.

Finally, we believe that NA problems and methods are worthy of treatment by physicists and engineers due to both its potential applications and, even more importantly, their fundamental theoretical nontrivialities. We hope that our book will be able to promote, at least to a degree, interest in this fruitful area of science.

Subject index

Printed in the United States
By Bookmasters